Lecture Notes in Geoinformation and Cartography

Series Editors: William Cartwright, Georg Gartner, Liqiu Meng,
Michael P. Peterson

For further volumes:
http://www.springer.com/series/7418

Michael P. Peterson
Editor

Online Maps with APIs and WebServices

 Springer

Editor
Prof. Michael P. Peterson
University of Nebraska, Omaha
Dept. Geography & Geology
Nebraska
USA

ISSN 1863-2246   ISSN 1863-2351 (electronic)
ISBN 978-3-642-27484-8   ISBN 978-3-642-27485-5 (eBook)
DOI 10.1007/978-3-642-27485-5
Springer Heidelberg New York Dordrecht London

Library of Congress Control Number: 2012935753

© Springer-Verlag Berlin Heidelberg 2012
This work is subject to copyright. All rights are reserved by the Publisher, whether the whole or part of the material is concerned, specifically the rights of translation, reprinting, reuse of illustrations, recitation, broadcasting, reproduction on microfilms or in any other physical way, and transmission or information storage and retrieval, electronic adaptation, computer software, or by similar or dissimilar methodology now known or hereafter developed. Exempted from this legal reservation are brief excerpts in connection with reviews or scholarly analysis or material supplied specifically for the purpose of being entered and executed on a computer system, for exclusive use by the purchaser of the work. Duplication of this publication or parts thereof is permitted only under the provisions of the Copyright Law of the Publisher's location, in its current version, and permission for use must always be obtained from Springer. Permissions for use may be obtained through RightsLink at the Copyright Clearance Center. Violations are liable to prosecution under the respective Copyright Law.
The use of general descriptive names, registered names, trademarks, service marks, etc. in this publication does not imply, even in the absence of a specific statement, that such names are exempt from the relevant protective laws and regulations and therefore free for general use.
While the advice and information in this book are believed to be true and accurate at the date of publication, neither the authors nor the editors nor the publisher can accept any legal responsibility for any errors or omissions that may be made. The publisher makes no warranty, express or implied, with respect to the material contained herein.

Printed on acid-free paper

Springer is part of Springer Science+Business Media (www.springer.com)

Contents

Part I Background

1 Online Mapping with APIs 3
  Michael P. Peterson

2 Web Mapping Services: Development and Trends 13
  Manuela Schmidt and Paul Weiser

3 Current Trends in Vector-Based Internet Mapping:
  A Technical Review 23
  Christophe Lienert, Bernhard Jenny, Olaf Schnabel, and Lorenz Hurni

4 Map Mashups and APIs in Education 37
  Emmanuel Stefanakis

Part II API Mashups

5 Multimedia Mapping on the Internet Using Commercial APIs 61
  Shunfu Hu

6 The GIS Behind *i*MapInvasives: The "Open Source Sandwich" 73
  Georgianna Strode

7 Towards a Dutch Mapping API 91
  Edward Mac Gillavry, Thijs Brentjens, and Haico van der Vegt

8 LatYourLife: Applying Multiple API Services for Task Planning ... 105
  Amin Abdalla

9 Guidelines for Implementing ArcGIS API for Flex Developers 123
  Georgianna Strode

Part III Symbolization

10 Web Services for Thematic Maps 141
Otakar Cerba and Jachym Cepicky

11 A Technical Survey on Decluttering of Icons in Online Map-Based Mashups 157
Haosheng Huang and Georg Gartner

12 Web Map Design for a Multipublishing Environment Based on Open APIs 177
Pyry Kettunen, L. Tiina Sarjakoski, Salu Yliriskų, and Tapani Sarjakoski

13 User Scalable Graduated Circles with Google Maps 195
Douglas Paziak

14 Webservices for Animated Mapping: The TimeMapper Prototype 205
Barend Köbben, Timothée Becker, and Connie Blok

15 The Possibilities of Globe Publishing on the Web 219
Mátyás Gede

Part IV Applications

16 Mapping Social-Network Interactions 241
James O'Brien and Kenneth Field

17 Online Map Service Using Google Maps API and Other JavaScript Libraries: An Open Source Method 265
Shunfu Hu

18 Online Information Dissemination at the Wisconsin State Cartographer's Office Using Map Services and APIs 279
Howard Veregin and Timothy Kennedy

19 WebGIS Systems for Planetary Data Access at the PDS Geosciences Node 299
J. Wang, D.M. Scholes, and K.J. Bennett

Index 315

Contributors

Amin Abdalla Research Group Geoinformation, Geoinformation and Cartography, Vienna University of Technology, Wien, Austria, abdalla@geoinfo.tuwien.ac.at

K. J. Bennett Department of Earth and Planetary Sciences, Washington University in St. Louis, St. Louis, MO, USA, bennett@wunder.wustl.edu

Thijs Brentjens Geonovum, Amersfoort, The Netherlands, t.brentjens@geonovum.nl

Otakar Cerba Section of Geomatics, Department of Mathematics, University of West Bohemia in Pilsen, Plzen, Czechia, ota.cerba@seznam.cz

Jachym Cepicky Section of Geomatics, Department of Mathematics, University of West Bohemia in Pilsen, Plzen, Czechia

Kenneth Field ESRI Inc, Redlands, CA, USA, j.field@kingston.ac.uk

Georg Gartner Research Group Cartography, Department of Geoinformation and Cartography, Vienna University of Technology, Vienna, Austria, georg.gartner@tuwien.ac.at

Edward Mac Gillavry Webmapper, Haarlem, The Netherlands, edward@webmapper.net

Mátyás Gede Department of Cartography and Geoinformatics, Eötvös Lornd University, Budapest, Hungary, saman@map.elte.hu

Shunfu Hu Department of Geography, Southern Illinois University, Edwardsville, IL, USA, shu@siue.edu

Haosheng Huang Research Group Cartography, Department of Geoinformation and Cartography, Vienna University of Technology, Vienna, Austria, haosheng.huang@tuwien.ac.at

Lorenz Hurni Institute of Cartography, ETH Zurich, Zurich, Switzerland, hurni@karto.baug.ethz.ch

Bernhard Jenny Department of Geosciences, Oregon State University, Corvallis, OR, USA, jenny@karto.baug.ethz.ch

Timothy Kennedy Wisconsin State Cartographer's Office, University of Wisconsin-Madison, Madison, WI, USA, ttkennedy@wisc.edu

Pyry Kettunen Department of Geoinformatics and Cartography, Finnish Geodetic Institute, Masala, Finland, Pyry.Kettunen@fgi.fi

Barend Köbben Faculty of Geo-Information Science and Earth Observation, ITC – University of Twente, Enschede, The Netherlands, kobben@itc.nl

Christophe Lienert Institute of Cartography, ETH Zurich, Zurich, Switzerland, lienert@karto.baug.ethz.ch

James O'Brien Kingston University London, Centre for GIS, London, United Kingdom, j.obrien@kingston.ac.uk

Doug Paziak Private Cartographic Contractor, 7528 Pinkney Street, Omaha, NE 68134, USA, dpaziak@hotmail.com

Manuela Schmidt Institute of Geoinformation and Cartography, Vienna University of Technology, Wien, Austria, manuela.schmidt@tuwien.ac.at

Olaf Schnabel Department for City Planning, Zurich, Switzerland, olaf.schnabel@zuerich.ch

D.M. Scholes Department of Earth and Planetary Sciences, Washington University in St. Louis, St. Louis, MO, USA, scholes@wunder.wustl.edu

Emmanuel Stefanakis Department of Geodesy and Geomatics Engineering, University of New Brunswick, Fredericton, NB, Canada, estef@unb.ca

Georgianna Strode Florida Resources and Environmental Analysis Center (FREAC), Florida State University (FSU), Tallahassee, FL, USA, GStrode@admin.fsu.edu

Contributors

Haico van der Vegt Kadaster, Apeldoorn, The Netherlands, Haico.Vegt@kadaster.nl

Howard Veregin Wisconsin State Cartographer, University of Wisconsin-Madison, Madison, WI, USA, veregin@wisc.edu

J. Wang Department of Earth and Planetary Sciences, Washington University in St. Louis, St. Louis, MO, USA, wang@wunder.wustl.edu

Paul Weiser Institute of Geoinformation and Cartography, Vienna University of Technology, Vienna, Austria, paul.weiser@tuwien.ac.at

Part I
Background

Chapter 1
Online Mapping with APIs

Michael P. Peterson

Abstract Bringing maps to users has been made much easier with the World Wide Web. Millions of maps now make their way through a world-wide network of computers. A major change occurred in 2005 in how those maps were delivered when Google Maps implemented a tile-based mapping system based on AJAX that facilitated interactive zooming and panning. The following year, an Application Programmer Interface (API) was released that gave programmers access to the underlying mapping functions. It was now possible to place data on top of the Google base map and make this map available to anyone. This system was created at tremendous expense. It is calculated that the number of tiles required at 20 zoom levels is nearly 1.5 trillion. At 15 KB per tile, this equates to 20 Petabytes or 20,480 TB and a data storage cost of between US $2 million and US $2 billion per data center. This expenditure indicates the level of importance that online companies place on maps. It also represents a shift in how maps of all kinds are delivered to users. Mobile devices are a further indication of this change in map delivery.

1.1 Introduction

This book is about new approaches for online mapping, a form of map presentation that can trace its origins to the introduction of the graphical World Wide Web in 1993. The Web drastically expanded the use of the Internet for the distribution of maps. Apps on mobile devices have since become a primary way that maps are delivered to users.

Since the introduction of Google Maps in 2005, online mapping has been defined by Application Programming Interfaces (APIs). These online software libraries provide the means to acquire, manipulate and display information from a variety

---

M.P. Peterson (✉)
Department of Geography/Geology, University of Nebraska at Omaha, Omaha, NE 68182, USA
e-mail: mpeterson@unomaha.edu

of sources. Although APIs are used for many different types of applications, the creation of maps is one of the major uses. The relative ease of overlaying all types of information with online mapping APIs has further transformed cartography from a passive to an active enterprise.

APIs are the basis of map mashups. The term mashup was first used for a movement in pop music that involved the digital mixing of songs from different artists and genres. In technology, the term is used for a melding of web resources and information. A mashup combines tools and data from multiple online sources. The most common mashup application is the mapping of data.

Mashups are an integral part of what is commonly referred to as Web 2.0. Beginning about 2004, the term Web 2.0 began to be used for a variety of innovative resources, and ways of interacting with, or combining web content. In addition to mashups, Web 2.0 also includes wikis, such as Wikipedia, blog pages, podcasts, RSS feeds, and AJAX. Social networking sites like MySpace and Facebook are also seen as Web 2.0 applications.

The advantage of using a major online mapping site is that the maps represent a common and recognizable representation of the world – a base map. Overlaying features on top of these maps provides a frame of reference for the map user. A particular advantage for thematic mapping is the ability to spatially reference thematic data. In the past, thematic maps have limited the display of spatial reference information such as cities and transportation networks partly to emphasize the distribution being mapped. The inclusion of these features provides valuable locational information to the thematic map user.

This chapter provides an overview to online mapping with APIs, and an overview of this volume.

1.2 The Online Base Map

Google Maps changed the online mapping landscape. Known for its search engine, Google effectively added a map-based search through Google Maps. In the process, they found a more effective way to indirectly make money from online maps by charging businesses to be found. In addition, by not including ads around the map, like MapQuest, they left more room for the map on the computer screen. More importantly, from a map user's perspective, Google Maps changed the way we interact with maps.

The delivery of a Google Map is based on image tiling. This technique had been used since the early days of the World Wide Web to speed the delivery of graphics. In comparison to text, images require more storage and therefore take longer to download. A solution is to divide the image into smaller segments, or tiles, and send each tile individually through the Internet. These smaller files often travel faster because each can take a different route to the destination computer. The tiles are reassembled on the receiving end in their proper location on the web page. With a moderately fast Internet connection, all of this occurs so quickly that the user rarely

1 Online Mapping with APIs　　　　　　　　　　　　　　　　　　　　　　5

Fig. 1.1 Individual map tiles from Google Map at six different levels of detail (zoom levels). In 2005, Google introduced a tiling system to deliver online maps. Over a trillion tiles are used for Google's 20 zoom levels

notices that the image is actually composed of square pieces. With slower connections, the individual tiles are clearly evident.

Figure 1.1 depicts a series of map tiles at different levels of detail (LOD). All tiles are 256 × 256 pixels and require about 15 KB a piece to store in the PNG format. Table 2.1 shows the number of tiles that are used in a tile-based mapping system for 20 levels of detail (LOD), or zoom levels, and the associated storage requirements and estimated storage costs. With 20 LODs, approximately one trillion tiles are needed for the whole world. At an average of 15 KB per tile, the total amount of memory required is 20 Petabytes, or 20,480 Terabytes. No single computer currently has this much storage capacity.

The cost of storing this much data has not been made public by Google or any other company. It is estimated in Table 1.1 based on a cost of about US $100 per Terabyte, the cost of a hard-drive in 2011 that does not include the housing or computer connection. As can be seen from Table 1.1, storing the entire one trillion tiles on disk drives would be about US $2 million ($100 × 20,480 TB). This assumes that all of the tiles are pre-made and stored. It is likely that many of the less popular tiles are 'made-on-the-fly' when they are requested.

In order to achieve faster response times, there is strong indication that data centers use faster random-access memory (RAM) to cache the most popular map tiles. At the current US $30 for 1 GB of RAM, storing the entire map of the world

Table 1.1 The number of tiles, storage requirements, and storage costs used by a tile-based online mapping system to represent the world at different levels of detail (LOD) or zoom levels

Levels of detail (LOD)	Number of tiles	Ground distance per pixel in meters	Storage requirements at 15 KB per tile	Disk storage costs at US $100 per Terabyte	RAM storage costs at US $30 per Gigabyte
1	4	78,272	60 Kilobytes (KB)	$0.000006	$0.002
2	16	39,136	240 KB	$0.00002	$0.007
3	64	19,568	968 KB	$0.0001	$0.03
4	256	9,784	3.75 Megabytes	$0.0004	$0.11
5	1,024	4,892	15 MB	$0.001	$0.44
6	4,096	2,446	60 MB	$0.006	$1.76
7	16,384	1,223	240 MB	$0.02	$7.03
8	65,536	611.50	960 MB	$0.09	$28.13
9	262,144	305.75	3.75 Gigabytes (GB)	$0.37	$112.50
10	1,048,576	152.88	15 GB	$1.46	$450.00
11	4,194,304	76.44	60 GB	$5.86	$1,800.00
12	16,777,216	38.22	240 GB	$23.44	$7,200.00
13	67,108,864	19.11	968 GB	$93.75	$28,800.00
14	268,435,456	9.55	3.75 Terabytes (TB)	$375	$115,200.00
15	1,073,741,824	4.78	15 TB	$1,500	$460,800.00
16	4,294,967,296	2.39	60 TB	$6,000	$1,843,200.00
17	17,179,869,184	1.19	240 TB	$24,000	$7,372,800.00
18	68,719,476,736	0.60	960 TB	$96,000	$29,491,200
19	274,877,906,944	0.30	3.75 Petabytes (PB)	$384,000	$117,964,800
20	1,099,511,627,776	0.15	15 PB	$1,536,000	$471,859,200
Total	**1,466,015,503,700**		**20,480 TB or 20 PB**	**$2,048,000**	**$629,145,600**

would be more than US $629 million (see Table 1.1). If all tiles are stored on either a disk drive or in RAM, we could estimate that the cost of map storage at each data center would be somewhere between $2 and $629 million. Google maintains more than 30 data centers. A still faster storage option would be to use a graphical processing unit (GPU). These devices are specifically designed to store and manipulate images and transfer image data much faster than computer memory. Map storage on GPUs would be at least twice as expensive as RAM, or about $1.3 billion for a map of the world at 20 levels of detail.

These data storage requirements and costs are only for a single map. The satellite view, with tiles in the JPEG format, requires approximately the same amount of storage space. Other maps provided by Google are the Terrain view (offered at only 15 levels of detail) and the bicycle map (12 larger scale levels of detail). All other maps are transparent overlays. Combining all of these data storage costs – perhaps as much as $2 billion, provides some indication of the importance placed on maps by Google and other companies.

1.3 Mapping APIs

Introduced in 2005, shortly after Google Maps, the Google Map API consists of a series of map-related functions (Google Maps JavaScript API V3 Basics 2011). These functions control the appearance of the map, including the scale, position, and any added information in the form of points, lines or areas. The purpose of the API is to make it possible to incorporate user-defined maps on websites, and to overlay information from other sources. The use of the Google Maps API is free, provided the site does not charge for access and does not generate more than 25,000 maps a day. Designed for business applications, a pay version of Google maps, called Google Maps API Premier, provides some additional functions dealing with geocoding and usage tracking.

Soon after the introduction of Google Maps in 2005, Microsoft, Yahoo!, and MapQuest changed their online mapping service to incorporate an AJAX-type interface. By 2006, Yahoo! had released its own API. The Yahoo! Maps API is much the same as Google's implementation but does not support polygons and still requires the use of an electronic key. While the key is made freely available, it limits the use of the API to the server that is specified when the key is requested. Other online map providers include OpenStreetMap, ESRI, and Nokia (OviMap).

In mid-2009, Microsoft re-labeled its Live Local web mapping service to Bing Maps, a part of the company's search engine services. Bing Maps includes a street map, an aerial view, Bird's-Eye view, StreetSide view, and 3D Maps. The oblique Bird's Eye view has more detail than Google's satellite view. In contrast to the Yahoo! Maps API, Bing Maps does support polygons. Most other online map providers include an API.

The development of APIs is still in an early stage and is progressing in a haphazard manner. While very similar, the function calls used by the major providers have slight differences and it is time-consuming to re-write the code for each. A standard set of functions has been developed that works with many online mapping systems. The open source Mapstraction API makes it possible to easily switch between each of the mapping APIs but implements only the common functions (Duvander 2010).

Google is still leading the development of mapping APIs with regular additions of new functions. The current Google Maps v3 was introduced specifically to meet the needs of online mapping through mobile devices. It reduces the amount of data communications overhead, thus increasing the speed of map display.

1.4 Behind the Online Base Map

While online maps can be based on any type of server they are usually associated with data centers – specialized buildings, usually without windows, that house a large number of computers. Figure 1.2 depicts the Google data center in Council

Fig. 1.2 Google data center in Council Bluffs, Iowa. Power generators, pictured on the *right*, are located behind the windowless main building

Bluffs, Iowa, and part of the associated power facilities behind the main building. Diesel generators are used to make certain that electricity is available in case of a power failure. A lead-acid battery back-up system is in place to power the computers until the generators are running. To reduce power demands, not all services are maintained during power outages. It has been reported that, in the case of a natural disaster such as an earthquake, Google data centers in California have contingencies to acquire diesel fuel by helicopter.

Power is a major concern for a data center. Each is estimated to use 10 MW of electricity, requiring about ten large diesel generators. Google has calculated the amount of energy used for each search done through its search engine. They estimate that each search requires 0.0003 kWh. In terms of greenhouse gases, 1,000 search requests generates the equivalent CO_2 of a car driven 1 km (0.61 miles) (search: *Powering a Google Search*). Partly to reduce costs and greenhouse emissions, companies operating data centers have invested in renewable energy.

The major innovation introduced by Google Maps is the incorporation of Asynchronous JavaScript and XML (*AJAX*) into the relationship between the server and client. This was the culmination of many years of effort to re-shape interaction through the Internet. Essentially, AJAX maintains a continuous connection with the server – exchanging small messages in the background even when the user has not made a specific request (Garrett 2005). This leads to faster server responses when the user does make a request. AJAX might be thought of as an application that works in the background of a browser to anticipate what the user might want and be ready to communicate with the server to respond to a request. Operations in Google Maps that are particularly assisted by AJAX include zooming and panning, the most common form of interaction with maps.

AJAX is not a programming language in itself but a term that refers to the combined use of a group of different technologies. The technique uses a mix of HTML, Cascading Style Sheets (CSS), Document Object Model (DOM), and the eXtensible Markup Language (XML). These are all freely available technologies. Asynchronous communication is used to exchange data with the server while the user is idle so that the entire web page does not need to be reloaded each time the

1 Online Mapping with APIs

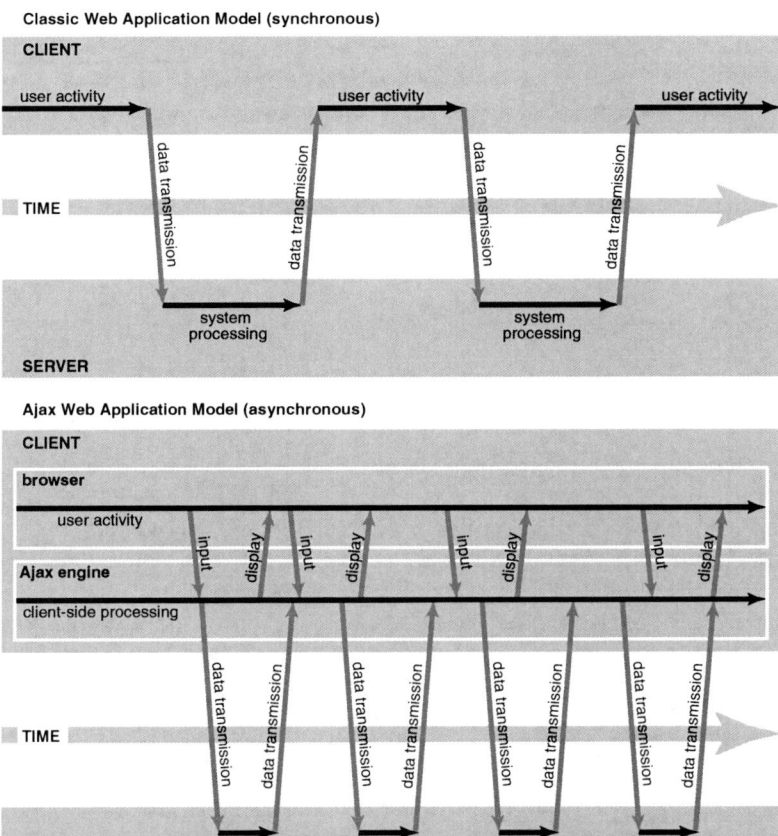

Fig. 1.3 The typical client–server communication is synchronous (*top illustration*). AJAX uses asynchronous communication between the client and the server. A connection is maintained to the server to speed interaction

user makes a change (see Fig. 1.3). The result is increased interactivity, speed, and an improved user interface.

AJAX eliminates the usual start-stop-start-stop type of interaction. When the map is scrolled, additional map tiles are automatically downloaded. The tiles are added almost instantly because a connection is maintained to the server so that additional tiles can be quickly loaded. As the user scrolls, more of the map or satellite image is downloaded from the server without the user specifically asking for the additional tiles.

Asynchronous communication is made possible by the AJAX engine, JavaScript code that resides between the user and the server. Instead of loading the webpage at the start of a web session, the AJAX engine is initially loaded in the background.

Once loaded, the XMLHttpRequest object begins its work. This JavaScript code downloads data from the server without refreshing the web page. A user action that normally would generate an HTTP request to the server becomes instead a JavaScript call to the AJAX engine. If the engine can respond to a user action, no response from the server is required. If the AJAX engine needs something from the server in order to respond to a user request – such as retrieving new data – the engine makes the request without interrupting the user's interaction with the application. AJAX has effectively transformed the online client/server experience.

1.5 Web Mapping Services

A considerable amount of geographic information has been placed into GIS databases since these systems came into widespread use in the 1980s. In order for this information to be useful to more people, a method was needed for "pulling" the information from the database. In 1999, the Open Geospatial Consortium defined a set of standards for distributing geographic data (OGC 2011). The purpose was to both facilitate the distribution of data and make layers of information easily available to Internet users. A series of standardized services were defined to supply geodata to any platform connected to the Internet. With this standard method of data access, a web mapping service is able to interact with and display maps through an Internet-based interface.

Initially, extracting information to a GIS database required interacting with a large and complicated database. The OGC streamlined the process by placing the burden for extracting data on the server. As defined by OGC, the web mapping service consists of two functions: (1) GetCapablites that defines the capabilities of the server such as the supported file formats, the available map layers, and the method of display; and (2) GetMap that tells the database what is needed. The database reads the request and creates the map-based data based on the requirements laid out by GetCapabilities. The data is then sent to the web mapping service.

Most web mapping services support a handful of other functions. For example, "GetFeatureInfo" sends specific information about locations on the map, such as the name of the road or the height of a location. "GetLegendGraphic" function deals with the symbols used on the map.

The OGC standard leads to the definition of a variety of services, including:

Web Map Service (WMS) – georeferenced map images typically in the form of raster tiles (PNG, GIF, or JPG), but they can also be in a vector format. Requests are made using a standard web URL address.

Web Coverage Service (WCS) – a geographical area that can be overlaid on a map but cannot be edited or analyzed. WCS is used to transfer coverages that consist of objects such as data points, pixels, or paths defined with vectors.

1   Online Mapping with APIs   11

Web Feature Service (WFS) – allows the request for geographical features, essentially the information behind the map. WFS web service allows features to be queried, updated, created, or deleted by the client. The data is usually provided in an XML format like GML.

The open source GeoServer application is the reference implementation of a server for the WMS, WFS, and WCS standards.

1.6 Mobile Mapping

Mobile devices have provided a new, portable medium for maps. Screens on cell phones have become larger, positioning technologies have improved, and all types of location-aware applications have been developed. Location Based Systems (LBS) emerged with the overall goal of providing information specific to the current location of the mobile user.

While the Internet and mobile phones developed during the same time period, there were major differences in how they developed. For example, the Internet originated through a government program while the mobile phone network was largely created by private interests. There was very little government involvement in building the mobile phone network. In Europe, governments mandated that mobile phone companies share cell phone towers. The more laissez-faire approach in the United States resulted in every company building their own towers. For a country as large in area as the US, this contributed to significantly greater costs for infrastructure development. These costs were passed to users, increasing subscription prices and slowing adoption.

Despite the added expense of owning and using mobile phones, the number of mobile phone subscribers quickly surpassed the number of Internet users. The Computer Industry Almanac (2005) reported in 2005 that the worldwide number of cellular subscribers surpassed 2 billion – exactly twice as many Internet users at that time and up from only 11 million in 1990 and 750 million in 2000. The use of cell phones expanded rapidly during the first decade of the twenty-first century with 4.6 billion subscribers by 2010.

While the US Federal Communications Agency maintained a laissez-faire relationship with the cell phone industry, it did mandate automatic location identification (ALI) on mobile phones. ALI stipulated positioning within 100 m or less to ensure that emergency workers could find cell phone callers. Wireless carriers were required to have 95% ALI-capable handsets among their subscriber bases by Dec. 31, 2005. The ALI mandate was the main impetus for the growth of location-aware cell phones, at least in the US (GPS World 2007).

In order to comply with ALI, most carriers initially decided to integrate GPS technology into cell phone handsets rather than overhaul the tower network used to triangulate the position of mobile phones. GPS does not work inside of buildings and is power hungry, quickly draining mobile phone batteries. For these reasons, cell towers were eventually upgraded to support position finding. Of the 3.3 billion cell phones in use in 2008, only 175 million had GPS (Bray 2008).

Initially, a location-aware mobile phone would only determine its location if an emergency call was made. It was not possible to get direct access to location data. Later, location data was provided continuously as an aid to navigation, transforming mobile phones into personal navigators. Many people now access maps primarily through their mobile devices.

1.7 Summary

The online map is a new entity. The first map to be incorporated within a web page was displayed by Mosaic web browser in 1993. Initially, scanned paper maps predominated as online maps. Database driven maps appeared in the latter part of the 1990s, along with the growth of data centers. Mobile devices began to be used extensively for map delivery beginning with the introduction of Apple's iPhone in 2007.

With all of these changes in the way maps are delivered to users, it is appropriate to examine the process by which this is done, the various applications, and how these maps can be improved. This is the overall purpose of this book. A culmination of many years of work by the Maps and the Internet commission of the International Cartographic Association, authors are from Austria, Czechia, Finland, Hungary, the Netherlands, Switzerland, United Kingdom, and the United States. The first part of the book examines the background of the online map. The second looks specifically at mash-ups. The third part examines methods of symbolization, and the last part examines applications.

References

Bray H (2008) GPS turns cell phones into powerful navigators. Boston Globe, April 17
Duvander, Adam (2010) Map Scripting 101: An example-driven guide to building interactive maps with Bing, Yahoo!, and Google Maps. San Francisco: No Starch Press.
Computer Industry Almanac (2005) China tops cellular subscriber top 15 ranking. http://www.c-i-a.com/pr0905.htm
Garrett JJ (2005) Ajax: a new approach to web applications. AdaptivePath.com, Feb. 18. http://www.adaptivepath.com/ideas/essays/archives/000385.php. Accessed 19 June 2008
Google Maps JavaScript API V3 Basics (2011) (search: *Google Maps Javascript API V3 Basics*)
GPS World (2007) FCC to require full E911 adherence by 2012. GPS World, 12 Sep 2007
Open Geospatial Consortium (2011) OGC standards and specifications. http://www.opengeospatial.org/standards

Chapter 2
Web Mapping Services: Development and Trends

Manuela Schmidt and Paul Weiser

Abstract Web mapping services like Google, introduced in 2005 have altered the online mapping experience. Not only could maps be viewed in a fast and simple way but there was also the possibility to create Mashups through APIs, leading some to proclaim the "democratization of mapping". Addressed here is the development of these mapping services, how they impacted the existing Web mapping environment and possible future areas of development. An emphasis is placed on the technical developments from desktop to mobile applications, as well as the development of base maps and map types from pre-rendered tiles to editable map styles in different viewing modes from bird eye view, 3D, and augmented reality. While the first maps produced with APIs were mostly static point maps, new features have enabled dynamic and interactive applications with "GIS-like" functionalities, often supported by third party implementations.

2.1 Introduction

In 2006, a year after the appearance of Google Map Mashups, the free software developer and activist Erle Schuyler summarized the state of map APIs as follows:

> At present, all that these map APIs offer is ultimately a way to put points on a map – what we've [...] referred to as "red dot fever". [...] Where is the broader palette for telling new and different stories on the Web with maps? Where is the bi-directionality, the interactivity, the Wiki nature? (Schuyler 2006)

Before proceeding, it is good to ask whether much has changed since Schuyler's assessment.

M. Schmidt (✉)
Institute of Geoinformation and Cartography, Vienna University of Technology,
1040 Vienna, Austria
e-mail: manuela.schmidt@tuwien.ac.at

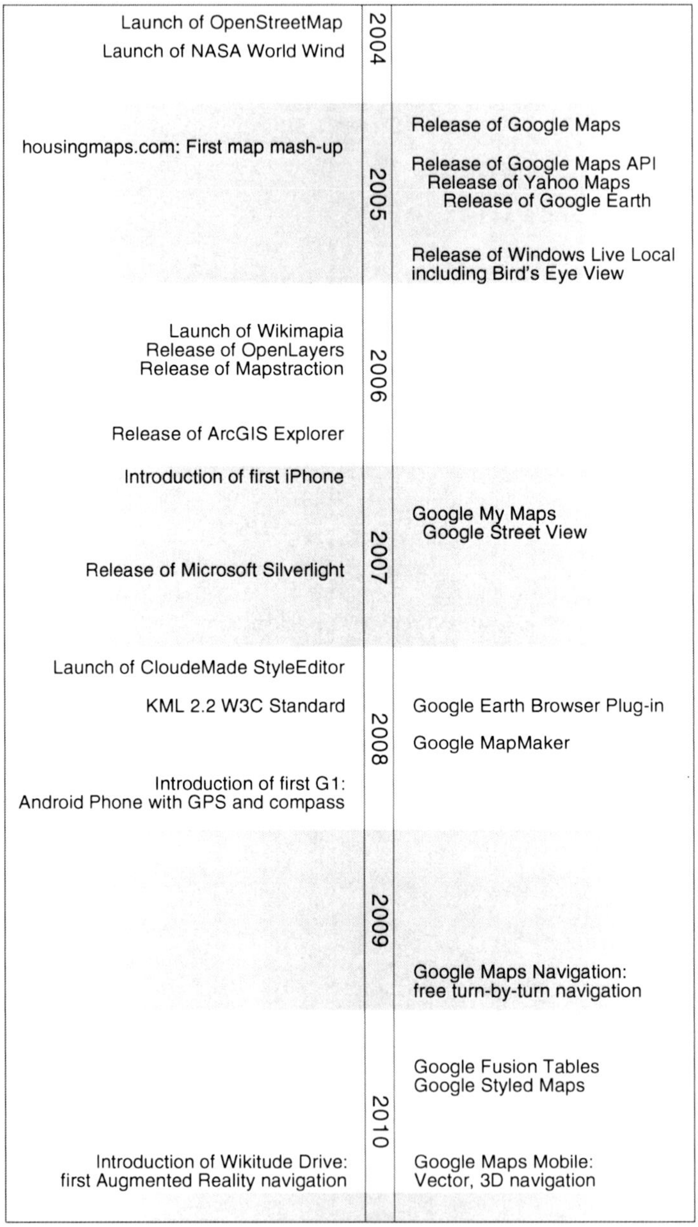

2004	Launch of OpenStreetMap Launch of NASA World Wind	
2005	housingmaps.com: First map mash-up	Release of Google Maps Release of Google Maps API Release of Yahoo Maps Release of Google Earth Release of Windows Live Local including Bird's Eye View
2006	Launch of Wikimapia Release of OpenLayers Release of Mapstraction Release of ArcGIS Explorer Introduction of first iPhone	
2007	Release of Microsoft Silverlight	Google My Maps Google Street View
2008	Launch of CloudeMade StyleEditor KML 2.2 W3C Standard Introduction of first G1: Android Phone with GPS and compass	Google Earth Browser Plug-in Google MapMaker
2009		Google Maps Navigation: free turn-by-turn navigation
2010	Introduction of Wikitude Drive: first Augmented Reality navigation	Google Fusion Tables Google Styled Maps Google Maps Mobile: Vector, 3D navigation

Fig. 2.1 Development of web mapping services depicted by some exemplary services and tools

One way to answer this question is to explore the stages of development of Web mapping services. Figure 2.1 shows a time-line depicting the release of important tools of Web mapping services on the one side and the introduction of related tools, projects, and products on the other. The following section will focus on cartographic aspects by discussing the different map and content types as well as map styles. Section 2.3 gives an overview on advances in API technology by highlighting the aspects of widespread usage, mobile usage and expert usage of Web mapping services. The final chapter gives a short summary and discussion.

2.2 Development of Online Maps

When Google published its Web mapping service in 2005, it was the first free service providing a global coverage of satellite map views (Purvis et al., 2006). Other companies followed offering satellite as well as road map views. They usually also provided a hybrid view, i.e., the combination of a road network and satellite views. Most road network data, however, was restricted to areas covered by commercial data providers. We argue that maps have changed considerably since that. The following paragraphs give an overview of the new types of maps and new approaches to map content and styles.

2.2.1 Map Types

Parallel to the launch of 2D web mapping services, 3D desktop applications like NASA World Wind and Google Earth were introduced. Shortly after, Microsoft integrated the 3D terrain view in the browser, at some places complemented with 3D buildings; however, a proprietary plug-in was needed to access this version. In 2008 also Google published a plug-in offering Google Earth's 3D capability in a browser (Google 2008a). In addition to 3D and different aerial views, another street map view created considerable controversy: panoramic, street-level imagery called "Street View" (Google) or "StreetSide View" (Bing).

An early innovation in addition to the basic map, satellite, hybrid view, was the "Bird's Eye View" integrated in Windows Live Local (now: Bing Maps) (CNet 2005), not only giving top-down aerial views, but images taken at an oblique, 45° angle, allowing for a view on the front and back sides of buildings. The drawback of this feature, however, is that only areas with a high population are covered.

Concerning the map itself, an improvement to the standard map view was the Google Maps "Terrain" view introduced in 2007 (Google 2007b), that displays physical features, i.e., shaded relief representation.

Most map services allow users to integrate new custom map types. This requires pre-rendering of the host titles and their storage on a server. Third party tools and open source scripts appeared that support the rendering of tiles from different data sources and their hosting on a cloud-server. Figure 2.2 compares the map type choices of Google Maps in 2005 to those in 2011. However, some of the options do not refer to map types, but to image or real-time data overlays, such as "Photos" or "Traffic".

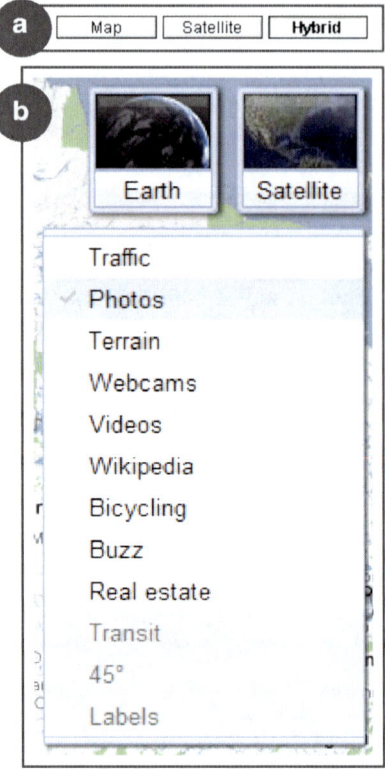

Fig. 2.2 Simple map type control in the first Google Maps release in 2005 for map, satellite and hybrid view (**a**). Extended map type control of Google Maps in January 2011 (**b**). Map types include real-time data overlays such as traffic

2.2.2 Map Content

Base maps are based on a mixture of public data commercial data. The two major data provider are NavTeq and TeleAtlas. However, these data are costly, quickly outdated and restricted to specific areas covered by the data acquiring companies. Large companies have invested large sums of money to purchase smaller companies to acquire their data (e.g., Nokia acquiring NavTeq (Nokia 2007), Microsoft acquiring the Imagery and Remote Sensing Company Vexcel (Microsoft 2006)). For an easier and cheaper data acquisition, Google introduced a tool called Map Maker in 2008, that enabled users to contribute data themselves (Google 2008b). This tool was only available for areas with no or little commercial data coverage, e.g., India, Pakistan, Iceland. Within a short time, large areas were mapped in this crowd-sourcing manner.

This user-generated approach is also used in a project called OpenStreetMap that tries to build a free map database of the world. Until now, many Web mapping service providers shied away from using OpenStreetMap data, because of unclear license terms and a claimed lack of quality assurance. However, studies quality of OpenStreetMap data in comparison to commercial data vendors (e.g., Haklay 2010;

Neis et al., 2010) show little difference. Applications using OSM data have proved successful with an increasing competition in the online map market. Big players like MapQuest (2010) and Microsoft (Bing 2010) are starting to use OpenStreetMap as a source for their base maps. In additions, they also offer tools to contribute data back to the OpenStreetMap project, showing the increasing relevance of open and crowd-sourced data also for commercial purposes.

Point and feature overlays also gained importance in web mapping services. Content overlays can be differentiated into those that can be displayed on request (as shown in Fig. 2.2) and content that is included in the base map, e.g., POIs, companies, restaurants, etc. For Google Maps this content is managed in Google Places. Other content overlays originate from sensors (e.g., traffic, webcams, weather), user-generated content platforms (e.g., photos, Wikipedia, video) or business directories (e.g., real estate). Custom content overlays are created through APIs (see Sect. 2.3).

2.2.3 Map Styles

One of the key elements that made web mapping services successful was their pre-rendered raster tiles (Smith 2008). While this ensured fast loading, it had the disadvantage of lacking flexibility in terms of style and content. In the beginning, the base maps always looked the same. A small change to that paradigm was discovered by developers who used hard-coded filters to colorize, de-saturate or blur map tiles. For more than optical modifications, custom map tiles had to be used. As described in Sect. 2.1 this usually implied that the user had to own the geo data, be able to configure and render the data and host the rendered tiles on their own servers.

In 2009 the tool "StyleEditor" by CloudMade (2009) was introduced that allowed the user to style the map. Consisting of a browser application with a GIS-like interface, the user can create individual styles by selecting map features from the OpenStreetMap database and assigning visibility and design. The resulting individualized map styles can be used in combination with a Cloudmade API or downloaded as raster tiles for use in combination with other mapping APIs.

It was expected that such a custom styling tool would be freely available for open geo data, as it accesses to underlying vector data. Interestingly, in late 2010, Google also published a similar tool called "Styled Maps" within their Google Maps API v3. With a so-called "Wizard", users could select feature types and determine the style in terms of visibility, hue, lightness and saturation. While this tool is still an advanced and rarely used option, it enables the programmer to not only change the look-and-feel of the base map, but also its features. Nevertheless, both mentioned tools are still restricted to fairly simple style changes within their range of geometric features.

2.3 Advances in API Technology and Features

2.3.1 Mapping for the Masses

The first map mash-up was created in April 2005 by Paul Rademacher, a 3-D graphic artist who combined housing data from Craigslist with the newly available Google Maps. This started a new chapter in Web mapping dominated by so-called "programmer-mapmakers" (Plewe 2007). Web mapping services delivered simple, attractive and intuitive interfaces; APIs allowed for data mash-ups. However, at this time, generating a map mash-up required the technical knowledge of creating a Web site and combining the public service with contributed data, thus limiting the mapmakers to web developers. This has changed, at least with the publication of Google My Maps announced as "Map-making: So easy a caveman could do it" (Google 2007a). Maps could be created with a simple drag and drop interface, allowing anyone to add placemarks, text, photos and videos to simple mash-ups, which could be saved, embedded and forwarded as a KML file. Also, Google Earth enabled users to create and contribute geo content, such as simple 3D buildings, easily. Also, CloudMade and other vendors started providing similar services. Plewe (2007) identified these new players in the mapping market as "consumer-mapmakers", who were enabled by consumer-oriented or citizen-oriented services to share their personal geography with the world, without having to have programming or design skills.

Mapping has become a daily routine for many people. For example, users can map their own location by "checking in" at geo-social networks such as Facebook, Twitter or Foursquare. Even though some of these services don't directly produce maps, they create a geo-footprint that can be queried through an API and visualized using most standard mapping APIs.

2.3.2 Maps Going Mobile

Until recently, a major drawback of web-based maps was their limited mobility compared to paper maps (Peterson 2003). Mapping applications were originally intended for desktop computers and fast local broadband connections. This, however, has changed with various recent developments.

Mobile computing hardware has changed significantly. Processor speeds and memory of mobile devices have increased dramatically and are now capable of rendering large vector data in reasonable time. High-end mobile devices are now on par with somewhat older gaming consoles like Nintendo's Wii (TechAutos 2010). Increased battery capacities and ultra-low voltage processors ensure that even more challenging computational tasks do not drain battery-life. Also, fast mobile broadband connections and cheap rates have made ubiquitous and affordable access to mobile mapping applications affordable.

The most important feature, in terms of hardware, is arguably the wide-spread use of reasonable large touchscreens in many mobile devices (Gartner Research 2010). Map interaction using a touchscreen is more natural and intuitive than any other method was. Touchscreen technology offers a direct way of manipulating content and facilitates better hand-eye coordination, thus making it superior over mice or keyboards (Shneiderman 1991).

Recent hardware developments have laid the foundation for the release of mapping APIs tailored to mobile devices offering new features such as car or pedestrian navigation, off-line use of maps, and various location based services. Modularization and optimization of API code enable mobile mapping application that, in terms of speed, can compare to common desktop APIs (Fox 2009).

With many people owning smart phones with a constant connection to the Internet, mobile mapping has become a ubiquitous tool for the masses. One of the most important mapping features in this context is navigation support. One important application Google Maps Navigation, introduced in 2009, was the first free application offering GPS turn-by-turn navigation, including live traffic data and search along the route. While this and all other versions of Google Maps before relied on a strong Internet connection, a new version of Google Maps for Mobile (Google 2010) changed that paradigm by offering vector maps, that were available offline and even allowed for offline rerouting. Applications like these will be strong competitors to traditional navigation systems, offered by Garmin, Nokia, and others.

Another application worth mentioning in this context is Wikitude Drive (2010), which combines the video captured through the phones' camera with driving instructions. In the future, more applications are expected to creatively extend the scope of Augmented Reality in combination with mobile mapping.

2.3.3 Mash-Ups for Experts

The number of features and services available through mapping APIs has increased considerably. While static point, polyline, or polygon overlays have become standard use, more and more dynamic information layers are added to the API's portfolios. Examples are traffic layers, showing real-time traffic information in urban areas, or bicycle layers, providing bike routes and overlays specific to bicycling usage. Other advanced capabilities include:

(1) Directions can be requested for different travel modes. (2) Support for elevation requests. (3) Geocoding, reverse geocoding and direction requests are usually restricted to a certain number of requests per day; otherwise a commercial license needs to be acquired. (4) Newer API versions also allow for customized base maps and for integrating custom map types, that can even use other projections than Mercator's, as long as they are rectilinear. (5) "Google Fusion Tables" allow for the online gathering, sharing and visualization of data tables. Even though this tool is still in development, it seems like a promising development for the

organization and visualization of thematic data. (6) The integration of third party tools has become easier and it is now possible to query databases for places, people, or businesses and display the results on the map.

With mapping tools performing tasks like geocoding, base map customization and data analysis, mash-ups are clearly moving towards specialized and expert GIS-like applications.

2.4 Summary and Discussion

The major mapping API providers continue to maintain and extend their mapping portfolio. However, some of the services have not survived. The existing map services have attracted millions of users – both desktop and mobile. The scope of web mapping applications has widened from easy-to-use consumer-oriented tools to highly specialized applications with GIS functionalities. We see new companies as well as old players trying to keep pace by using new technological concepts, different licenses and revised business models.

From a cartographic perspective, we can expect interesting developments from tools like the above mentioned. Map mash-ups until now assemble data on top of base maps, which are usually pre-designed by professional cartographers or designers. Tools like "Styled Maps" allow everybody to design base maps. On the one hand, this will enable unique and exciting new mash-ups – on the other hand, this might lead to bad and unreadable maps in some cases. Cartographic tools like ColorBrewer (www.colorbrewer2.org, Brewer 2010), which supports (non-expert) map makers to select suitable color schemes for maps, will be increasingly important.

As for Schuyler's last question: Despite the advances of web mapping within the last few years, there is still a lot of potential to collaborate, to elaborate, to tell stories with creative new methods and to use the data in exciting and useful new ways.

References

Bing (2010) Bing engages open maps community. Online: http://www.bing.com/community/site_blogs/b/maps/archive/2010/11/23/bing-engages-open-maps-community.aspx. Accessed 24 Jan 2011
Brewer CA (2010) http://www.ColorBrewer.org. Accessed 24 Jan 2011
CloudMade (2009) Cloudmade releases style editor. Online: http://blog.cloudmade.com/2009/07/16/cloudmade-releases-style-editor-11/. Accessed 24 Jan 2011
CNet (2005) Microsoft offers a new angle on maps. Online: http://news.cnet.com/Microsoft-offers-a-new-angle-on-maps/2100-1032_3-5986057.html. Accessed 24 Jan 2011
Fox P (2009) Google I/O 2009 – Performance tips for Geo API Mashups. Video: http://www.youtube.com/watch? v = zI8at1EmJjA&feature = player_embedded. Accessed 24 Jan 2011

Gartner Research (2010) Gartner says touchscreen mobile device sales will grow 97 percent in 2010. Online: http://www.gartner.com/it/page.jsp?id=1313415. Accessed 24 Jan 2011

Google (2007a) Map-making: so easy a caveman could do it. Online: http://googleblog.blogspot.com/2007/04/map-making-so-easy-caveman-could-do-it.html. Accessed 24 Jan 2011

Google (2007b) http://googlemapsapi.blogspot.com/2007/12/v294-terrain-and-new-maptypecontrol-to.html. Accessed 24 Jan 2011

Google (2008a) http://google-latlong.blogspot.com/2008/05/google-earth-meet-browser.html. Accessed 24 Jan 2011

Google (2008b) http://groups.google.com/group/google-map-maker/browse_thread/thread/7ba81462f965c1dd. Accessed 24 Jan 2011

Google (2010) http://google-latlong.blogspot.com/2010/12/next-generation-of-mobile-maps.html. Accessed 24 Jan 2011

Haklay M (2010) How good is volunteered geographical information? A comparative study of openstreetmap and ordnance survey datasets. Environ Plann B 37(4):682–703

MapQuest (2010) http://blog.mapquest.com/2010/07/09/mapquest-opens-up/. Accessed 24 Jan 2011

Microsoft (2006) http://www.microsoft.com/presspass/press/2006/may06/05-04VexcelPR.mspx. Accessed 24 Jan 2011

Neis P, Zielstra D, Zipf A, Struck A (2010) Empirische Untersuchungen zur Datenqualität von OpenStreetMap - Erfahrungen aus zwei Jahren Betrieb mehrerer OSM-Online-Dienste. AGIT 2010. Symposium für Angewandte Geoinformatik, Salzburg, Austria

Nokia (2007) http://www.nokia.com/press/press-releases/showpressrelease?newsid=1157198. Retrived Jan 12 2011

Peterson MP (2003) Maps and the Internet. Elsevier, Amsterdam

Plewe B (2007) Web cartography in the United States. Cartogr Geogr Inf Sci 34(2):133–136

Purvis M, Sambells J, Turner C (2006) Beginning google maps applications with PHP and Ajax: from novice to professional. Apress, Berkeley

Schuyler E (2006) Web Map API Roundup. Online: http://mappinghacks.com/2006/04/07/web-map-api-roundup/. Accessed 24 Jan 2011

Shneiderman B (1991) Touch screens now offer compelling uses. IEEE Software 8(2):93–94. doi:10.1109/52.73754, 107

Smith P (2008) Take control of your maps. Online: http://www.alistapart.com/articles/takecontrolofyourmaps. Accessed 24 Jan 2011

TechAutos (2010) Making sense of smartphone processors: the mobile CPU/GPU guide. Online: http://www.techautos.com/2010/03/14/smartphone-processor-guide/. Accessed 12 Jan 2011

Wikitude (2010) Wikitude drive: never take your eyes off the road again. Online: http://www.wikitude.org/en/drive. Accessed 24 Jan 2011

Chapter 3
Current Trends in Vector-Based Internet Mapping: A Technical Review

Christophe Lienert, Bernhard Jenny, Olaf Schnabel, and Lorenz Hurni

Abstract Possibilities and limitations of Internet cartography software largely depend on the pace set by the software industry. The variety of commercial and non-commercial software caters for the needs of a continuously growing mapping community, including both professional and amateur cartographers. This chapter provides an overview of state-of-the-art technologies for vector-based Web-mapping as of the beginning of 2011. Both proprietary and open format technologies are discussed for vector data rendering in browsers, highlighting their advantages and disadvantages. The discussed technologies are Adobe Flash, Microsoft Silverlight, Scalable Vector Graphics (SVG), JavaFX, Canvas, and WebGL. The chapter also discusses client and server side frameworks which provide Application Programming Interfaces (APIs) for creating custom interactive maps, mainly by overlaying raster images with vector data.

3.1 Introduction

Internet maps are the major form of spatial information delivery, as the Internet is today the primary medium for the transmission and dissemination of maps (Peterson 2008). For map authors, the maze of available techniques for creating and distributing Web maps is overwhelming, while authoring tools for Web-maps meeting the demands of high-quality cartography are difficult to find. Map authors may choose between GIS and graphics software products to create maps for the Internet, but these off-the-shelf maps oftentimes fall short of effectively conveying information. There are three main reasons for this shortcoming: (a) the design of these maps sometimes does not take into account the specific limitations of digital displays (Jenny et al., 2008); (b) the maps are often restricted in using standard

---

C. Lienert (✉)
Landscape and Waters, Canton of Aargau, 5001 Aarau, Switzerland
e-mail: christophe.lienert@ag.ch

functionality provided by the authoring software; and (c) they do not take full advantage of interactive features available in modern Web-browsers. Not only is the situation of available products confusing and overwhelming, there are also out-of-date technologies, which are not developed further. Likewise, some of the new technologies are characterized by short life cycles: they have disappeared as fast as they have arrived on the scene.

In comparison to raster-based maps, vector graphics formats offer a series of advantages for interactive mapping: (a) They are scalable without loss of information or graphical artefacts; (b) the symbolization is adjustable on-the-fly (e.g., line width, transparency, fill color); (c) the geometry and symbolization can be animated; (d) map features can be shown and hidden without regenerating and reloading the entire map; (e) attributes can be attached to each individual map feature; (f) map features, such as diagrams, can be generated on-the-fly; and (g) the geometry can be changed, allowing for lossless projection to other coordinate systems (Schnabel and Hurni 2009).

In this chapter, we confine ourselves to the description and assessment of current technologies for vector-based mapping on the Internet. Raster-based Web-mapping is not part of this chapter, and only cross-platform and cross-browser technologies are treated. The chapter refers to the state-of-the-art as of the beginning of 2011 and the discussed technologies relate to the most current browser versions, i.e., Mozilla Firefox 4, Internet Explorer 9, Apple Safari 5, Opera 11, and Google Chrome 8.

3.2 Browser Technology for Vector Data Rendering

Vector mapping is based on vector graphics which use geometrical primitives such as points, lines, curves, or polygons. These primitives, in turn, are all based on mathematical equations. Base technologies for vector mapping may be defined as technology, or software, or even entire application programming interfaces (API), which are capable of creating, editing, and extending such vector-based graphics for the Web. The graphic objects may be changed by editing the geometry information and the graphical attributes. Affine transformation operators allow for stretching, twisting, and rotating the objects.

The most current and established vector technologies for the Web are presented in this section. Both the associated authoring tools and the way vector content is provided to the user are discussed. A distinction in proprietary and open-source software is made since considerable differences exist as to business and development models. An open-source approach allows for the extraction and further modification of vector graphics objects, or even for the technical advancement of authoring and rendering software. Proprietary source code, in contrast, is mostly delivered in a compiled binary form and is therefore non human-readable. Table 3.1 shows an overview of base technologies for vector mapping.

Java Applets and Vector Markup Language (VML) technology are not discussed in this article. Due to their complex programming environment, Java Applets are

3 Current Trends in Vector-Based Internet Mapping: A Technical Review

Table 3.1 Base technologies for Web-based vector mapping

Technology/software	Company/consortium	Authoring tools	Format
Flash/Flex	Adobe	Flash Builder, Flash Professional	Proprietary
XAML/Silverlight	Microsoft	Expression Blend, Visual Studio	Proprietary
SVG	W3C	Illustrator, Corel Draw, Inkscape, XML Editors	Open
JavaFX	Oracle/Sun	NetBeans, Eclipse	Open
Canvas	WHATWG/W3C	–	Open
WebGL	Khronos Group	–	Open

comparatively little used to produce vector-based maps (Byrne et al., 2010). JavaFX is a modern alternative for the Java environment providing similar graphical capabilities, and discussed in this chapter. VML is deemed out-dated since it is a rejected World Wide Web Consortium (W3C) standard and is only supported by Internet Explorer (Zaslavsky 2003).

3.2.1 Proprietary Technology

The business models of software companies producing proprietary technologies and software are, by and large, based on licensing. Customers purchase a number of licenses which have to be renewed annually, or for each update. Usually, not the vector rendering technology itself is licensed, but auxiliary tools for creating the content.

Proprietary technologies and associated authoring tools are geared towards the designer community creating Web-based content, as well as programmers using various frameworks and code libraries for the development of Web applications. The two most widely used proprietary products are Adobe Flash and Microsoft Silverlight. They both provide high-performance authoring tools for graphic designers and programmers.

3.2.1.1 Adobe Flash

Originally developed by Macromedia, Adobe Flash was designed for animated Web-based vector graphics. Adobe's marketing targets graphic designers and authors of Rich Internet Applications (RIA). This orientation is reflected in the development of new tools, with Flash/Flex being the most well-known for application development (Noble and Anderson 2008). Currently, the cross-platform and cross-browser framework Flex comprises MXML (an XML-based vector graphics description language), ActionScript (a JavaScript-related language) and, for rendering, either the Flash browser plug-in, or the Adobe Integrated Runtime (AIR) for desktop applications. MXML is capable of describing various graphical user interface (GUI) components and vector objects. In addition, raster graphics, filter effects, videos, sound, animations etc. can be defined with MXML. User interaction can be realized with custom MXML ActionScript code.

Various authoring programs allow for the generation of Adobe Flash content. For designers, Adobe provides the visual design environment Flash Professional, while programmers draw on the tools from Adobe Flash Builder. In a typical workflow, either an MXML file is created using Adobe Flash Builder, or FLA files and/or ActionScript classes are created using Adobe Flash Professional. The resulting files are then compiled to a binary SWF file and presented with the Flash plug-in or the Adobe Integrated Runtime (AIR).

Advantages of the Flash framework include the performant rendering engine, the integration of multimedia content (e.g., video, sound and animation), a wide range of auxiliary tools for designers, and the wide-spread dissemination of the Flash Player for rendering Flash content. Adobe claims Flash Player is installed on more than 98% of Internet-enabled desktops worldwide (source: adobe.com/products/flashplayer/faq). Among the disadvantages of Adobe Flash, there are the dependency on one software vendor who may arbitrarily change the code base or the functionality of tools and plug-ins. Also, security concerns are raised when using a plug-in, particularly in regard to arbitrary, remote code execution and passing on of cached user information (Bradbury 2010).

Yet, Flash remains popular in the graphics industry. Typical use cases include games and multimedia graphics with animation or video, advertisement banners, and RIAs of varying complexity. Due to its wide dissemination, GIS and Web-mapping applications feature built-in map export functionalities compatible with Adobe Flash Player. An example is ESRI's ArcGIS API for Flex on top of the ArcGIS Server, which allows map authors to design customized interactive Web-maps, with options to edit or query data, and integrate temporal data.

3.2.1.2 Microsoft Silverlight

Microsoft's counterpart of Adobe Flash, the cross-platform and cross-browser Silverlight framework, consists of an XML-based vector graphics description language, known as XAML, which may be manipulated by various programming languages, such as C#, VB.NET, or JavaScript. Silverlight uses a subset of the Microsoft .NET framework, particularly the Windows Presentation Foundation (WPF). The necessary browser plug-in is available for Windows and Mac OS X and is installed on 50% of desktop computers worldwide (source: riastats.com).

Two authoring tools are available from Microsoft for generating Silverlight content: The visual authoring environment Microsoft Expression Blend for designers and the code-based Microsoft Visual Studio for programmers. In a typical workflow, a XAML file is created with Expression Blend or Visual Studio, compiled to a binary XAP file, and then presented in the browser by means of the Silverlight plug-in.

The performant rendering engine, the integration of multimedia content, and the availability of auxiliary tools for programmers are the main advantages of the Silverlight framework. The disadvantage in terms of the dependency on one single software vendor is similar to Adobe Flash.

Silverlight is suitable for programmers experienced with the Microsoft Windows .NET framework. It is supported by various development tools and Microsoft's dominant market position adds to its successful diffusion. Typical use cases include business applications. ESRI, traditionally closely connected to Microsoft, supports Silverlight with a separate API for creating interactive maps.

3.2.2 Open Standards

Open-source software is freely available and users may directly contribute to its enhancement by extending specific functionalities and publishing new code. In this section, four open-source technologies are discussed: Scalable Vector Graphic (SVG), Oracle/Sun JavaFX, WHATWG/W3C Canvas, and WebGL.

3.2.2.1 Scalable Vector Graphics (SVG)

SVG is a XML format for vector graphics. SVG is a recommended standard of the W3C consortium that all modern Web-browsers draw without the use of a plug-in, including Chrome, Firefox, Internet Explorer (as from version 9), Opera and Safari. However, the level of SVG support considerably varies between the different browsers. The SVG specification includes vector and raster graphics, filter effects, point symbols, masking, animation, and many other features (Neumann and Winter 2003). SVG is extendable with JavaScript, allowing for the creation of interactive graphics and graphical user interfaces.

Among the applications capable of creating SVG content are Adobe Illustrator, Inkscape, Xara Extreme, and Open Office Draw for designers, or different XML editors for programmers. In a typical workflow, an SVG file is created containing geometry data, and a separate JavaScript file with the application logic (e.g., the interactive functions).

The main advantages of SVG are the direct support in browsers, and the large variety of vector elements and visual effects. Another major advantage is the possibility to use multiple coordinate systems in a single drawing, which makes the SVG standard attractive for mapping applications: map features are based on native geographic coordinates, while user interface elements use screen coordinates.

The disadvantages of SVG are the sub-optimal support for multimedia, and the slow rendering. This issue is currently addressed by browser authors. Internet Explorer 9, for example, will introduce hardware accelerated SVG rendering.

3.2.2.2 JavaFX

JavaFX, now developed by Oracle, is a cross-platform and cross-browser framework for the development of Rich Internet Applications (RIA). It is based on the Java Runtime Environment, which is installed on about 75% of desktop computers worldwide (source: http://riastats.com). The tools needed for generating JavaFX content are NetBeans or Eclipse, both Integrated Development Environments (IDE) for experienced programmers. In a typical workflow, JavaFX code is compiled to Java bytecode and saved to a JNLP or JAR file. These files are then passed to the browser and executed using the Java Runtime Environment or Java Micro Edition on mobile phones.

Among the advantages of the JavaFX framework are the integration of Java drawing classes, and the thorough security concept. However, starting up the JavaFX plug-in is slower than starting up Silverlight or Flash. Another major disadvantages are missing tools for designers. Integrating multimedia elements, such as video and sound, is possible; but owing to the lack of authoring tools, JavaFX is mainly used by experienced programmers. Being an open-source framework with a thorough security concept, JavaFX comes into play for developing large business applications in which maps may be an integral part. However, for Web applications, it is currently not as widely used as is Flash or Silverlight. It should also be noted that with the release of JavaFX 2, the hitherto recommended JavaFX scripting language will not be developed any further.

3.2.2.3 Canvas

Canvas is a HTML element, which uses JavaScript commands for drawing graphic primitives (e.g., rectangles, paths, text). The Canvas version for drawing 2D graphics is standardized by the Web Hypertext Application Technology Working Group (WHATWG) and will be part of the upcoming HTLM5 specification. HTML5 is the next major revision of the HTML standard, which is currently under development by the W3C (Mansfield-Devine 2010). No browser plug-in is required to render Canvas elements, as it is already implemented in Chrome, Firefox, Opera, Safari, and Internet Explorer. Canvas is combinable with other Web standards, but it represents a lower conceptual protocol level than, for example, SVG, as it is not based on a built-in scene graph or a Document Object Model (DOM). Drawing commands are not converted to graphical features for later access or manipulation. Instead, each JavaScript drawing command immediately changes the pixels of the generated image. After rendering vector data, only the individual image pixels may be manipulated using JavaScript.

Currently, no mature graphic authoring tools exist for Web designers to create Canvas drawings. Content is therefore mainly created by programmers using text editors and custom-made code. In a typical workflow, a HTML file is extended with JavaScript code drawing the Canvas graphics. The JavaScript code might be embedded into the HTML file or stored in separate files. The browser automatically loads the JavaScript when rendering the Canvas element.

In the future, Canvas has a considerable potential to compete established vector data rendering technologies, such as Adobe Flash. The main advantages of the Canvas element are the support by all browsers, the fast rendering, and its options for raster data manipulation. The major disadvantage is the missing scene graph, which complicates linking with event handlers for interactive graphics, and which may considerably increase complexity when dealing with a large number of complex graphical primitives.

3.2.2.4 WebGL

WebGL is a 3D graphics API for Web applications that extends the HTML Canvas element. The specification is currently a work in progress, and implementations are not yet finalized. WebGL is specified by the non-profit technology consortium Khronos Group, which controls various open standards, for example, the OpenGL standard for rendering 3D graphics.

Similar to the 2D variant of Canvas, three-dimensional drawing with WebGL is controlled by JavaScript code without using a built-in scene graph. It accesses the computer's graphics card via the platform-independent OpenGL API which entails a very high rendering performance. WebGL uses the OpenGL ES 2.0 standard, a subset of OpenGL, which is also supported by devices with limited computing power, such as smartphones, tablet computers and other mobile devices. WebGL rendering is based on shader programs that calculate rendering effects on graphics hardware with a high degree of flexibility.

WebGL is not yet part of Web-browsers for end users. However, developer versions of Chrome, Firefox, and Safari contain experimental implementations. As a consequence, WebGL is currently mainly applied by programmers and early adopters for experimental applications.

Various scripting libraries are available to create WebGL content or for loading 3D objects that are designed with 3D modeling software (e.g., Autodesk 3ds Max). JSON (JavaScript Object Notation) is often used to describe and load 3D objects.

The advantages of WebGL include very fast rendering, and the versatility offered by shader programs for graphical special effects. Due to its early development status and owing to the lack of authoring tools, expert programmers are the exclusive user group of WebGL. Another disadvantage is the lack of support by Internet Explorer. However, WebGL has a considerable potential for both 2D and 3D map visualization.

3.3 Vector Overlay for Client-Side Mapping

In the previous section, independent general-purpose vector-based Internet standards are discussed. The standards are implemented in Web-browsers, or require additional plug-ins. The discussion is now moving towards a higher abstraction level, i.e., frameworks and APIs for mapping which build on these standards. Such client-side frameworks and interfaces offer additional functionality for cartographic applications, and encapsulate and further abstract the underlying visualization standards.

Client-side frameworks are widely used, since they greatly facilitate the creation of vector-based Web maps allowing cartographers to focus on their core competency in design and data visualization. The concept of such toolkits is to provide an API that allows map authors to create so-called mash-ups. Such maps usually combine a raster map in the background with custom, overlaid vector data. Often, the default graphical user interface provided by the API for manipulating the map is also customized, either by using functionality of the mapping framework, or by integrating specialized external libraries. The list below shows the most popular toolkits for generating map mash-ups.

The *Google Maps API*, the *Microsoft Virtual Earth/Bing Maps API,* and the *Yahoo! Maps API* offer similar functionality. They provide access to a multi-scale worldwide raster map, and a specialized graphical user interface for navigating the map. Authors work with a JavaScript based or a Flash-based API to embed their map contents. A wide range of functionalities and services are available for data integration and map drawing.

OpenLayers is a GUI and a customization tool for combining raster and vector data sources. It consists of a JavaScript library for displaying map data in Web-browsers, without any server-side dependencies. Unlike Google's, Microsoft's or Yahoo!'s APIs, the OpenLayers API is entirely free and open-source. It is often used in combination with OpenStreetMap, a freely editable map of the world.

The *carto.net* framework is for SVG based maps. By means of a programming language and the DOM (Document Object Model), SVG documents are manipulated. The DOM allows any Web-enabled programming language to create, manipulate, and delete map elements. In most cases, JavaScript is used for these manipulations.

CartoWeb (www.cartoweb.org) and *p.Mapper* are two frameworks running on MapServer, which is discussed in the next section. A graphical user interface and customization tools for web maps are provided, using JavaScript on the client-side and PHP MapScript on the server-side.

3.4 Vector Overlay for Server-Side Mapping

Web-map servers used to be restricted to raster-based output. Raster functionality includes the tiling of data, the conversion of data to various formats, or the resampling of raster images using different down- or up-scaling operators.

3 Current Trends in Vector-Based Internet Mapping: A Technical Review   31

Nowadays, Web-map servers are increasingly able to produce vector graphics formats. The concept is much the same as for client-side mapping: vector data are handled as individual, addressable objects and often overlay background raster data. Different Web-map servers share common characteristics, such as their cross-platform and cross-browser capabilities, and their more or less strict support of Open Geospatial Consortium (OGC) standards. In this section, a selection of Web map servers are discussed in more details, with emphasis laid on vector formats.

3.4.1 MapServer

Formerly known as UMN MapServer, MapServer is the most widely used open-source map server worldwide. It is very popular with a large user community and with numerous programmers who further develop functionalities and features. MapServer is able to read data from a variety of enterprise geodatabases, such as Oracle, IBM DB2, or PostgreSQL via ESRI ArcSDE. It can also read data from spatial databases, such as Oracle Spatial, PostgreSQL/PostGIS, and from several GIS file formats, such as ESRI shapefiles. The main cartographic features include data filtering operations, anti-aliasing, on-the-fly projection, and visualization of data in form of pie and bar charts. Beside its capability to output raster data according to Web Map Service (WMS) versions 1.0, 1.1.1, 1.3.0, and Web Coverage Service (WCS) versions 1.0, 1.1.x, it also supports vector-based output standards, such as Geography Markup Language (GML), SVG, PDF, and Web Feature Service (WFS) version 1.0.0. In order to visualize vector output, the W3C standard Styled Layer Descriptor (SLD) 1.0.0 may be applied, but usually, users define the cartographic symbolization in a so-called mapfile. The advantages of MapServer are its active user community, on-the-fly map projection, and its easy integration in different Web servers, such as Apache and IIS. The fact that MapServer does currently not fully support SLD may be viewed as a disadvantage.

3.4.2 QGIS Mapserver

This open-source map server is based on Quantum GIS (QGIS), which is a free and open-source desktop GIS. It has a rather small, but very active user and developer community, mainly based in Europe. QGIS mapserver is able to read various data sources, ranging from ESRI Shapefiles to GML, or spatial databases like PostgreSQL/PostGIS. It features anti-aliasing and visualizes geodata by means of patterns, point symbols, or pie and bar charts. Beside the raster-based WMS 1.1.1 and 1.3.1 output, QGIS mapserver supports the vector-based output standards GML (Geography Markup Language), and WFS (Web Feature Service) in combination with SLD (Styled Layer Descriptor) 1.0.0. The SLD symbolization description file is typically generated using the Quantum GIS desktop application, or the associated PublishToWeb plug-in.

Among the advantages of QGIS mapserver are the fast rendering, and the possibility to visualize geodata as diagrams, patterns and point symbols, together with the full support for SLD. Another advantage is the integration in the Quantum GIS desktop application, making the functionalities of QGIS mapserver accessible to a wide user group. A disadvantage remains the small number of active developers.

3.4.3 GeoServer

GeoSever is an entirely Java-based open-source map server. Its worldwide community is large and active. GeoServer can handle data directly from most common spatial databases. PostgreSQL/PostGIS, IBM DB2 with Spatial Extender, Oracle Spatial, and ArcSDE, as well as standard GIS file formats such as ESRI Shapefiles are manipulable through GeoServer. Some of the advantageous features are the ability for anti-aliasing, versioning, and its security concept. Beside the most current raster-based standard outputs (see above for QGIS mapserver), GeoServer also exports to the vector formats WFS 1.0 and 1.1, PDF, SVG, KML, GeoRSS (geocoded Web feeds), as well as GML 2.1.2 and 3.1.1. It also fully supports SLD to create cartographic symbolization. GeoServer runs predominantly on the Apache Web server. The support by its active community and by major software companies is an additional advantage.

3.4.4 ESRI ArcGIS Server

This proprietary and popular map server allows users to link to the ESRI GIS software portfolio. Using ArcSDE, it handles spatial databases, such as Oracle Spatial, IBM DB2, and PostgreSQL/PostGIS, as well as a range of GIS file formats. Beside anti-aliasing, filtering, and 3D output, it also offers a variety of geo-processing functionality. ArcGIS Server is able to export raster-based data according to the most current standards, and also provides vector-based output, such as WFS 1.0 and 1.1, GML 2.0 and 3.1, KML 2.1 and 2.2. For cartographic symbolization, SLD 1.0 is supported. ESRI ArcGIS Server offers different Web-services, which are, however, typically conceptualized using the Desktop ArcGIS. The same applies to the definition of the cartographic symbolization. Among the major advantages are the geo-processing functionalities and 3D output.

3.4.5 Intergraph Geomedia WebMap

This software is another proprietary map server which handles spatial databases such as Oracle Spatial and Microsoft SQL Server. Geomedia WebMap is able to export WMS and different raster formats, as well as vector-based standards, such as WFS, GML, or SVG. Geomedia WebMap is mainly used in the business sector.

3.5 Web-Based Vector Map Examples

The following two map examples illustrate how geographic data may be visualized by vector-based Internet maps. The example in Fig. 3.1 contains hydrological real-time data which are automatically edited, processed, and visualized in an interactive vector map, along with interactive time series graphs. The map is based on data stored in a real-time PostgreSQL/PostGIS database which are automatically converted to SVG. The real-time visualizations have been created on the basis of the carto.net SVG framework and have been extended with specific interactive GUI components.

The example in Fig. 3.2 shows a city plan accessible to administrative officials as well as to the general public. The map is based on Microsoft Silverlight technology. Most of its content is delivered as WMS raster data using the ArcGIS Server via a

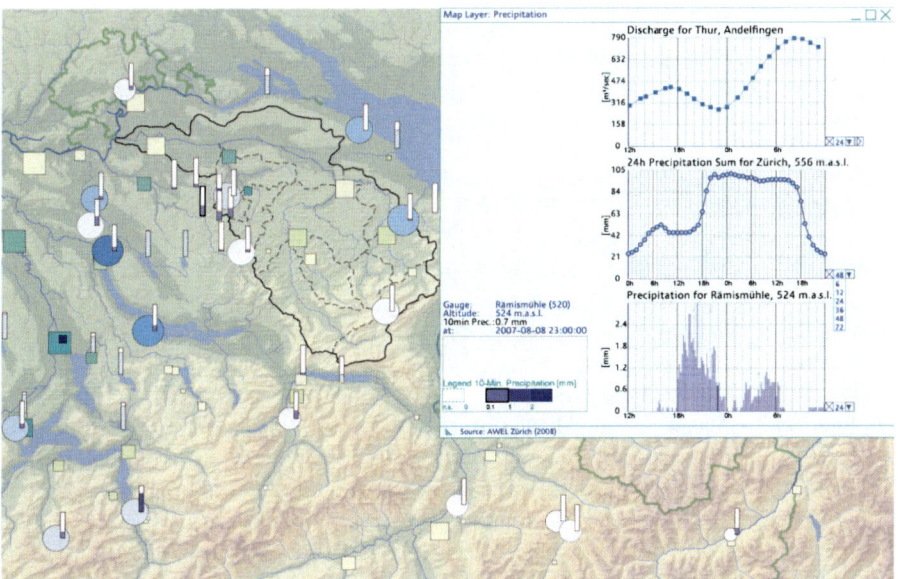

Fig. 3.1 SVG real-time map and time series graphs (Lienert et al., 2010)

Fig. 3.2 Silverlight city map (www.stadtplan.stadt-zuerich.ch)

REST API and integrated in a Silverlight GUI. Vector-based interactive objects, such as points of interest and borders, are placed over the raster background. The vector data in this example are stored in an Oracle Spatial database.

3.6 Conclusions

Vector-based Internet technologies are continuously developing and changing, with new standards and formats appearing, and old ones disappearing. The number of technologies may seem confusing, but at a technical level, similarities between the different technologies prevail: the use of equal graphical primitives (e.g., SVG and Silverlight), the integration of multimedia (e.g., Flash and Silverlight), and the use of JavaScript, or derivates thereof, outweigh the differences between the technologies, which are mainly rendering speed and the underlying authoring and programming environments.

The anticipated move from desktop applications to purely Web-based applications and services may drive browser vendors to further adopt Web-mapping standards and workflows (Jolma et al., 2008). Currently with self-contained mobile applications (so-called Apps) on the rise, however, the Web-browsers are not the ubiquitous user interface for which many were hoping (Kennard and Leaney 2010). Thus, a standardization of formats for vector-based Internet graphics is not to be expected soon.

Table 3.2 shows application domains and the disseminations of vector-based Internet mapping technologies by beginning of 2011, for different user groups and for different use cases. This table is surely subject to rapid change. A wide range of

Table 3.2 Application and dissemination of vector-based Internet mapping technologies

Technology/software	Targeted authors	Typical use cases	Authoring tools	Dissemination
Flash/Flex	Designers and programmers	Multimedia, advertisement, games	++	Very high
XAML/Silverlight	Designers and programmers	Business applications	++	Medium
SVG	Designers and programmers	Infographics (diagrams, maps)	+	High
JavaFX	Programmers	Business applications	+	High
Canvas	Programmers	RIAs, infographics, games	–	High
WebGL	Programmers	3D graphics (experimental)	–	Low

mapping frameworks and authoring tools cover various skill levels, some straightforward, others guided, yet others still experimental. However, the cartographers' choice of vector mapping technology certainly also correlates to their technical and programming skills.

Since authoring tools facilitate the map making process, technologies such as Adobe Flash and Flex, or Microsoft Silverlight may better meet the needs of design-oriented cartographers. Their popularity and diffusion is accordingly high. Due to their code-based environment and required programming skill, the features offered by software applications such as JavaFX, Canvas or WebGL reach a smaller number of cartographers. These technologies, however, enjoy growing popularity and may be complemented by some authoring tools in the future.

Client-side and server-side frameworks allowing for vector data overlay are becoming increasingly popular and are found on numerous commercial and non-commercial websites. There is also a trend for mapping software on mobile devices, such as cell phones, to adopt this overlay concept. For many map authors, client-side frameworks constitute a widely used basis for building customized maps. These frameworks provide free raster-based, multi-scale background world maps as well as functionality for custom vector data overlays. Technically more adept cartographers with more computer science expertise, in turn, may set up a Web map server for generating custom base maps combined with vector data overlays.

References

Bradbury D (2010) The battle of the Internet browsers. J Infosecurity 7(2):34–37
Byrne J, Heavey C, Byrne PJ (2010) A review of Web-based simulation and supporting tools. J Simul Model Pract Theory 18(3):253–276
Jenny B, Jenny H, Räber S (2008) Map design for the Internet. In: Peterson MP (ed) International perspectives on maps and the Internet. Springer, Berlin, pp 31–48
Jolma A, Ames DP, Horning N, Mitasova H, Neteler M, Racicot A, Sutton T (2008) Free and open source geospatial tools for environmental modelling and management. In: Jakeman AJ, Voinov AA, Rizzoli AE, Chen SH (eds) Developments in integrated environmental assessment. Elsevier, Amsterdam, pp 163–180

Kennard R, Leaney J (2010) Towards a general purpose architecture for UI generation. J Syst Software 83(10):1896–1906

Lienert C, Kunz M, Weingartner R, Hurni L (2010) Monitoring and comparing: a cartographic web application for real-time visualization of hydrological data. In: Konecny M, Zlatanova S, Bandrova TL (eds) Geographic information and cartography for risk and crisis management. Springer, Berlin, pp 409–424

Mansfield-Devine S (2010) Divide and conquer: the threats posed by hybrid apps and HTML 5. J Netw Security 2010(3):4–6

Neumann A, Winter AM (2003) Webmapping with Scalable Vector Graphics (SVG): delivering the promise of high quality and interactive web maps. In: Peterson MP (ed) Maps and the Internet. Elsevier, Amsterdam, pp 197–220

Noble J, Anderson T (2008) Flex 3 cookbook. O'Reilly, Sebastopol

Peterson MP (2008) International perspectives on maps and the Internet. Springer, Berlin

Schnabel O, Hurni L (2009) Cartographic web applications – developments and trends. In: Proceedings of the 24th international cartography conference, Santiago, 2009

Zaslavsky I (2003) Online cartography with XML. In: Peterson MP (ed) Maps and the Internet. Elsevier, Amsterdam, pp 171–196

Chapter 4
Map Mashups and APIs in Education

Emmanuel Stefanakis

Abstract The rapid technological evolutions in map mashups and APIs during the last few years offer the possibility to build effective educational tools with limited efforts and programming skills. Many subjects, in particular geography, social studies, science, even math and Greek/English studies may be supported by mashup technologies at all levels of education. This chapter summarizes several attempts made recently in cooperation with educational organizations (schools and museums) to develop a series of prototype educational frameworks for the subject of history. A wide range of tools and APIs have been adopted to meet the educational needs as well as to evaluate their efficiency and completeness. Special attention has been given in involving the users (the pupils and teachers) at all phases of development, from design to the actual implementation of both the content and application scenario/interface.

4.1 Introduction

Map mashups are currently so popular that a lot of people incorrectly assume that all mashups use maps. Since early 2005, when Google Maps were released, most of the major online map providers have offered a mapping service for users and a mapping API for developers (over 50 powerful and reliable APIs related to mapping and geolocation are already available).

Recently, the online mapping space has exploded. A vast range of applications to all geographical domains (such as marketing and ecology) are available and many of them are rather popular to web users. The educational uses of map mashups are emerging rapidly. Maps are useful in any subject, in particular geography, social

E. Stefanakis (✉)
Department of Geodesy and Geomatics Engineering, University of New Brunswick,
P.O. Box 4400, Fredericton, NB E3B 5A3, Canada
e-mail: estef@unb.ca

studies, science, and even math and Greek/English studies. Obviously, map mashup technology is able to play an important role in many subjects and at all levels of education.

It is widely recognized that school pupils cannot easily perceive the geographical space. On the other hand, school books are missing maps and other geovisualization content. As a consequence, it is quite hard for the pupils to locate events into space and recognize any geographical interactions and relations.

The last couple of years, at Harokopio University of Athens (Greece) and in cooperation with several educational organizations in Athens (schools and museums), the use of map mashups in education have been examined. Specifically, several prototype applications to assist the subject of history, from mythology to classical ages and recent history, have been developed and evaluated. Various tools and APIs have been adopted for this reason, such as, Google Maps API, OpenLayers API, and Geonetwork Opensource.

The scope of this Chapter is to present these prototypes, focussing on their technical aspects and functionality as well as on the feedback of the users (the pupils and teachers), during a precious evaluation process. The discussion is organized as follows. Section 4.2 presents a prototype educational framework for modelling, analyzing and visualizing the Ancient Greek Myths. Section 4.3 presents a prototype application to assist the teaching of Roman and Byzantine history in the Primary School. Section 4.4 shows how the last prototype has been customized to assist an educational program of the National Archaeological Museum of Athens, regarding the Antikythera Wreck and Mechanism dated in 150 BC. Section 4.5 presents a set of tools developed to assist the study of the Greek historical battleship "G. Averof"; acted in three wars during the first half of the twentieth century. Finally, Section 4.6 concludes the discussion.

4.2 A Journey to the Ancient Greek Myths

The first prototype is an educational framework for modeling, analyzing and visualizing the Ancient Greek Mythology (Stefanakis 2008a). This prototype aspires to help school pupils and other general users in getting a clear geo-spatial perception of the mythology events. This is accomplished by making use of the recent technological developments in geovisualization, web mapping and geographic mashups (Brown 2006; Mitchell 2005; Young 2008; Andrienko and Andrienko 2005; Stefanakis 2008b).

The Greek Myths involve many heroes, characters and gods and consist of numerous parallel events. The events occur in the geographical space and, although the absolute time is missing, they are related to each other by temporal (topological) relations, inside a single myth or across myths. Additionally, apart from the myth descriptions, there is a series of multimedia items that accompany each myth, such as ancient statues and figurines, representations on the ancient vases and amphorae, as well as a series of modern paintings or poems.

The scope of this prototype was to model a set of Greek Myths as they are presented in the official school book compiled by the Pedagogical Institute of the Greek Ministry of Education (PI-Schools 2010). The modeling takes into consideration the spatial, temporal and thematic peculiarities of the events comprising the myths as well as the educational requirements for the analysis and visualization of the myths content in the school class.

An appropriate data repository has been set up to accommodate the myths content, while a Web Client Interface has been developed to serve the myths to the users on the web. End-users are able to browse in the myths content and have the events visualized on top of Google Maps.

4.2.1 Software Systems and Tools

The educational framework has been developed using merely Free and Open Source Software Systems. It is also compliant with the specifications of the W3C Consortium, i.e., it makes use of XML-based languages and tools.

Specifically, PostgreSQL DBMS and its PostGIS extension have been used in the role of the data repository. XML (eXtensible Markup Language) has been adopted as the basic format (language) for expressing the data extracted from the repository. XSL (eXtensible Style Language) and Google Maps API have been used to transform the XML documents extracted from the data repository to appropriate HTML documents and Javascripts; served to the end-users through the Web Client Interface.

4.2.2 Data Repository

Figure 4.1 presents the basic classes of the prototype in UML. The spatial dimension is compatible with the OGC proposal ("Simple Features" Proposal). As for the temporal dimension, the individual events are interrelated via a reflective association with temporal topological relationships (Allen 1983).

As shown in the UML diagram (Fig. 4.1a) a Myth (Myth) is organized into Sections (MythSection). The prototype follows the structure of the content in the school book; i.e., the Myths and the Sections included in the prototype have been extracted from the book.

Each MythSection is composed of Events (Event). These Events have been discovered from the corresponding Sections of the school book. In other words, an attempt has been made to split the text in the book into a series of Events. This task was not always straightforward, provided that the book content is a free (not a formal) text. At the current version, the prototype accommodates the most important events of each book Section and focuses on the events with spatio-temporal descriptions.

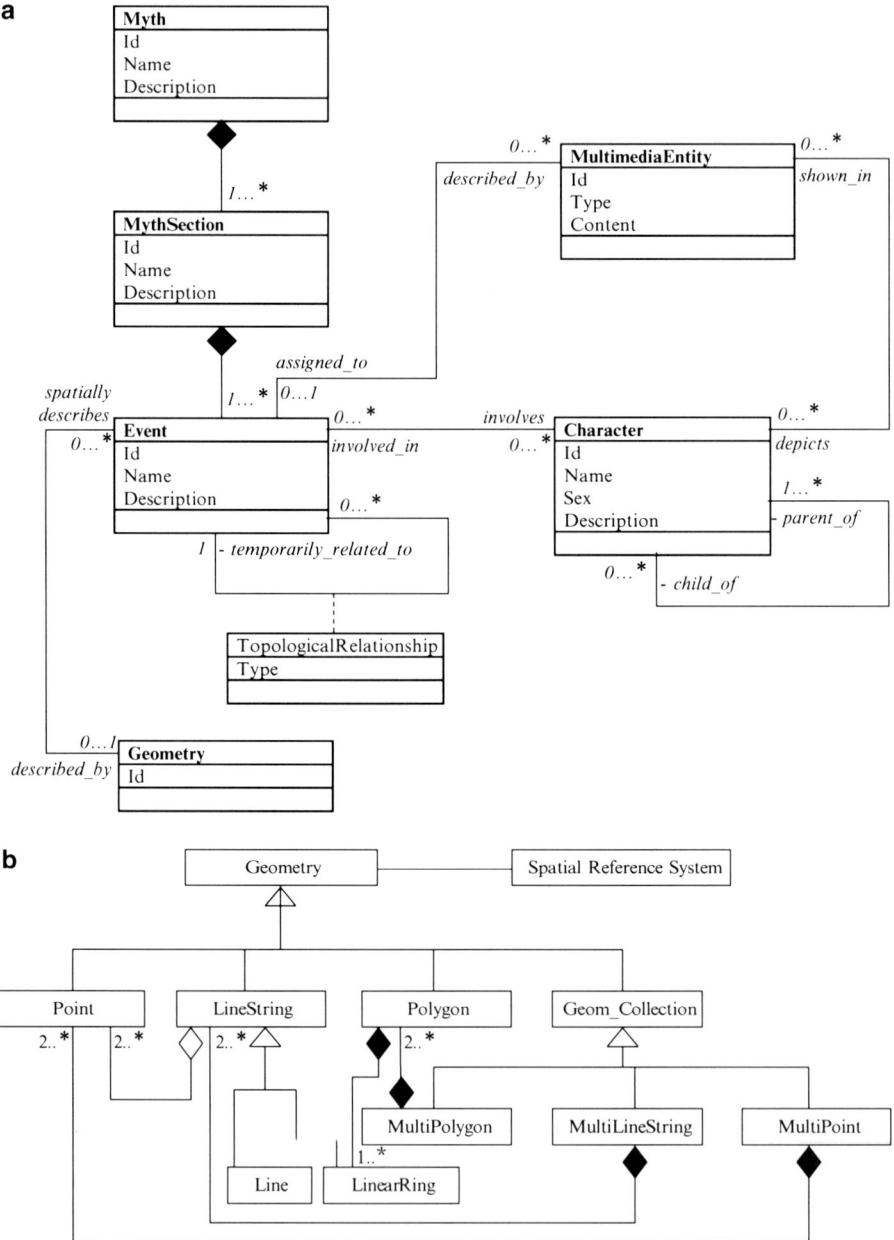

Fig. 4.1 (a) The basic classes of the prototype (UML diagram), (b) the geometry class according to the OGC proposal (part of)

4 Map Mashups and APIs in Education   41

Each Event involves several characters (Character); it is spatially described (OGC Geometry); and assigned a set of multimedia items (MultimediaEntity). Events are interrelated each other through temporal relations (Allen 1983). Figure 4.1b shows part of the Geometry class according to the OGC proposal.

4.2.3 The Web Client Interface

The data residing into the repository are served to the users through a Web Client Interface, specifically developed for this reason. This has been accomplished by the workflow presented in Fig. 4.2.

As shown in Fig. 4.1, each Myth consists of Sections and each Section comprises a sequence of Events. All these items have been modeled and stored in the data repository (database) and are served to the user through the frames of the Web Client Interface. As shown in Fig. 4.2, two types of XML documents are extracted from the data repository. The top XML document contains the thematic (non-spatial) information related to the Myth Events, while the bottom XML document contains the spatial information, i.e., the geographic places where each event occurs as well as the temporal sequence of the events composing a Myth Section.

Figures 4.3 and 4.4 show two sample XML documents (part of) as extracted from the data repository. The document in Fig. 4.3 contains the thematic information. Specifically, it shows the Section "Birth of Hercules", which is part of the Myth "Hercules". Two Events are shown in the figure. Each Event has three nested elements, i.e., <description> (event description), <image> (image file path), and <img_descr> (image description).

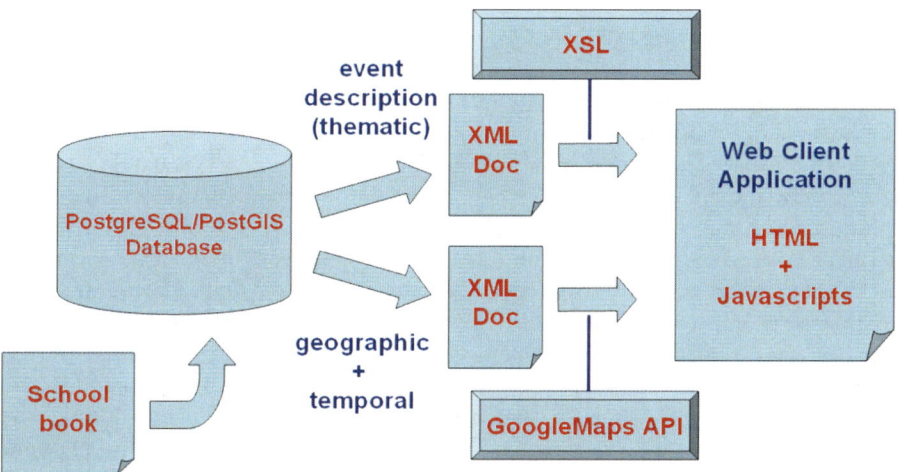

Fig. 4.2 The workflow to the generation of the content served through the Web Client Interface

```
<?xml version="1.0" encoding="UTF-8" ?>
- <section>
   <title>Η γέννηση του Ηρακλή.</title>
 - <event>
      <seq>1</seq>
      <description>Ο Αμφιτρύωνας και η Αλκμήνη, κόρη του βασιλιά των Μυκηνών, βρήκαν καταφύγιο
         στη Θήβα. Η Αλκμήνη γέννησε τον Ηρακλή, που ήταν γιος του Δία, και τον Ιφικλή.</description>
      <image>../images/math_1_50_img_64.jpg</image>
      <img_descr>Ο Ηρακλής - γιός του Δία</img_descr>
   </event>
 - <event>
      <seq>2</seq>
      <description>Ο Δίας έστειλε τον Ερμή να φέρει τον Ηρακλή στον Όλυμπο και να πιει το γάλα της
         Ήρας ώστε να γίνει αθάνατος.</description>
      <image>../images/math_1_50_img_59.jpg</image>
      <img_descr>Η δημιουργία του Γαλαξία</img_descr>
   </event>
 - <event>
```

Fig. 4.3 Description of the events and images in XML

```
- <H>
  - <H1>
    - <markers>
        <marker name="ΟΛΥΜΠΟΣ" descr="Το βουνό του Ολύμπου" lat="39.9883" lng="22.3479" type="H1" />
        <marker name="ΘΗΒΑ" descr="Η πόλη της Θήβας" lat="38.3158" lng="23.3196" type="H1" />
        <marker name="ΔΕΛΦΟΙ" descr="Το μαντείο των Δελφών" lat="38.4721" lng="22.4924" type="H1" />
        <marker name="ΜΥΚΗΝΕΣ" descr="Η πόλη των Μυκηνών" lat="37.7288" lng="22.7588" type="H1" />
    </markers>
    - <courses>
        <course suc="1" from="ΘΗΒΑ" to="ΟΛΥΜΠΟΣ" type="H1" />
        <course suc="2" from="ΟΛΥΜΠΟΣ" to="ΘΗΒΑ" type="H1" />
        <course suc="3" from="ΘΗΒΑ" to="ΔΕΛΦΟΙ" type="H1" />
        <course suc="4" from="ΔΕΛΦΟΙ" to="ΘΗΒΑ" type="H1" />
        <course suc="5" from="ΘΗΒΑ" to="ΜΥΚΗΝΕΣ" type="H1" />
    </courses>
  </H1>
```

Fig. 4.4 Description of the placemarks and movements (courses) in XML

The document in Fig. 4.4 contains the spatial and temporal information. Specifically, it spatially describes the places on earth, where the myth events take place, through a series of markers. Each <marker> element has a list of attributes, i.e., the name, a description, the longitude and the latitude. The latter two are expressed in WGS'84 spatial reference system as they have been stored in the data repository. The coordinates have been derived from the GeoNames geographical database. Additionally, this XML document describes the course followed through a sequence of movements from marker to marker (<course> element).

The top (thematic) XML document is transformed through an XSL transformation (see Fig. 4.2) to an HTML frame included in the Web Client Interface. On the other hand, the bottom (spatio-temporal) XML document is transformed through another XSL transformation (see Fig. 4.2) to an HTML with Google API Javascripts; also included in the Web Client Interface.

The Web Client Interface consists of a set of frames which accommodate the appropriate content, according to the user's selections. Table 4.1 presents the configuration of the frames and their content. Figure 4.5 presents an example screen

4 Map Mashups and APIs in Education

Table 4.1 The frames of the Web Client Interface and their content

Header frame...		
List of myths... (the user selects here the myth of interest)	**Myth sections...** (the user selects here the section of interest) **Myth section events...** (the user reads here the events composing the selected myth section)	**Map interface...** (the Google map enriched with myth section details, i.e., placemarks and paths, is visualized here; the user may interact with the map content through a set of map controls)
Footer frame...		

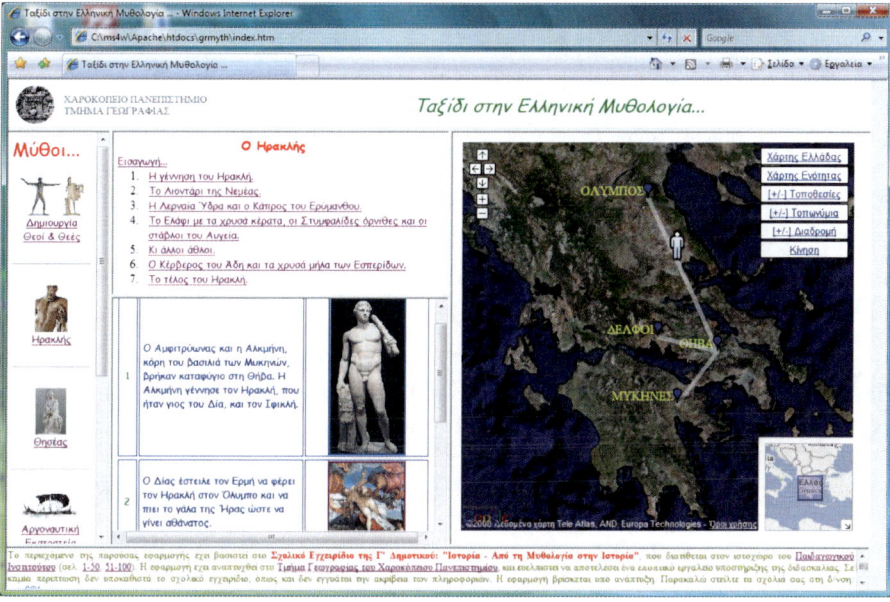

Fig. 4.5 An example screen of the web client application

of the Web Client Interface. In this example, the user has selected the following: (a) Myth: Hercules; and (b) Myth Section: The birth of Hercules.

As a result, the Myth Section Events frame shows the details of the events composing the selected Section, while the Map Interface frame highlights on top of Google Maps (map type: Satellite) the places where the events took place, i.e., Thebes, Mycenae, Delphi and Olympus. Additionally, the courses of Hercules are drawn on the map frame (white lines). The user may interact with the map through a set of control buttons as shown in Fig. 4.5, such as zoom-in/out and pan, switch on/off the placemarks/paths/labels, etc. S/he may also make an icon (here the little white figure) move along the drawn paths and simulate the spatio-temporal movements as described in the corresponding myth section.

4.2.4 Educational Aspects

Special attention should be given in the educational aspects of the prototype. Currently, there is a close interaction with experts in the primary education community, in order to meet the educational requirements and make this prototype a useful tool in the school class. The main concerns are: (a) making the presentation of the myths attractive and pleasant to the pupils; (b) helping them recognize the geographical space, focusing on the physical geography and encouraging them interrelate the past with the present infrastructures; and (c) compiling attractive and educational quizzes and game scenarios to help pupils understand the myths and recognize their interactions and relations in space and time (in the "chorochronos").

One question that the reader is expected to pose is: why Google maps? The answer is rather simple and naïve: because they are available for free! Google Maps API is very handy. Developers may easily built up tools with geo-visualization capabilities. On the other hand, the users may locate the events into the contemporaneous earth map and see for instance that: "Hercules on the way to Mycenae has passed near my house!" Obviously, the prototype needs to incorporate as an optional background image a digital map of the Greek Mythology compiled for the application needs. This is one future extension.

Although the first reactions from various user communities, i.e., pupils, parents and teachers are positive; the interface must be elaborated and enriched as regards to its functionality (Stefanakis 2008c). Some extensions are: (a) the overlay of diagrams, 2-D and 3-D maquettes and plans of the archaeological places on top of Google Maps; (b) the enrichment of the content with other multimedia items, such as sound and videos; (c) the development of quizzes and game scenarios for pupils to both test and assist their understanding/knowledge; and (d) the translation of the content in other languages (e.g., English, German, Spanish, etc.).

4.3 The Roman and Byzantine Empires

In the second prototype a map mashup prototype application to assist the subject of history at the fifth Grade of the Greek Primary School has been developed and evaluated (Stefanakis 2010). The textbook is unique to all pupils, according to the directive of the Greek Ministry of Education. It is a modern book (Glentis et al., 2008), with simple language and many illustrations. On the other hand, the book provides a limited number of printed maps. Additionally, most of them are rather abstractive and low-quality maps.

The Scope of the mashup application was to enrich the map content of the book and to familiarize the pupils with the geographic space of the Mediterranean Sea, the Middle East, and Central Europe. In other words, we have attempted to

4 Map Mashups and APIs in Education

highlight the geographical aspects of the Roman and Byzantine Empires and show how the geographical space affects the people's life and history.

4.3.1 Software Systems and Tools

The prototype application has been built in pure HTML using OpenLayers. OpenLayers is a JavaScript library for displaying map data in web browsers, with no server-side dependencies. It is a Free Software, developed for and by the Open Source software community (OSGeo), and implements a JavaScript API for building rich web-based geographic applications (such as the commercial Google Maps and MSN Bing Maps APIs). Furthermore, OpenLayers implements industry-standard methods for geographic data access, such as the OpenGIS Consortium's Web Mapping Service (WMS) and Web Feature Service (WFS) protocols.

4.3.2 Functionality

Figure 4.6 presents the user interface of the prototype application. The interface consists of a map window (where the satellite image and the current political map of the Mediterranean Sea and the surrounding areas are drawn) and a series of buttons, which implement the application functionality. The user through these buttons may have the main towns of the era and the main rivers shown on the background map. S/he can also choose the main historical events and have them drawn on the map (e.g., boundaries, invasions, etc.). Finally, on floating windows s/he can read more content (such as maps and illustrations available on the web and/or the textbook) as

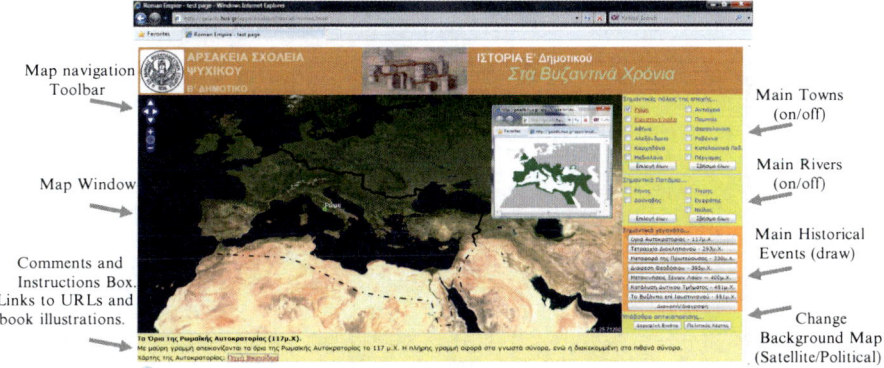

Fig. 4.6 The user interface of the prototype (here the Roman Empire in 117 AD). Available at http://gaiadb.hua.gr/apps/arsakeio/istoriaE/

Table 4.2 Geographic features and historical events visualized in the prototype

Towns	Rivers	Events
Rome	Danube	The Roman Empire (117 AD)
Constantinople	Rhine	Diocletian's Tetrarchy (293 AD)
Athens	Nile	The capital of the Byzantine Empire in Constantinople (330 AD)
Alexandria	Euphrates	Theodosius Division (395 AD)
Carthage	Tigris	Invasions of the Roman Empire (Goths, Huns, etc.)
Mediolanum		The Western Empire Collapses (451 AD)
Antioch		The Byzantine Expansion during Justinian's reign (551f AD)
Pompeii		
Thessaloniki		
Ravenna		
Chalons		
Pergamos		

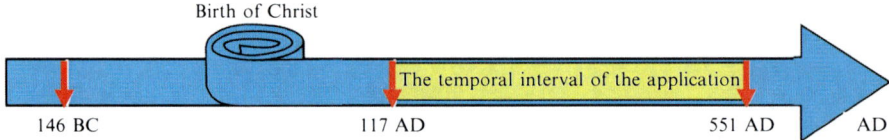

Fig. 4.7 The time axis and the period the prototype currently refers to

being accessible though appropriate links in the comments box (at the bottom of the interface).

In the current version, the application visualizes the towns, the rivers and the events listed in Table 4.2. These events refer to the temporal interval shown in Fig. 4.7. Notice that, the application content is strictly limited to what is included in the textbook.

The default background image map is the MODIS (Moderate Resolution Imaging Spectroradiometer) layer, retrieved from NASA Jet Propulsion Laboratory Server (http://jpl.nasa.gov) by posing the following GetMap request, according to the OGC's WMS specification:

```
http://wms.jpl.nasa.gov/wms.cgi?
SERVICE=WMS&VERSION=1.1.1&
REQUEST=GetMap&
LAYERS=modis&
BBOX=20,20,64,62&STYLES=&SRS=EPSG:4326&
WIDTH=2000&HEIGHT=1000&FORMAT=image/png
```

This image resolution (2,000 × 1,000 pixels) is adequate, provided that the zoom buttons in the toolbox (Fig. 4.6) allows a limited number of zoom levels (up to three times, X3). The user may switch the MODIS layer (satellite image) to a political map of the same region, showing the contemporary country boundaries

(Fig. 4.8d). The latter has been compiled for the needs of the prototype and is stored locally at HUA Server, along with the application.

The geometric features (towns and rivers) and the geometries assigned to the historical events in Table 4.2 are drawn using the OpenLayers library constructs. The source data have been enriched by Wikipedia (Wikipedia 2010) and the Digital Atlas of Roman and Medieval Civilization developed by the Centre for Geographic Analysis at Harvard University (Darmc 2010).

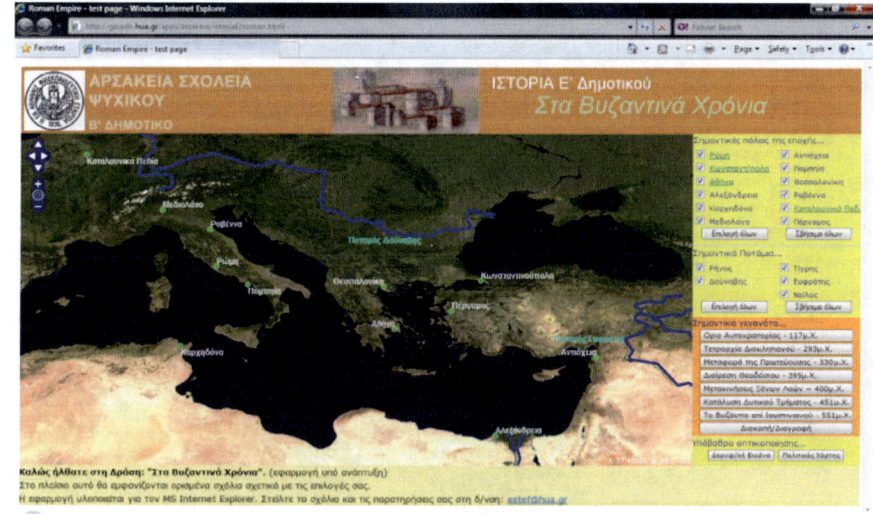

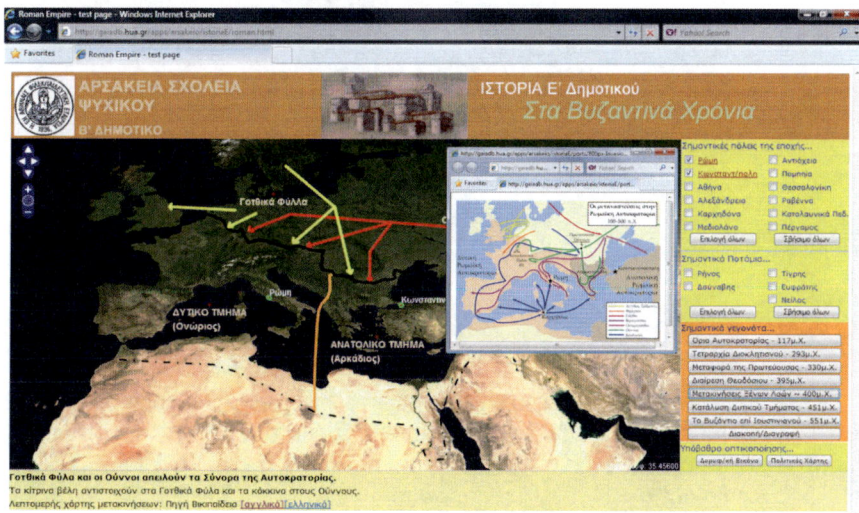

Fig. 4.8 (continued)

c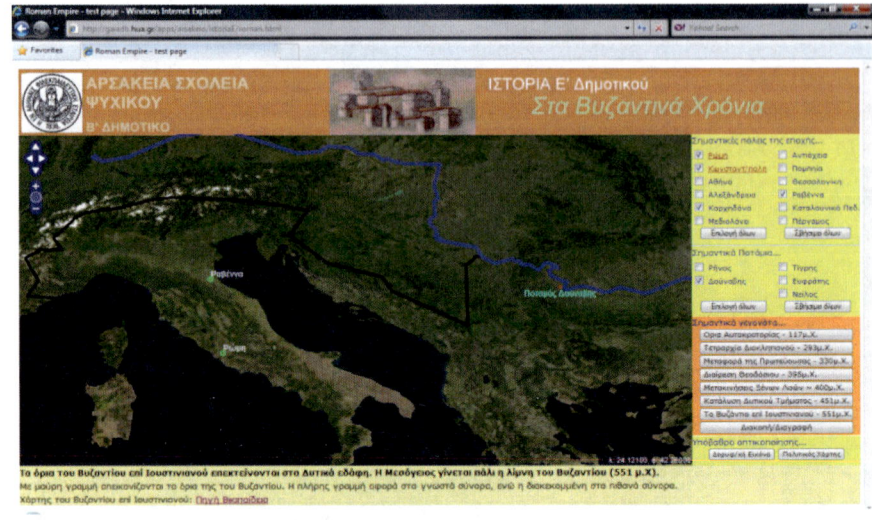

d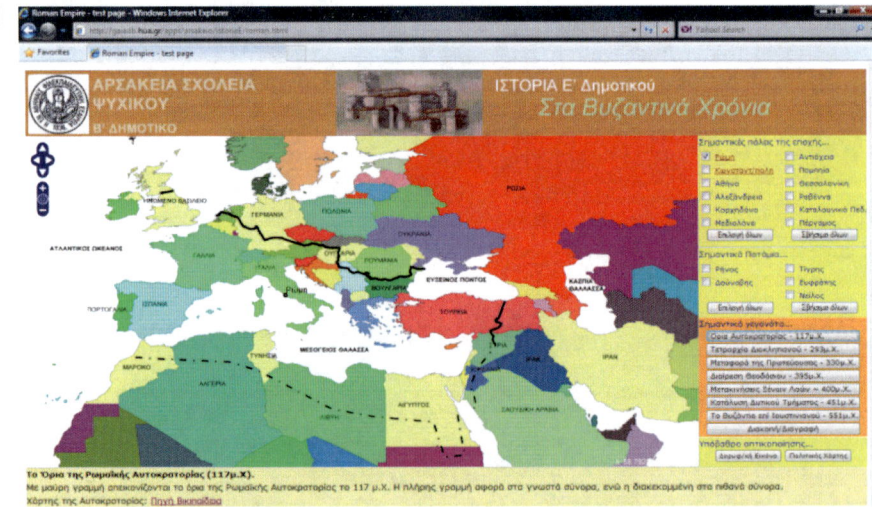

Fig. 4.8 Example screens of the prototype

4.3.3 Educational Aspects

Both the content and functionality of the application have been evaluated preciously and at multiple levels. First, the author has cooperated closely with the history teachers of the biggest private school in Athens, Greece, during the development phase, to clarify the application scope and content.

The initial prototype has been presented to a limited number of school pupils. Their feeling and comments have been decoded and leaded to significant changes in the initial version. Then the revised version has been presented to two regular school classes of 33 pupils each (66 pupils in total). Their feedback was really positive. Their comments have also been collected and leaded to minor revisions of the prototype. This revised version of the prototype is currently available at http://gaiadb.hua.gr/apps/arsakeio/istoriaE/.

The whole process, including the development, the presentations and the revisions, helped us understand the significance of the application in education, while raised several useful guidelines for future developments. As for the significance, the pupils understood (as stated by themselves) the following issues, which were not clear after reading the textbook: (a) the distribution of the main towns during the Roman and Byzantine Empire was along the coastline of the Mediterranean Sea (due to commercial and climatic reasons, etc.; see Fig. 4.8a); (b) the physical boundaries (such as rivers and mountains) have been crucial since the roman era (Fig. 4.8c Part of the Byzantine Empire borders during Justinian's reign; the Danube and the Alps); and (c) the weakness to administrate and preserve large territories forced the Emperors to the division of the Empire several times (Diocletian, Theodosius, etc.; Fig. 4.8b shows the Theodosius division and the main invasions of the Empire, e.g., Goths, Huns, etc.).

As guidelines for future developments, it is worthy to mention the following. Firstly, the content of the application must be limited to what is included in the textbook. The pupils want to see in a more attractive way (e.g., though map visualizations), what they have already read in the book. Any additional content has been proved confusing and/or boring to them. Secondly, a series of simple but illustrated activities using either paper or screen maps are very attractive to them. Example activities include: (a) the re-design of boundaries on top of clear map backgrounds, (b) the comparisons with the contemporary status, and (c) the measurement and re-allocation of various figures and entities.

4.4 The Antikythera Wreck

The educational framework presented in the previous Section has also been customized and used in another context. The Antikythera Wreck is a shipwreck from about 150 BC. It was discovered by sponge divers on the Greek island of Antikythera in the early 1900s. The wreck produced numerous statues dating back to the fourth century BC, as well as the world's oldest known analogue computer, the Antikythera Mechanism (Fig. 4.9). The device is displayed in the National Archaeological Museum of Athens.

The origin of the mechanism is still unknown. There are several opinions expressed by archaeologists in the past. At Harokopio University of Athens and in cooperation with the Department of Education at the Museum, we have developed a web application, which shows the alternative origins and courses of the ship. This application is based on the educational framework presented in Sect. 4.3.

Figure 4.10a presents the interface of the application. The user may select, read and visualize the various opinions of the scientists as regards to the origin and the

Fig. 4.9 The Antikythera mechanism (Source: Wikipedia)

a

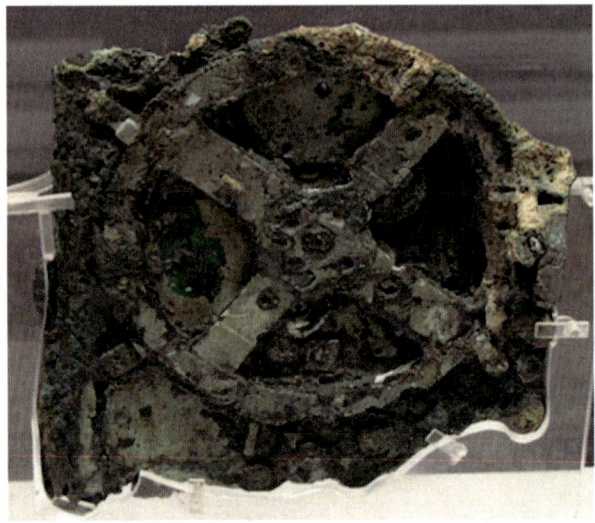

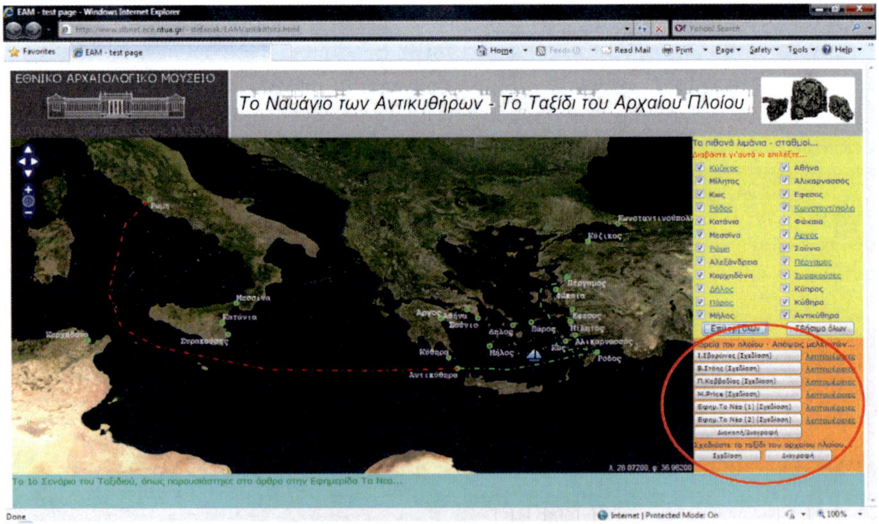

b

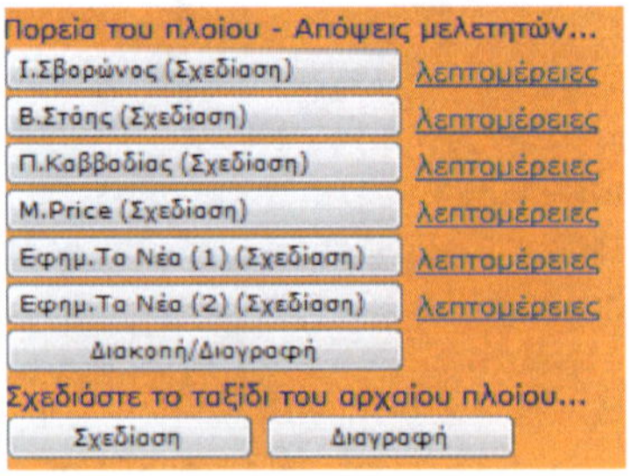

Alternative opinions expressed by various scientists.

Fig. 4.10 The interface of the Antikythera Wreck application (**a**). The opinions of the scientists regarding the origin and the course of the ship before the wreck (**b**)

course of the ship before the wreck (Fig. 4.10b). A moving icon shows an alternative course from the origin to the spot of the wreck, while the user may read the documentation as presented by the corresponding scientist.

4.5 The Battleship "G. Averof"

The last prototype has been developed as part of a research project for the promotion and enrichment of the Battleship G. Averof s Museum (Stefanakis and Kritikos 2008). The battleship "G. Averof" (Fig. 4.11) was a legendary warship of the Greek fleet during the first half of the twentieth century with an active history in three wars (Balkans, World War I and II). The ship has been acquired by the Greek Navy due to the funds offered by the wealthy Greek patriot and benefactor Georgios Averof. The ship-building has been concluded in 1911 in Italy.

"G. Averoff" is the world's only surviving heavily armoured cruiser of the early twentieth century. Today, it is a Museum operated by the Greek Navy. In the Department of Geography at Harokopio University of Athens – in cooperation with the Admiral in duty and the personnel at the Museum Battleship "G. Averof" and the permission of the Hellenic Navy – we have launched in Spring 2007 a research project with subject: "the promotion and enrichment of the Museum Archives".

Specifically, the research project is focused on: (a) the "life" in the battleship during the war periods (with presentation of the interior space in digital form, accompanied with photos and documents – during war periods the ship crew was 1,200 persons); (b) the movement of the ship in space-time and detailed mapping of the naval engagements and other operations; (c) the development of the ship during

Fig. 4.11 The battleship "G. Averof" (Source: Museum Archives)

the last century (armaments, breakages, demodernizations); and (d) the virtual visits in the Museum (with interactive multimedia and web applications).

4.5.1 Geographic Mashups

A pilot application to present the most important events of the battleship using *mashups* has been implemented in Google Maps API (Fig. 4.12). This application highlights the powerful visualizations that may be readily obtained by combining the content of the Project repositories/databases and the geographic mashups.

The application consists of an interface generated in HTML and visualizes the content of a KML (or *KMZ*) file on top of Google Maps. The latter has been generated by a *Java script*, which takes as input the result of a *SELECT* query in the historical photographs (image) database (Fig. 4.13).

4.5.2 The Geographic Catalogue Server

A rich digital content of various types and formats, ranging from digital image files to web mapping applications has been produced during the project. In order to make this content available and increase usability, we have generated a *geographic catalog server*. This server is able to diffuse all the digital content on the web, using widely accepted standards for services and protocols (such as those proposed by ISO TC211 and OGC). In other word, the catalog server transforms the digital repository to a *Spatial Data Infrastructure* (SDI; Williamson et al., 2004) of the battleship "G. Averof" history.

The software package adopted to implement the geographic catalog server is the *GeoNetwork Open Source Server* (Fig. 4.14). We have stored in this server appropriate *metadata* for the digital content (i.e., digital photographs, deck plans and battle plans) and the applications generated in the project. The metadata are in XML format and follow the specification of ISO19139.

The metadata items have been generated in an automated procedure, similar to the one in Fig. 4.13 and they are subsequently imported (*batch import mode*) to the GeoNetwork Catalog Server. Figure 4.15 presents graphically the procedure for the generation of the photographs metadata items.

Figure 4.16 presents the metadata of a historical photograph as presented to the end-user. As mentioned above, this metadata is encoded based on an ISO19139 template (XML format). Should Web users be authorized, they are capable to connect to the GeoNetwork server, access, browse the metadata and retrieve the geospatial content and/or applications. The server provides them an efficient interface to pose queries on metadata items with thematic, spatial and temporal predicates.

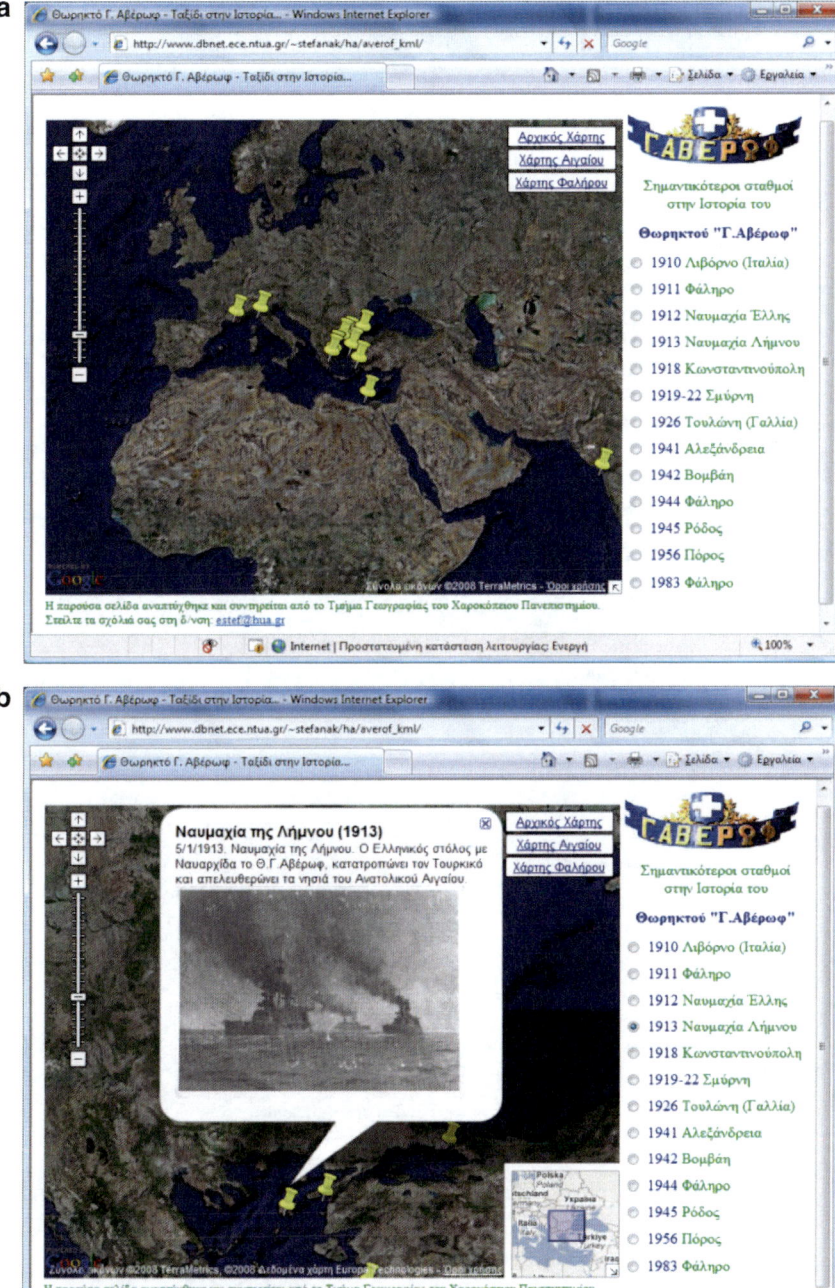

Fig. 4.12 The Google Maps API application presenting the most important events of the battleship history: (**a**) all the placemarks, (**b**) focus on the naval battle of Limnos. Available at: http://averof.dynalias.net/averof_kml

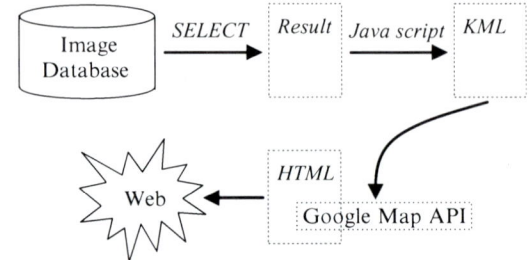

Fig. 4.13 The procedure towards the generation of the Google Maps API Application

Fig. 4.14 The initial screen of the catalog server for the battleship "G. Averof"

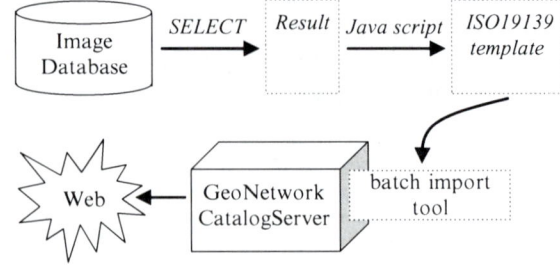

Fig. 4.15 The procedure towards the generation of the metadata items for the historical photographs

Fig. 4.16 The metadata items for a historical photograph (damages from the naval battle of Elli 12/Mar/1912) as presented to the end-user. This metadata is encoded based on an ISO19139 template (XML format)

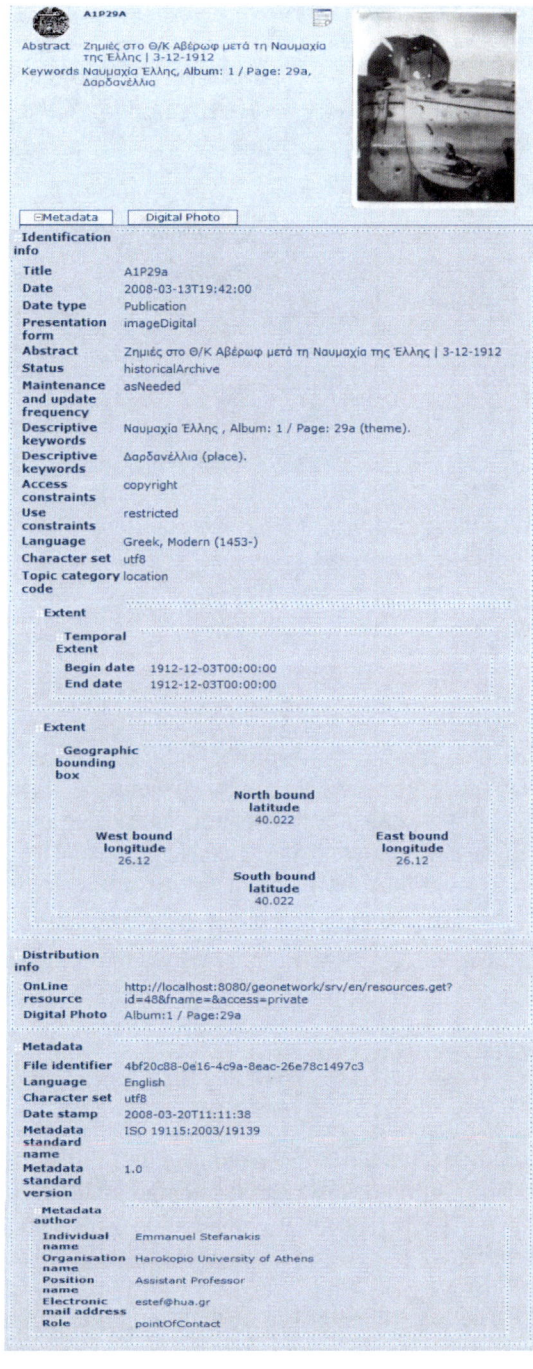

Fig. 4.17 An example query to the metadata items of the catalog server for the battleship "G. Averof"

Figure 4.17 presents the customized query interface in GeoNetwork Catalog Server for the battleship "G. Averof". End-users may enter: (a) the *thematic predicates* at the left side of the screen (e.g., a Keyword = "Naval battle of Elli") and/or at the right side of the screen (e.g., Category = "Image file"); (b) the *spatial predicate* on map frame at the middle of the screen (e.g., Spatial Query Window = [min_long, min_lat, max_long, max_lat]); and/or (c) the *temporal predicate* at the top-right side of the screen (e.g., Temporal Interval = [1911–1915]).

4.6 Conclusion

Current technological evolutions in map mashups and APIs offer the possibility to build effective educational tools with limited efforts and programming skills. This chapter summarizes several attempts made the last couple of years at Harokopio University of Athens in cooperation with educational organizations (schools and museums), to develop a series of prototype educational frameworks for the subject of history.

The experimental use of the prototypes by pupils, students and teachers in the school classes is rather encouraging. The prototypes can help users locate the historical events into space and perceive the existing geographical relations and interactions. The success of these prototypes is their customization to the pupils' needs, general experience and knowledge. This may be accomplished by involving the pupils and teachers at all phases of development, from design to the actual implementation of both the content and application scenario/interface.

Acknowledgments The author wishes to thank the Hellenic Fellowship Foundation (IKY), the Arsakeio Schools of Athens, the Museum Battleship "G. Averof" and the National Archaeological Museum of Athens for their contribution in the development of the prototype applications.

References

Allen JF (1983) Maintaining knowledge about temporal intervals. Commun ACM 26:832–843
Andrienko N, Andrienko G (2005) Exploratory analysis of spatial and temporal data: a systematic approach. Springer, Berlin
Antikythera Mechanism. http://en.wikipedia.org/wiki/Antikythera_mechanism. Accessed 1 July 2012
Antikythera Wreck. http://en.wikipedia.org/wiki/Antikythera_wreck. Accessed 1 July 2012
Brown MC (2006) Hacking GoogleMaps and GoogleEarth. Wiley, Indianapolis
Darmc (2010) Digital atlas of Roman and Medieval civilization. http://darmc.harvard.edu/. Accessed 1 July 2012
G. Averof's Museum. http://www.bsaverof.com/. Accessed 1 July 2012
GeoNames. http://www.geonames.org/. Accessed 1 July 2012
Geonetwork Opensource. http://geonetwork-opensource.org. Accessed 1 July 2012
Glentis S, Maragkoudakis E, Nikolopoulos N, Nikolopoulou M (2008) The Byzantine era. PI-Schools, Athens. http://www.pi-schools.gr/books/dimotiko/. Accessed 1 July 2012
Google Maps API. http://code.google.com/apis/maps/. Accessed 1 July 2012
Google Maps. http://maps.google.com/. Accessed 1 July 2012
Hellenic Navy. http://www.hellenicnavy.gr/. Accessed 1 July 2012
HTML. http://www.w3.org/html/. Accessed 1 July 2012
ISO/TC211. http://www.isotc211.org/. Accessed 1 July 2012
KML, Keyhole Markup Language. http://code.google.com/apis/kml/. Accessed 1 July 2012
Mitchell T (2005) Web mapping. O'Reilly, Beijing
OGC, Open Geospatial Consortium. http://www.opengeospatial.org/. Accessed 1 July 2012
OSGeo, Geographic Open Source. http://www.osgeo.org/. Accessed 1 July 2012
PI-Schools (2010) A cross thematic curriculum framework for compulsory education. The Greek Ministry of Education, http://www.pi-schools.gr/download/programs/depps/english/10th.pdf. Accessed 1 July 2012
PostGIS. http://postgis.refractions.net/. Accessed 1 July 2012
PostgreSQL. http://www.postgresql.org/. Accessed 1 July 2012
Stefanakis E (2008a) A journey to the ancient Greek myths: an enhanced educational framework to story-telling with geo-visualization capabilities. In: The Proceedings of the first international workshop on story-telling and educational games (STEG'08), Maastricht, 16 Sep 2008
Stefanakis E (2008b) Web services for mapping. Tutorial. In: The 3rd international conference on internet and web applications and services (ICIW 2008) Athens. http://www.dbnet.ece.ntua.gr/~stefanak/WebServMap_Stefanakis.pdf. Accessed 1 July 2012
Stefanakis E (2008c) Interactive visualization and analysis of the ancient Greek myths. In: GeoVisualization of dynamics, movement and change workshop, AGILE 2008 conference, Girona, May 2008. http://geoanalytics.net/GeoVis08/. Accessed 1 July 2012
Stefanakis E (2010) Introducing map mashups in primary school. In: Proceedings of the 6th international conference on geographic information science (GIScience 2010), Zurich, 14–17 Sept 2010

Stefanakis E, Kritikos G (2008) The battleship "G. Averof" promotion and enrichment of the museum archives. In: Proceedings of the XXI ISPRS congress, Commission IV, WG IV/1, Beijing, July 2008, pp 67–72
W3C. http://www.w3.org/. Accessed 1 July 2012
Wikipedia (2010) Wikipedia: the free encyclopaedia. http://wikipedia.org/. Accessed 1 July 2012
Williamson IP, Rajabifard A, Feeney MEF (2004) Developing spatial data infrastructures: from concept to reality. Taylor & Francis, London
XML. http://www.w3.org/XML/. Accessed 1 July 2012
XSL. http://www.w3.org/Style/XSL/. Accessed 1 July 2012
Young M (2008) Practical Google maps mashups with Google mapplets, GeoRSS and KML. Apress

Part II
API Mashups

Chapter 5
Multimedia Mapping on the Internet Using Commercial APIs

Shunfu Hu

Abstract Multimedia mapping provides a unique approach to integrating geospatial information in digital map format and multimedia information (e.g., text, photographs, sound, and video). Multimedia mapping on the Internet is the direct result of advancement of Web mapping techniques, Internet technology, and Web standards (e.g., HTML, XML, Ajax). However, like traditional Internet GIS applications, multimedia mapping on the Internet is suffered from slow response time, limitation of data sizes, and the slow client/server communications. In addition, the development of multimedia mapping on the Internet requires a huge investment of computer hardware and software, and a steep learning curve for the application developer to become knowledgeable about all of the components involved. It is imperative that a better solution is needed to implement multimedia mapping projects on the Internet. This chapter discusses a new approach of multimedia mapping that utilizes commercial Application Programming Interface (APIs) such as Google Maps API, Yahoo! Flickr API, and YouTube API that can provide faster response time, greater user interaction, as well as higher quality multimedia presentation. A case study of an online visitor guide for the campus of Southern Illinois University Edwardsville is demonstrated.

5.1 Introduction

Multimedia mapping refers to the integration of computer-assisted mapping system and multimedia technologies that allows the incorporation not only of geospatial information in digital map format, but also multimedia information (Hu 2010). The term "multimedia" implies the use of a personal computer (PC) with information presented through the following media: (1) text (descriptive text, narrative and

S. Hu (✉)
Department of Geography, Southern Illinois University, Edwardsville, IL, USA
e-mail: shu@siue.edu

labels); (2) graphics (drawings, diagrams, charts or photographs); (3) digital video (television-style material in digital format); (4) digital audio sound (music and oral narration); (5) computer animation (changing maps, objects and images); and virtual reality (VR). The development of multimedia mapping techniques has gone through several stages, including the development of interactive maps and electronic atlases during the 1980s (Peterson 1995; Openshaw and Monnsey 1987; Rhind et al., 1988; Shepherd 1991), the development of "hypermap" in the early 1990s (Wallin 1990; Laurini and Milleret-Raffort 1990; Cotton and Oliver 1994; Cartwright 1999), the integration of hypermedia system (which features hypertext, hyperlinks and multimedia) and geographic information system (GIS) in the late 1990s and early 2000s (Shiffer 1998; Bill 1998; Hu 1999; Soomro et al., 1999; Chong 1999; Hu et al., 2003; Yagoub 2003; Goryachko and Chernyshev 2004; Belsis et al., 2004), and more recently the integration of Web-based mapping tools and multimedia-rich Web applications using Hypertext Markup Language (HTML), Extensible Markup Language (XML), and Asynchronous JavaScript and XML (Ajax) (Hu 2006, 2009).

Hu (2006) described multimedia mapping on the Internet as a Web application, which is based upon the interactions between client and server computer systems through network technology. The client side allows Internet users to access remote computers on the Internet by providing requests through standard Web browser software such as Microsoft Internet Explorer, Netscape Navigator, or other custom-generated software such as ESRI ArcExplorer. The server side consists of at least three components, including the web server, the map server and the data server. Web server software such as Netscape's FastTrack Server or Microsoft's Internet Information Server provides the capability to manage and respond to requests from the client side. The map server, interacting with the web server, implements data processing in a GIS application. Examples of such map servers include ESRI ArcView Internet Map Server (IMS), MapObjects IMS, and ArcIMS, MapInfo MapX and MapXtreme. The data server provides various data sets such as ESRI ArcView shapefiles, ArcInfo coverages, remotely sensed data, and/or other statistical data. Typically, the components on the server side can be placed on more than one computer. The network technology provides Internet software components that communicate with each other on various computers connected by the network. Those components include HTTP (i.e., hypertext transfer protocol), and TCP/IP (transmission control protocol/Internet protocol). Protocols are the languages that make Internet communication possible. In Hu's work, the multimedia mapping was based upon interactions between three components: (1) a web-based mapping application developed to manipulate the cartographic features and their attributes; (2) a web-based hypermedia system designed to manipulate multimedia information including hypertext, hyperlinks, graphics, photographs, digital video, and sound; and (3) a mechanism linking the web-based mapping application and the hypermedia system. Such implementation of the web-based multimedia mapping, just like traditional Internet GIS applications, is suffered from slow response time, limitation of data sizes, and the slow client/server communications. In addition, it requires the huge investment of computer hardware (i.e., dedicated computer server) and

Table 5.1 Comparison of various multimedia mapping techniques

	Multimedia mapping technique		
	Desktop	Web-based hypermedia GIS	Commercial APIs
Programming (scripting)	Visual basic	Visual basic and HTML	JavaScript and XHTML
Digital mapping	Arcview, ArcGIS or MapObjects	MapObjects IMS or ArcGIS IMS	Google Maps API
Data storage	Local (computer hard drive or CD-ROM)	Remote (server)	Remote (multiple professional services)
Response time	Short (local access)	Long (through network cable)	Fast (through Ajax)
Media format	Window-based (.tif, .avi, .wav)	Web-based (.gif, .mov, .wav)	Any type
User	Single user	Internet user	Internet user

software (i.e., web server, map server), and a steep learning curve for the developer to be knowledgeable about the coding in native language of the map server.

The launch of Google Maps in 2005 has revolutionized web mapping service applications on the Internet. Based on Asynchronous JavaScript and XML (AJAX), Google Maps introduced a new type of server/client interaction that maintained a constant connection to the server for more immediate downloading of additional map information (Peterson 2008). In addition to implementing a better client/server interaction, Google also provides programmers free access to its code in the form of an Application Programming Interface (API). In other words, the API consists of a set of routines or functions that can be called by a programmer using JavaScript, php, or similar scripting language (Udell 2009). For example, HousingMaps (www.housingmaps.com) is a combination of Google Maps with the realty listings from Craigslist (www.craigslist.org). On the other hand, commercially available APIs for digital photograph sharing service such as Yahoo! Flickr and for digital video sharing service such as YouTube provide unique way to deliver multimedia information on the Internet. These APIs do most of the heavy lifting, and the developer can combine functionality easily to create a simple but useful application in a relatively short amount of time. They allow the developer to quickly build a web application without a need of "reinvent the wheel". Table 5.1 provides a comparison between the method used in the author's earlier multimedia mapping work (Hu 1999, 2006; Hu et al., 2003) and the new method proposed in this chapter.

The objective of this chapter is to discuss a new approach of multimedia mapping that utilizes Google Maps API, Yahoo! Flickr API, and YouTube API to incorporate descriptive text, photographs, and video clips in a Google digital map environment that can bring vital information to the Internet user. A case study of an online visitor guide that features hypertext, hyperlinks and multimedia for the campus of the Southern Illinois University Edwardsville (SIUE) is demonstrated.

5.2 Methodology

5.2.1 Study Area

The SIUE campus was selected as the study area to develop the online visitor guide. Situated on 2,660 acres (11 km^2) of beautiful rolling hills, and forested woodlands atop the bluffs overlooking the natural beauty of the Mississippi River's rich bottom land, the SIUE campus is the second largest campus landwise in the United States. The brick, slate and granite of the modern buildings complement the terrain and are softened by a carefully designed garden landscape that attracts visitors by its physical beauty. These characteristics make the campus an ideal location for the development of an online multimedia mapping project.

5.2.2 Data Sources

The data sets for this project included digital maps, digital photographs, and digital video clips, and descriptive text about points of interest (e.g., academic buildings) on the SIUE campus. The digital map and satellite imagery were provided by the Google Maps (via Google Maps API). A series of photographs for the points of interest were taken using Cannon SLR 40-D digital camera and were saved as JPEG format. A few samples of digital video clips were acquired using Panasonic digital video camcorder and saved as AVI format. Descriptive text about academic departments was provided through web links from the points of interests to those academic departments' web sites.

5.2.3 "Mashup" of Google Maps API, Yahoo! Flickr API, and Youtube API Through JavaScript and XHTML

Since the online visitor guide for the SIUE campus is a web application, every API implementation is based on a web page. JavaScript is the native language of Google Maps. In addition, Google Maps is built of XHTML (Extensible HTML), formatted with CSS (Cascading Style Sheet) (Udell 2009). Therefore, both JavaScript and XHTML are used for the multimedia "mashup" of the Google Maps API, Yahoo! Flickr API, and YouTube API for developing the online visitor guide of the SIUE campus. Figure 5.1 illustrates a conceptual framework for the integration of the three APIs in the World Wide Web environment.

First of all, in order to use the Google Maps API, a programmer has to set up a Google account and apply for a Google Maps API key (http://code.google.com/

5 Multimedia Mapping on the Internet Using Commercial APIs

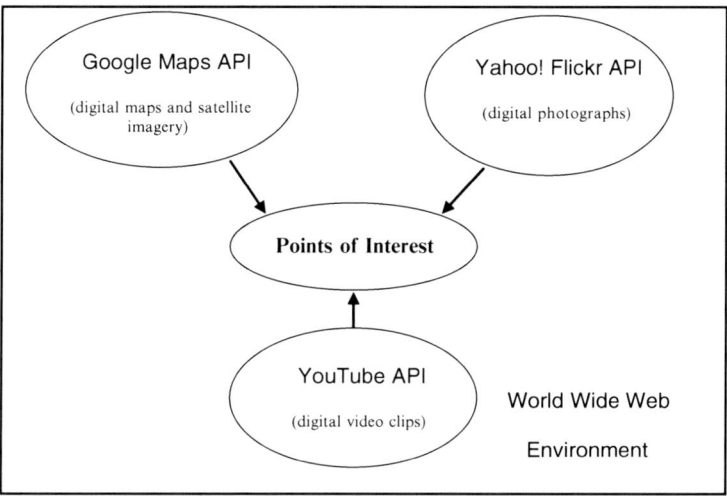

Fig. 5.1 Web-based multimedia mapping "mashups" using commercial APIs to provide multimedia information for points of interest in a digital map environment

apis/maps/signup.html). This is essentially the authorization from Google to use Google Maps on the programmer's web site. Version 3 of Google Maps, released in mid-2010, no longer requires a key.

In the design of the web application, a two-column layout design was adopted, including a column for a map container and the other column for a sidebar. In the map container, the Google map in various types for the campus is displayed, and points of interest (e.g., academic buildings) are marked with the default Maps API marker icons – the reddish-balloon-with-a dot. The sidebar contains a list of the same points of interest marked on the map. Such content is handy to the page visitor because it enables him/her to the points of interest at a glance. If the user is interested in a particular academic building, for example, but does not know where it is on campus, the sidebar list lets him/her go straight there without having to hunt around the map for it. Therefore, it is beneficial to establish a real linkage between the sidebar and the map.

In order to provide the user with interaction with the map, a few standard Google Maps controls are added, such as Pan and Zoom controls (GLargeMapControl); Map Scale control (GScaleControl); Overview control (GOverviewMapControl); and Map Type control (GMapTypeControl): Map, Satellite, Hybrid, Terrain and Earth. In addition, tooltips to the markers are provided, along with clickable marker icons with Google Maps API's standard Infowindow, which can attach multimedia content to a marker. Three tabs associated with the Infowindow are provided: Details tab for the descriptive text, hypertext and hyperlinks; Photo tab for photographs or Flickr slideshow; and Video tab for digital videos or YouTube movies.

Flickr (www.flickr.com) is an photo hosting and video hosting website created by Ludicorp and later acquired by Yahoo!. In addition to being a popular website for users to share and embed personal photographs, the service is widely used by bloggers to host images that they embed in blogs and social media. **SlideFlickr** (www.slideflickr.com) is an online Flickr slideshow generator that can help create Flickr slideshows, which can be later embedded on a website using a few lines of JavaScript code. For this project, all the photographs were pre-loaded onto a Yahoo! Flickr account (free of charge) and organized into sets based on the academic buildings on the SIUE campus. Then, a SlideFlickr slideshow for each academic building was generated and its SlideFlickr IDs were obtained to embed those slideshows into the SIUE campus visitor Web page.

YouTube (www.youtube.com) **is a free online video streaming service** that allows the user to upload, share, and view videos. It is a place for people to engage in new ways by sharing and viewing videos. YouTube accepts many different video file formats such as MPEG, AVI, MOV, WMV and converts it to Flash video file format or FLV format. Note that FLV files cannot be played back directly by the Flash Player (a free player by Adobe) – they must be embedded in an SWF file. SWF files are Flash files which end users see. Whenever the end user sees a Flash file on a web page, he/she is looking at an SWF file. Technically speaking, an SWF file is a compressed version of the FLA file which is optimized for viewing in a web browser, the standalone Flash Player, or any other program which supports Flash. The play window of YouTube videos is normally at 340×240 in size. The YouTube videos can also be integrated on websites using video embeds or APIs. For this project, video clips about various aspects of the campus activities initially saved in AVI format were loaded onto a YouTube account and the links were embedded into the web pages of the online campus visitor guide.

5.3 Results

The use of Google Maps API provides very efficient mechanism to deliver digital cartographic information to the Internet user with fast response time and user-friendly interaction. With Google Maps standard Map Type control, the user is able to choose any of the map types: map, satellite imagery, hybrid (map and imagery), or terrain (topography) about the environment under investigation. The addition of Google Earth can add more functionality for the user to explore the environment. Figure 5.2 shows an outlook of the online campus visitor guide for the SIUE campus. On the left side is the map container that displays the digital map environment for the campus with the high-resolution satellite imagery. Points of interest are marked on the map with the default Maps API marker icons. The sidebar on the right side displays a list of the same points of interest. The user can click on an icon inside the map container or a name of the points of interest in the sidebar to find out more information about it. For example, if the user moves the cursor over the Peck Hall building (Note: tooltips will tell the user the name of a

5 Multimedia Mapping on the Internet Using Commercial APIs

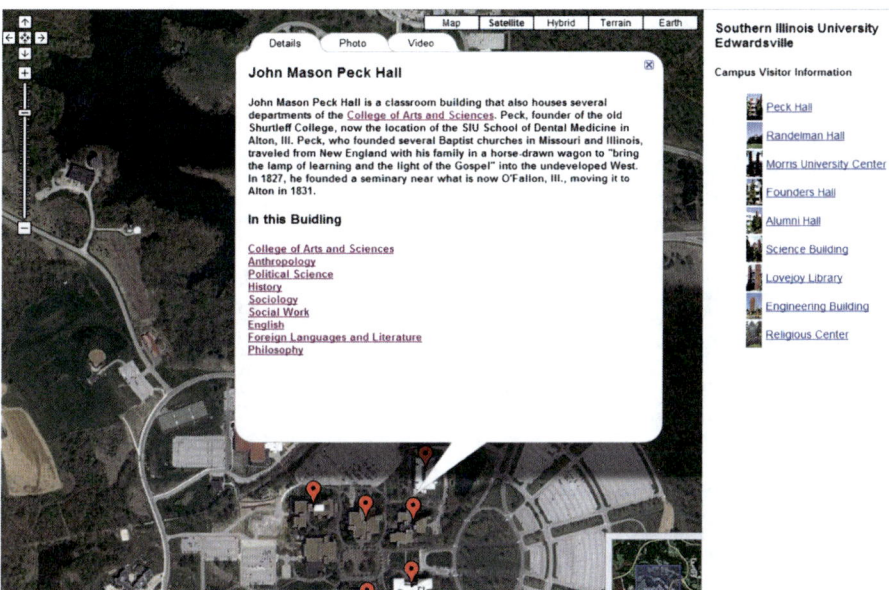

Fig. 5.2 Infowindow's details tab provides descriptive information about an academic building (e.g., Peck Hall)

point of interest) inside the map container and then clicks on the icon, the Google Infowindow will pop up. There are three tabs in the Infowindow. The Details tab describes the history of this building and provides a list of the departments (in hypertext format) housed in this building. The department hypertext provides hyperlink to the Web page of respective department. Similarly, the user can click on the name of a point of interest from the sidebar to launch the Infowindow associated with it.

Figure 5.3 demonstrates a Google Infowindow's Photo tab that provides access to digital photographs organized in Flickr slideshow about the Peck Hall building. The user can simply watch the slideshow or interact with the slideshow by clicking the forward or backward button.

Figure 5.4 demonstrates a Google Infowindow's Video tab that provides access to digital videos. One method for the video to be played is to launch a new video play window. In this case, a hyperlink can be provided for the user to click on. Another way is to embed a YouTube video into the Web page so the user can simply click on the Play button to watch the movie.

Finally, the Google Earth map type enables the Web page visitor to see the SIUE campus in the full three-dimensional splendor of Google Earth. Figure 5.5 demonstrates the Google Earth in the Internet browser. The user can click the name of a point of interest in the sidebar and the browser will fly into that point of interest. Then, the user can click on the marker to explore further information about it from the Google Infowindow tabs mentioned above.

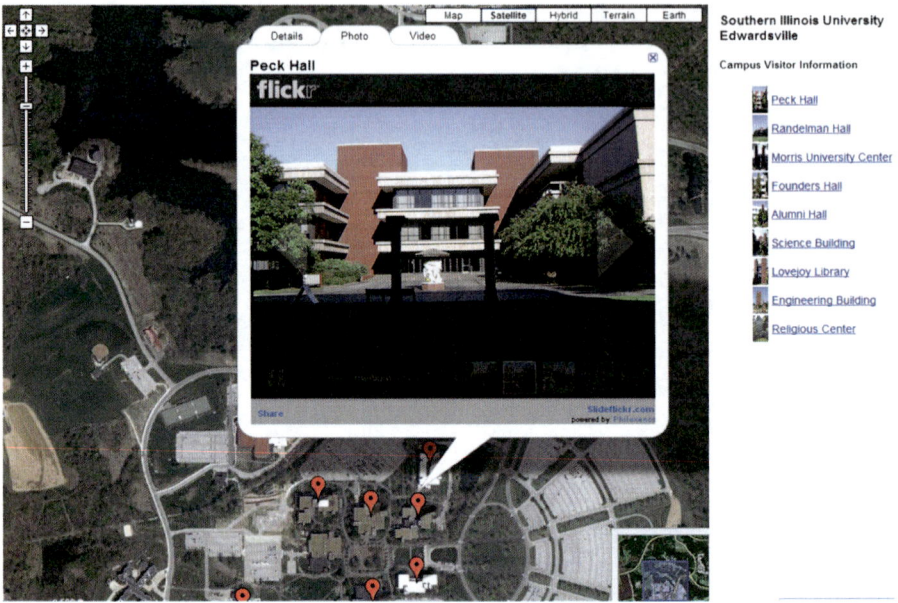

Fig. 5.3 Photo tab provides access to a Flickr slideshow. The user is able to control the slideshow by clicking forward or backward button

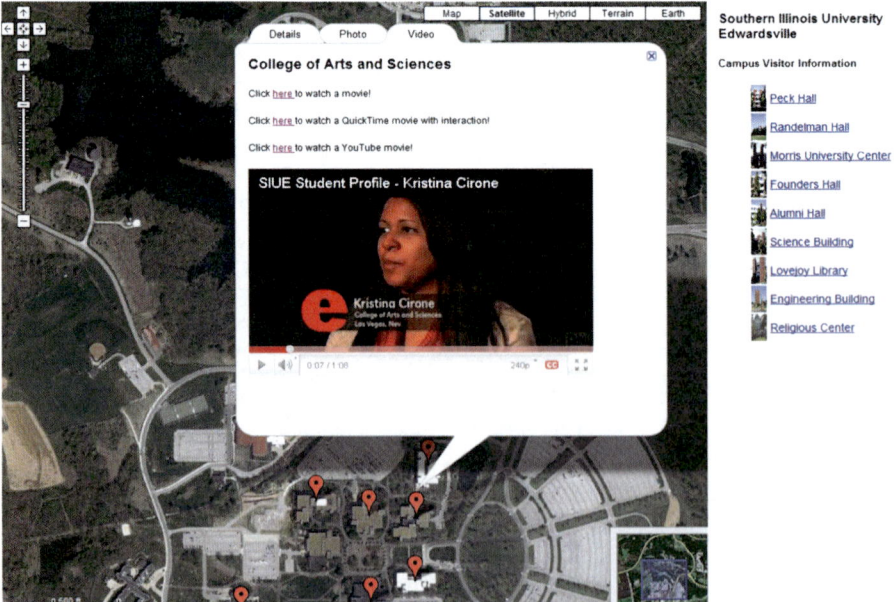

Fig. 5.4 Video tab provides access to digital videos. Shown here are a few hypertext and hyperlinks to digital video clips as well as an embedded YouTube movie

5 Multimedia Mapping on the Internet Using Commercial APIs

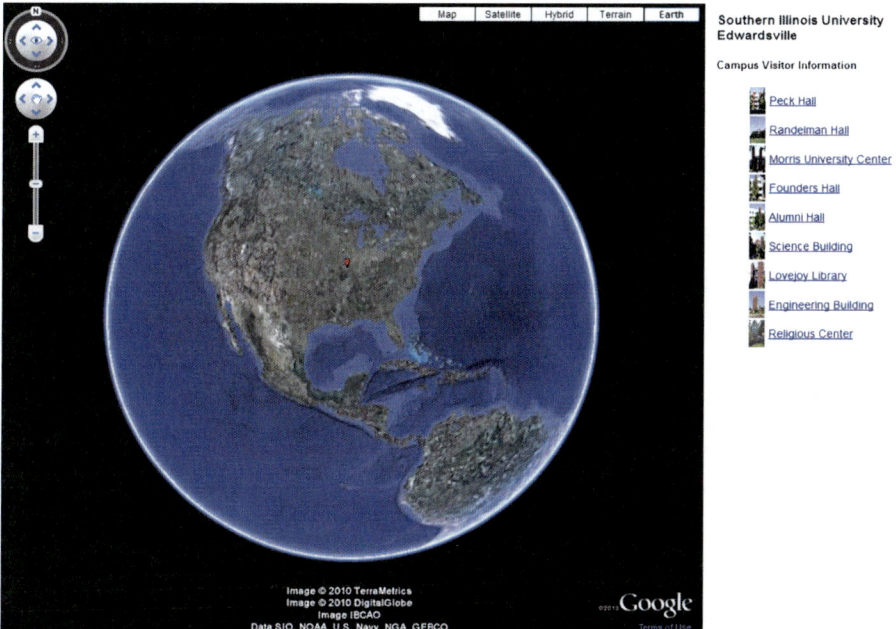

Fig. 5.5 Google earth in the internet browser

5.4 Conclusions

This paper has demonstrated a relatively easy approach to multimedia mapping on the Internet. The new approach combines spatial data, multimedia information and functionality from multiple sources such as Google Maps, SlideFlickr, and YouTube to create a new service such as the SIUE online visitor guide. This new approach clearly has some advantages over the previously documented multimedia mapping techniques on the Internet. First, the programmer can employ easy-to-use commercial APIs to develop multimedia mapping applications. With no investment on dedicated computer hardware (i.e., server) and specialized software (i.e., web server, map server), the developer just needs to have a web account, a Google account, a YouTube account, and a Yahoo Flickr account so he/she is able to develop a web page and deliver the multimedia mapping service to the Internet user. Second, with much faster response time and greater user interaction, the Internet user just needs a standard Internet browser to navigate (zoom in, zoom out, pan, etc.) from place to place within the Google Maps digital map environment. Third, the multimedia content, especially digital photographs, digital sounds and videos, can be delivered to the Internet user through free-of-charge professional services provided by third parties (e.g., Yahoo! Flickr or YouTube). Such professional services usually provide the state-of-the-art technology to deliver the multimedia content with the highest quality at the fastest speed (e.g., video streaming). The apparent benefit of using existing APIs may be of value to those who do not have the time or energy to invest in more customized solutions. It is also beneficial for college students to easily start off with the multimedia mapping technology and for organizations, such as the university in this case study, to see the benefits of the technology without having to make a huge investment.

References

Belsis P, Gritzalis S, Malatras A, Skourlas C, Chalaris I (2004) Enhancing knowledge management through the use of GIS and multimedia. Lect Notes Artif Int (Subseries Lect Notes Comput Sci) 3336:319–329

Bill R (1998) Multimedia-GIS concepts and applications. Geo-Inf Syst 11:21–24

Cartwright W (1999) Development of multimedia. In: Peterson MP, Gartner G (eds) Multimedia cartography. Springer, New York, pp 11–30

Chong AK (1999) Orthoimage mapping bases for hybrid, multimedia, and virtual environment GIS. Cartography 28:33–41

Cotton B, Oliver R (1994) The cyberspace lexicon – an illustrated dictionary of terms from multimedia to virtual reality. Phaidon Press, London

Goryachko VV, Chernyshev AV (2004) Multimedia and GIS-technologies in atlas mapping. Vestn Mosk Univ 5(2):16–20

Hu S (1999) Integrated multimedia approach to the utilization of an Everglades vegetation database. Photogram Eng Rem Sens 65(2):193–198

Hu S (2006) Design issues associated with discrete and distributed hypermedia GIS. In: Stefanakis E, Peterson MP, Armenakis C, Delis V (eds) Geographic hypermedia: concepts and systems. Springer, Berlin, pp 37–52

Hu S (2009) Advancement of web standards and techniques for developing hypermedia GIS on the Internet. In: Madden M (ed) ASPRS manual of GIS. American Society of Photogrammetry and Remote Sensing (ASPRS), Bethesda, MD, pp 975–983

Hu S (2010) Multimedia mapping. In: Warf B (ed) Encyclopedia of geography. SAGE Reference Publication, Thousand Oaks

Hu S, Gabriel AO, Bodensteiner LR (2003) Inventory and characteristics of wetland habitat on the Winnebago upper pool lakes, Wisconsin, USA: an integrated multimedia-GIS approach. Wetlands 23(1):82–94

Laurini R, Milleret-Raffort F (1990) Principles of geomatic hypermaps. In: Proceedings of the 4th international symposium on spatial data handling, Zurich, Switzerland, vol 2, pp 642–655

Openshaw S, Mounsey H (1987) Geographic information systems and the BBC's Domesday interactive videodisk. Int J Geogr Inf Syst 1(2):173–179

Peterson M (1995) Interactive and animated cartography. Prentice Hall, Upper Saddle River

Peterson MP (2008) International perspectives on maps and the internet: an introduction. In: Peterson MP (ed) International perspectives on maps and the internet. Springer, Berlin, pp 3–10

Rhind DP, Armstrong P, Openshaw S (1988) The Domesday machine: a nationwide geographical information system. Geogr J 154(1):56–58

Shepherd ID (1991) Information integration and GIS. In: Magurie DJ, Goodchild MF, Rhind DW (eds) Geographical information systems: principles and applications, vol 1. Longman Scientific and Technical Publications, Essex, pp 337–357

Shiffer MJ (1998) Multimedia GIS for planning support and public discourse. Cartogr Geogr Inf 25(2):89–94

Soomro TR, Zheng K, Turay S, Pan Y (1999) Capabilities of multimedia GIS. Chin Geogr Sci 9(2):159–165

Udell S (2009) Beginning Google maps mashups with mapplets, KML, and GeoRSS. Apress, New York

Wallin E (1990) The map as hypertext – on knowledge support systems for the territorial concern. In: Proceedings of the first European conference on geographical information system, EGIS' 90. EGIS Foundation, Munich, pp 1125–1134

Yagoub MM (2003) Building an historical remote sensing atlas and multimedia GIS for Al Ain. GEO: Connexion 2(7):54–55

Chapter 6
The GIS Behind *i*MapInvasives: The "Open Source Sandwich"

Georgianna Strode

Abstract Invasive species can be considered an ecological "time bomb" that contributes to a number of human and environmental problems, including reduced crop and livestock production and biodiversity loss. It is estimated that 5,000 acres of western lands become infested each day (North American Weed Management Association 2002). Westbrooks assesses the financial costs to Americans at $138 billion per year. Of the current methods for treating invasive species, Early Detection and Rapid Response (ED/RR) is highly ranked because it is cost-effective and environmentally sound and it increases the possibility of successful eradication (ibid).

For ED/RR to be effective, land managers must have access to comprehensive and accurate data. Tracking invasive species is challenging because multiple agencies and organizations collect data, often resulting in isolated datasets. In addition, each agency uses its own collection format. Sensitive and unverified data add further complications. Land managers need to see the "big picture," which often crosses political boundaries and results in differing ideas concerning collection and dissemination techniques.

To facilitate collaboration, *i*MapInvasives.org was created as a versatile mapping tool to serve the needs of land managers, conservation planners, and agency decision makers. Internet Geographical Information Systems (IGIS) provide collaboration and networking capabilities that were previously unavailable with desktop GIS. Data are collected from multiple agencies and Users can access data from all sources, regardless of the original data format. There are other useful features such as customizable Early Detection notifications and restriction of sensitive data from users with less system authorization.

G. Strode (✉)
Florida Resources and Environmental Analysis Center (FREAC), Florida State University (FSU), Tallahassee, FL, USA
e-mail: GStrode@admin.fsu.edu

*i*MapInvasives uses an "open source sandwich" approach to IGIS. The "sandwich" uses mostly open source technologies, but the middle layer, or map, may contain images produced from commercial products. Together, these complementary components provide the robustness required for tracking invasive species with ample flexibility for customization.

While the challenges facing *i*MapInvasives are specific to invasive species management, the conceptual and technological issues presented here apply to any area of data collection and dissemination. This chapter gives an overview of the internal workings of the system and details some of the features that are unique to invasive species tracking.

6.1 Importance of Early Detection and Rapid Response

When water chestnuts were discovered in a New Jersey lake, organizers acted quickly to remove them. Water chestnuts can choke out other species, deplete oxygen, and kill fish and other species. The species reproduces rapidly and can grow so densely that it can result in large floating mats that cover the water. The eradication effort was considered successful because the plants were removed before they could become established (Bouchal 2010).

Biodiversity loss caused by invasive species is increasing and could soon surpass habitat destruction and fragmentation (Crooks and Soule 1999) as dangers to the natural environment. Invasive species are considered to be ecological time bombs because, unlike a chemical spill, the damage replicates itself and can spread exponentially. Cornell University estimates the economic impact of exotic species in 1994 at $20 billion per year and at $138 billion per year in 1999 (Westbrooks 2004).

Often little is done about invasive species because there is conflict between the need to minimize environmental damages and control efforts. If public welfare is not at risk, invasive species are often ignored. Lack of awareness concerning treatment of invasive species contributes to the problem. Landowners, land managers, and gardeners are in a position to provide leadership to combat this environmental threat (ibid).

There are several methods used to help reduce the spread of invasive species. They include screening imports, border inspections, Early Detection and Rapid Response, and long-term management of large land areas. Of these four, Early Detection and Rapid Response is the most cost-effective and environmentally sound approach. It has the highest chances of successful eradication (ibid). Remote sensing has been used successfully for early detection (Lass et al., 2005) along with other methods.

Early detection strategies can be tailored for specific goals. A "weed-led" (sometimes called "weed-driven" or "species-driven") approach helps establish a targeted list of species, which is important because species can be new to one part of an area and be considered a serious problem in another area. A "site-led" (or "site-driven") program is designed to protect specific areas: vulnerable and valuable. Vulnerable areas can be the first places a species colonizes but may have little conservation value. Vulnerable areas can include areas along roads or railways, rubbish dumps, areas in or near water, and places modified by human activity. Valuable areas have high conservation value. It is possible to use a "site-led" strategy to monitor the vulnerable areas within a valuable area (Owen 1998).

Early detection of invasive species is becoming more important as policy makers and land managers realize the threat that invasive species can pose to the natural environment and to property values. While some invasive species databases exist, they usually are inventory-based and do not have sufficient Geographic Information System (GIS) capabilities to be useful for decision making. John Randall of The Nature Conservancy describes the role of *i*MapInvasives as:

> *i*MapInvasives is filling a need for a database system that will allow information on the distribution of invasive plant species across large areas like states to be compiled, mapped and analyzed. This will allow land managers, conservation planners and agency decision-makers to set priorities for prevention and control programs and budgets. Better yet, as the number of states using *i*MapInvasives increases, the possibility of mapping distribution and spread at even larger scales such as multi-state regions opens up. On the other end of the scale, some land managers may use *i*MapInvasives to better understand the distribution of invasive plants on their lands and surrounding properties which will give them an edge in developing control and action plans. (Randall 2010)

6.2 The Conceptual Sandwich

The term "open source sandwich" is a slang term used by map developers to describe an internet map made mostly with open source products and a map made from one or more commercial products. It is an open source technology stack with a proprietary map in the middle.

The term "map sandwich" is defined as several map layers pressed together to form one map. An example would be a demographic layer "sandwiched" between a terrain layer and reference information. The purpose of a map sandwich is to better control the final look of the map by choosing and arranging individual map layers. The resulting map should look professional (Frye 2010). James Fee carries the definition further to suggest that publishing map layers in easy-to-use formats helps map consumers integrate published map data layers in their maps. Making data layers easily accessible increases the chances that the public will use them (Fee 2010).

A "technology stack" is a suite of products that work together. An "open source technology stack" is a technology stack that uses only open source products. An example could be PHP or Python, Javascript, Apache, and PostgreSQL. It is important to remember that "open source" is defined by the license(s) that govern the use, distribution, and rights to the software's source code. Open source software is not defined by programming language or operating environment (Tiemann 2010), nor does it imply any indication of quality.

The "sandwich" model is a loose interpretation of the traditional architectural model. Even so, the "sandwich" model is a simple way to understand the three parts of internet mapping: map navigation tools (top), a static map image (middle), and supporting spatial databases (bottom). Stated another way, the top can be considered client-side, the bottom is considered server-side, and the map is in the middle.

6.3 *i*MapInvasives Product Choices

GIS is increasingly used to support environmental modeling as GIS technologies have improved. However, many environmental modeling frameworks fail to represent or describe geographic information in a rigorous way. Environmental models can become locked into a specific GIS platform that is difficult to change. Logic that is based on science can become intertwined with technology and it is difficult to separate the two. As a result, there is increased interest in new ways to integrate GIS and environmental modeling. There is increasing popularity of ad hoc approaches to integrate GIS and modeling, leading to leveraging GIS functionality from traditionally non-GIS software (Viger et al., 2004).

In choosing a technical solution for this project, several considerations were taken into account:

- Follow computer industry and GIS standards whenever possible
- Minimize use of plug-ins and vendor-specific extensions and APIs
- Use proven products
- Have complete programming control

The solution chosen is a hybrid comprised of open source and commercial products. Non-GIS software is used wherever possible to keep science logic and technology separated to avoid being locked into a particular GIS platform.

6.3.1 Products by Type

The main components in the technology stack include OpenLayers, PostgreSQL/PostGIS, Django, and Python. The map sandwich uses a combination of Environmental Systems Research Institute's (ESRI) Representational State Transfer (REST) and third-party map products such as Google or Bing. See Fig. 6.1 shows the products used by architectural position and product type. Figure 6.2 shows the products listed by sandwich position and by product type.

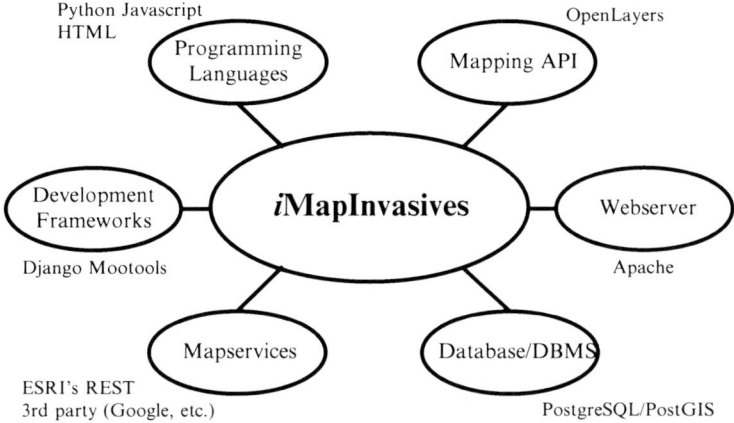

Fig. 6.1 Conceptual products by type

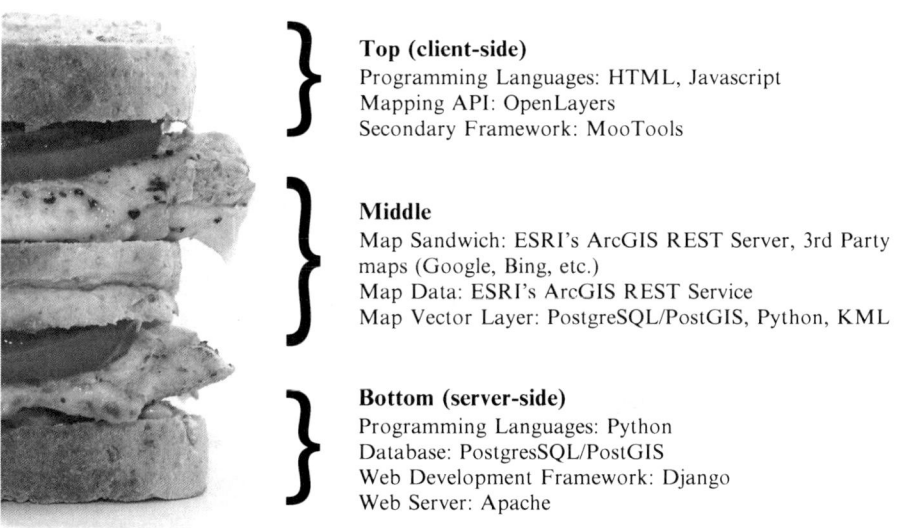

Top (client-side)
Programming Languages: HTML, Javascript
Mapping API: OpenLayers
Secondary Framework: MooTools

Middle
Map Sandwich: ESRI's ArcGIS REST Server, 3rd Party maps (Google, Bing, etc.)
Map Data: ESRI's ArcGIS REST Service
Map Vector Layer: PostgreSQL/PostGIS, Python, KML

Bottom (server-side)
Programming Languages: Python
Database: PostgresSQL/PostGIS
Web Development Framework: Django
Web Server: Apache

Fig. 6.2 Products used by architectural position

6.4 The Product Details

Ten products and formats are used for the *i*MapInvasives project. They are listed below in order of the sandwich position and by functionality.

6.4.1 Top (Client-Side)

Programming Languages (Client-Side)

HTML, Javascript
These web tools are common and need no further explanation.

Mapping API

OpenLayers
OpenLayers is a powerful open source, Javascript-based client-side library for creating interactive maps in the browser with no server-side dependencies. There are numerous built-in controls, such as the layer switcher, scalebar, overview map, permalinks, and others. Multiple layers are allowed and each layer object can be individually manipulated. As a result, developers have complete control of everything: controls, map events, layers, etc. (OpenLayers 2010).

Secondary Development Framework

Mootools
This is an object-oriented Javascript framework designed to help easily develop clean, cross-browser, elegant code (MooTools 2010). *i*MapInvasives uses this framework for automatic sorting and pagination of information from database queries and to show the results of a map identify. MooTools also offers "growls" (also known as "roars") to notify users of which actions they have just taken. A "growl" is one or more small windows that appear briefly (1–2 s) to give users a status update after they have made a significant change.

6.4.2 Middle ("Map Sandwich")

Map Background Images

ESRI's ArcGIS REST Server
The Representational State Transfer (REST) is a method for distributing media across systems such as the internet in the client/server environment (REST 2010). *i*MapInvasives uses REST to obtain a map image. A Web Mapping Service (WMS) image could be used in place of REST.
Below is a code sample to show how the request is made.
The ArcGIS Server REST interface allows us to request parts of the map from a URL: http://server/ArcGIS/rest/services/World_Shaded_Relief_esri/MapServer/export?bbox=72679.8759250002,4305685.42989042,797063.251575,5162872.42440958&bboxSR=&layers=&layerdefs=&size=&imageSR=&format=png&transparent=false&dpi=&f=image

Third Party Maps
   This could be any map source such as Google, Bing, Virtual Earth, or other mapping service.

Map Data

ESRI's ArcGIS REST Service
   The REST server was described previously for use as a background image.

Map Vector Layer

Keyhole Markup Language (KML)
   KML is an XML-based format for displaying geographic data. Originally used with Google Earth, KML is now used with many GIS products (KML 2010). *i*MapInvasives uses KML in the vector layer to easily show geographic boundaries in vector format.
   Other vector layer functions are controlled through Python and PostgreSQL/PostGIS. Each of these is discussed in a later section.

6.4.3 Bottom (Server-Side)

Programming Language (Server-Side)

Python
   Python is a widely used open source programming language, named after *Monty Python and the Flying Circus*. It runs on major operating systems and can integrate well with other programming objects (Python 2010). *i*MapInvasives uses Python in conjunction with the Django framework to serve web pages and to control User Levels and database functions.

Web Development Framework

Django
   Django was named after the famous 1930s jazz musician, Reinhart Django. The software was developed by an online news organization in order to meet the intensive deadlines of newsrooms. Django supports the "DRY Principle," which translates to "Don't Repeat Yourself." The focus is to automate as many tasks as possible so that they only need to be programmed once (Django 2010). The learning curve for Django is higher than with other products, but is well worth the effort. Django code is very compact, with some simple *i*MapInvasives HTML pages containing as little as four lines of code.
   The concept of a server-side web framework can be confusing partly because it is hard to imagine why a framework would be needed in the first place. Below is a hypothetical web page that illustrates some of the functionality that

Django offers. This hypothetical page displays a list of invasive species that meet a certain criteria and offers a link to users with a level five or higher. Let's assume that the database query has already been performed in Python and the results are stored in a container named *searchResults* that is sent as a parameter to this web page.

1. {% extends "imiadmin/base.html"%}
2. {% load smart_if%}
3. {% if user.iMapUserLevel.userlevel_id > 5%}
4. <a href="showMoreDetail.html"> Show More Detail</a>
5. {% endif%}
6. {{searchResults}}

Line 1 loads the page styling (header, logo, etc.). Line 2 loads a module making it possible to perform logic (if) statements. Lines 3, 4, and 5 show a hyperlink to users with User Levels higher than 5. Line 6 shows the results of the Python search.

To understand the value of Django, imagine the programming needed to perform this same task without using a framework.

Database/DBMS

PostgresSQL/PostGIS

PostgreSQL is an enterprise-class open source database with over 15 years of active development. PostgreSQL by itself can perform standard relational queries. PostGIS is an add-on that allows spatial queries. The queries use standard Structured Query Language (SQL) on spatial data that have been loaded into the database. The output of the spatial query is in alphanumeric (text) format (PostgreSQL 2010). For example, a question such as "In which school zone is this house located?" would be translated into SQL statement and sent to PostgreSQL/PostGIS. The answer returned would be in text format. *i*MapInvasives uses PostgreSQL/PostGIS extensively because it provides the ability to query any piece of information at any time and from a spatial standpoint if needed. Results can be passed on to another part of the system for further processing.

Web Server

Apache

The Apache HTTP Server Project has the goal of creating a commercial-grade and robust HTTP open source server. Apache is standard for many internet servers and the developers adhere to strict standards in their development (McMillan 2000). Apache is comparable to Windows Internet Information Services (IIS).

6.5 Customized Features

*i*MapInvasives contains the standard mapping features found in many online maps, plus many customized features that are needed to meet the legal and conceptual requirements required by the project. Several features are described below, including why they are needed and their implementation methods. Common features, such as general and spatial queries, are not included in this chapter. Although not a complete list of customized features, the following descriptions provide a good idea of the challenges the project faces and the flexibility that this technical model delivers.

6.5.1 User Level and Login

Much of the functionality of *i*MapInvasives is dependent on a User's security level (referred to as User Level). Users that do not log in are considered to be User Level zero and can only view a map designed for the general public. Users that log in are shown more detailed data. Administrators have the highest User Level and have full access.

The rationale behind user level access restrictions is the Users require differing levels of functionality. Citizen scientists are provided basic functionality while land managers are given more robust capabilities. Many species look alike and occasionally there are misidentifications among lesser trained users. Some data are sensitive and should not be displayed or should be generalized for users who may misinterpret the data. Some invasive species are harmful to humans and can damage private property (e.g., crack a house's foundation or destroy all other plants in the area). To avoid the possibility of false alarms and to protect privacy, the *i*MapInvasives Executive Committee (representatives from participating states and the Development Team) has implemented a policy to display data according to User level and/or zoom scale. Users can attain a higher access level by attending training classes and fulfilling educational requirements (iMapInvasives 2010).

6.5.1.1 Map for the General Public

A thematic map is available to the general public and requires no login. The data are shown thematically by county, watershed, or other polygon area. State Administrators may select species to feature on the public map page.

6.5.1.2 Map for Registered Users

This map requires that a user log in. Lower-level users are able to view confirmed, non-sensitive data. Higher-level Users are allowed to see full point data, both

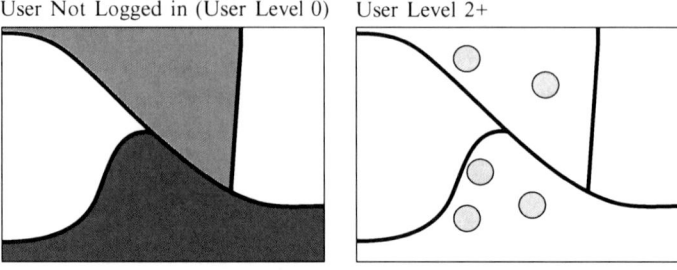

Fig. 6.3 The map display can change according to User Level

confirmed and unconfirmed. In some cases, the map can automatically switch from point data to thematic data based on user level and zoom scale, as shown in Fig. 6.3. For example, a map can be thematic for a low-level user and show detailed point data for a high-level user. The map might also show point data at a wide zoom level and change to thematic if a low-level user zooms in too close. Django is the tool used to manage the User Level information. The map image request is changed according to the User Level and/or zoom scale.

6.5.1.3 Minimum Viewing Level

State Administrators have the option of setting a minimum viewing level for each species. The effect is that Users must be of a certain User Level in order to view information about that species. The purpose is to maintain privacy both for sensitive species and their locations. This functionality might be useful in the following situations:

- For a politically sensitive species that is not yet declared to be an invasive, land managers can track the species without causing unnecessary alarm.
- For sensitive species, such as the feral pig, information on their location could lead to unauthorized hunting.

6.5.2 Invasive Species Database Structures

Invasive species data are gathered and stored in a variety of formats, from stick-pins in wall maps to computerized GIS inventories (NAWMA 2002). *i*MapInvasives models its database structure after the Weed Information Management System (WIMS) originally developed the Bureau of Land Management and then further developed by The Nature Conservancy (TNC). WIMS uses three relational tables to collect data for Observation, Assessment, and Treatment tables (WIMS 2010). *i*MapInvasives uses a variation of this model.

6 The GIS Behind *i*MapInvasives: The "Open Source Sandwich"

- Observation – the minimal data needed (who, what, when, where) stored in point format. Based upon the North American Weed Management Association's recommendation for minimum standards (North American Weed Management Association 2002).
- Assessment – an assessment made by a trained researcher or volunteer. There is an associated field of text or percentages, such as "the field is 80% covered." There can be many Assessments in an area. Before GIS, Assessment data was stored as point data, but *i*MapInvasives uses polygons.
- Treatment – a history of how the area was treated. There are fields for descriptions and percentages. There can be multiple Treatments in an area because there are many eradication methods: chemicals, grazing, manual extraction, or no treatment. Before GIS, Treatment data was stored as point data, but *i*MapInvasives uses polygons.

6.5.2.1 Data Conversion: The Crosswalk

Most agencies have established data collection procedures and data formats, and *i*MapInvasives is not attempting to change existing policies. However, it is necessary to have all data in a common format. To accomplish this, "crosswalks" were created to establish relationships between agency databases and *i*MapInvasives. Participating states can create a customized crosswalk that correlates existing data sets from their state to the *i*MapInvasives databases.

6.5.2.2 Surveys (Presence/Absence Data, or "Not Detected")

A survey is a report of an area that has been inspected for presence of a certain species. For example, "this area has been examined for species X and here is what we found." It is possible that the report shows no presence of a species, called a "negative survey" or "absence data." A negative survey is different from an area that has never been inspected. To be effective, negative surveys need to be visualized differently from areas where no survey has been conducted. Surveys will require polygons, and specific implementation plans for this are under development.

6.5.2.3 Mobile Applications

*i*MapInvasives is working on the following strategies to provide functionality for mobile applications:

- *What's Invasive!* – Users can use the *What's Invasive!* (What's Invasive 2010) application for seven common species. These data will be sent to the State Administrator, checked for quality control, and included in a bulk upload. No login required.

- Mobile Phone – A phone application will be available from inside *i*MapInvasives website. Users can add Observation data using their own phone. Login required.
- Global Positioning System (GPS) – In collaboration with NatureServe, *i*MapInvasives is developing an application for GPS use. This will capture point data and can be expanded to include polygons. Login required.

6.5.3 Supporting Non-GIS Databases

There are numerous support databases that work with the map to create a robust system which provides extensive functionality for the end-user. For instance, individual Users can control their User Profile and register for Early Detection Email Alerts. *i*MapInvasives also allows data to be entered as part of a "Project": a class Project, a 4-H Project, or a Grant Project. Project Leaders have access to functionality to view and manage their Project data. Also *i*MapInvasives provides extensive functionality for the *i*MapInvasives Database Administrator, including viewing the full list of Users, adding new Users, adding new species to the tracking list and setting which species are included on the Public Map.

6.5.4 Early Detection

One of the goals of the *i*MapInvasives Project is to support land managers and volunteers involved in Early Detection and Rapid Response (ED/RR) work. The life cycle of an invasive species in a given geographic area is described below. See Fig. 6.4 for diagrams of an invasive species' entry into a geographic region.

- Approaching Region (AR) – the species is reported in a neighboring region but is not yet reported in the area of interest
- Early Detection (ED) – the species has been reported in the area of interest at least once but fewer than four times
- Presence Established (PE) – the species has been reported four or more times and has probably established its presence in the area

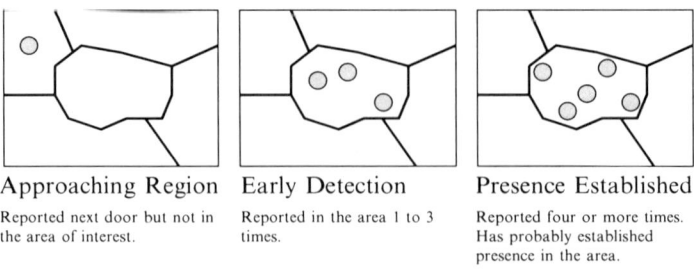

Approaching Region
Reported next door but not in the area of interest.

Early Detection
Reported in the area 1 to 3 times.

Presence Established
Reported four or more times. Has probably established presence in the area.

Fig. 6.4 Diagrams of the three stages of an invasive species' entry into a region

The late Leslie Mehrhoff, a noted biologist from the University of Connecticut, was instrumental in advocating this terminology. In 2006–2007, the New York State Invasive Pest Council, with support from the U.S. Forest Service, developed AR, ED, and PE lists for the eight Partnerships for Regional Invasive Species Management (PRISM) in the state of New York. Several sources were utilized to produce the lists, including the *New York Flora Atlas* and information from local experts, a process that took months. Now, with *i*MapInvasives, the AR and ED lists can be created with the click of a button. Users can choose their geographic region of interest: a county, a management area, a watershed, or other. In addition, Users can sign up to receive an AR or ED email alert for their region.

6.5.4.1 Distance-Based Alert

In some instances, a User is interested in receiving an alert based on distance instead of by county, or watershed, or other management area. For instance preserve managers might be interested in knowing when a new species is reported within 50 miles of their preserve, regardless of which county or watershed the new species falls in. In this case the User can register for a Distance-based Email Alert by simply specifying the land preserve and a distance to monitor surrounding the preserve, essentially establishing a buffer within which any new species reported triggers a Distance-based Email Alert. See Fig. 6.5 for a diagram of a Distance-based Alert.

6.5.4.2 Suspicious Distance Alert

Another important automated functionality of the *i*MapInvasives programming is the Suspicious Distance Alert. This alert is designed specifically to help the *i*MapInvasives Database Administrator notice and evaluate new records reported suspiciously far away from other records for that species. The Database

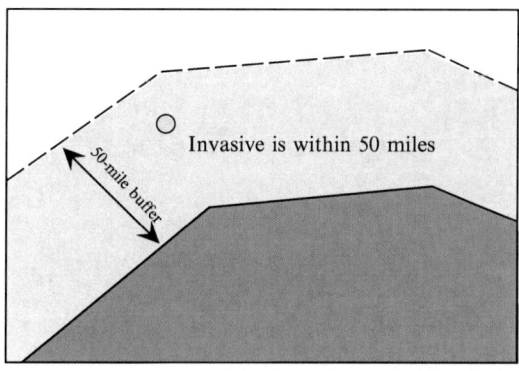

Fig. 6.5 A situation that triggers a Distance-based Email Alert

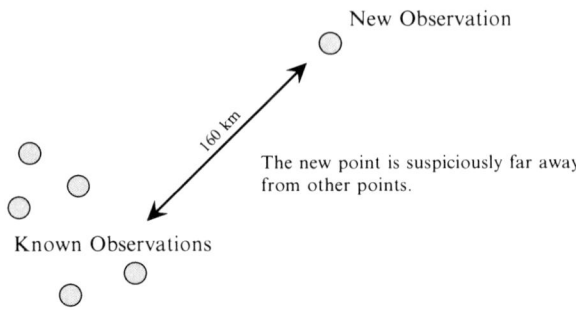

Fig. 6.6 A situation that triggers a Suspicious-distance Email Alert

Administrator can set a "suspicious distance" for each species in the database, depending on the range and biology of the species. For each new record entered into the *i*MapInvasives database, a query is automatically run to test to see if the new species is within the suspicious distance or beyond the suspicious distance threshold. If the new record is beyond the suspicious distance, then a Suspicious Distance Alert is sent to the *i*MapInvasives Database Administrator. In cases where a new record is "suspiciously" far from other known reports for that species, the record is most likely either 1: significant or 2: wrong. It is important to evaluate the information quickly and either 1: submit the information to land managers quickly or 2: in the case of a misidentification, delete the record from the database as quickly as possible. Figure 6.6 shows a scenario where a new observation is a suspicious distance from existing observations.

6.5.4.3 Significant Record Notice

An important aspect of ED/RR is to document species which are new to a county. In the world of biology it is important to document the existence of a new species in a county with a "voucher specimen." To support this effort, a GIS test is run when records are entered online to check if this record is a new species for the county. If it is a new species, the record is considered a "Significant" record, and the User sees the message: "This is a significant record. Please submit a voucher specimen."

6.5.5 User-Friendly Functionality

The Development Team actively seeks feedback from users in many different ways including "Usability Days" where invited users share their experiences. So far these meetings have been productive and have led to the redesign of portions of the user interface.

6.5.5.1 Maximizing the Map Display for the User

Edward Tufte recommends reserving 92% of the screen for the user (Tufte 2010). To this end, *i*MapInvasives users have the option of collapsing both the layer list and the title bars in order to maximize the map on their screen space.

6.5.5.2 Friendly Notices for the User

On the map, "Growls" appear on the screen briefly to remind the User of the results of their recent actions, such as toggling map layers or updating their User profile. If several actions occurred, the "Growls" are stacked vertically on the screen. The "Growls" are semitransparent and disappear after a few seconds. MooTools software contains this functionality.

6.5.5.3 Collapsible Lists

On the map, the elements inside the Table of Contents: zoom, query and layers, can each be collapsed or expanded. In addition, on the User and Administrative pages, sections within those tabs can also be collapsed or expanded to allow the User to more easily view the choices.

6.5.6 Vector Layer

The vector layer, sometimes called the "graphics layer," is an area where graphics can be drawn on top of the map. The vector layer is useful for visualizing data that is not stored in a database such as clustering or other types of GIS analysis. With OpenLayers, it is possible to have an unlimited number of vector layers because they can be created and destroyed as needed (Hazzard 2011, p. 50).

The Public Map contains only thematic species data. There are a number of different methods to create this type of map and *i*MapInvasives has utilized several versions before settling on the current methodology. The first strategy utilized a Web Map Service (WMS) and a Stylized Layer Definition (SLD) file. This option pre-processed the species counts and wrote them to an external file, which was time-consuming. This worked well but was slow. The second option uses Keyhole Markup Language (KML) as vector boundary definitions. This solution is very fast, does not require extra processing effort, and gives added interactivity with the mouseover effect.

Cluster analysis involves assigning points into groups, or clusters, in order to enhance visualization. The code for cluster analysis was developed after examining several options. The first experimental attempt used a Web Feature Service (WFS) but this option ran too slowly. The second attempt used a direct call to the PostgreSQL/PostGIS databases with a Python program that runs a clustering algorithm that generates a KML file. This method is working very well.

6.6 Business Model: Long-Term Planning

Development funds for the *i*MapInvasives Project have been provided by public and private sources, including The Nature Conservancy and the New York State Environmental Protection Fund through the Department of Environmental Conservation.

The cost of developing the core programming for *i*MapInvasives' robust functionality is expensive. However, because the project was designed for easy replication, making *i*MapInvasives available for a new state or province simply requires making a new state/province module, which is extremely reasonable.

6.6.1 Core Functionality and State/Province-Specific Programming

The programming for the *i*MapInvasives Project has been designed to facilitate adding new states or provinces: all the core functionality for *i*MapInvasives uses variables for state-specific aspects of the software. All the state-specific variables and layers are in state/province modules (or configuration file). Participating states/provinces pay an annual *i*MapInvasives service fee which covers the cost of maintaining the state-specific programming module for their state/province. This state/province module allows their state/province to have access to all the core functionality for the User map as well as the User and Administration capabilities. Figure 6.7 shows a diagram of the state/province access to the main programming code.

6.7 Conclusions and Future Work

*i*MapInvasives is proving to be a successful strategy for dealing with invasive species on a state, national, and perhaps international scale. The project has a rich variety of functionality and has been robustly tested by the steadily growing user base (as of this writing, seven states are participating). There is a monthly Executive Committee conference call to discuss current topics, and there are plans to start a User Group this year.

However, technological challenges lie ahead. At the time of this writing, there are several significant tasks under development. User-entered polygons for

Fig. 6.7 Participating states have access to the core programming functionality

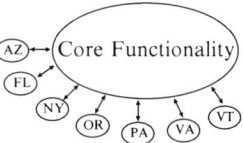

Assessments, Treatments and Surveys (potentially "negative surveys") appear to be the most challenging. Given that some functionality has been re-written (the thematic map portion was re-written), it is expected that more difficult tasks may require several attempts before determining the strategy that works best. This is an ongoing project continuously incorporating user feedback, and revisions are expected and welcomed. The project will continue to evolve over time.

On the conceptual front, there are opportunities to explore new ideas. Geovisualization techniques will be needed to display new concepts such as negative surveys. Tracking animals and insects are new concepts, as most tracking to date has involved plants. IGIS advancements suggest ideas that are new to the invasive species field and land managers need time to explore these concepts. There are opportunities for IGIS analysis that have yet to be determined.

Open source software has proven to be reliable and easy to work with. The software and functionalities detailed here were specifically designed for viewing, recording, and analyzing invasive species data. However, the design for this project, both the "open source sandwich" and the Core-modular programming, would work well for any mapping application that requires flexibility and customization.

Acknowledgements Thanks to Meg Wilkinson of the New York Natural Heritage Program (NYNHP) for her vision of this project and her skill in managing its direction. Thanks also to the members of the *i*MapInvasives Executive Committee for their guidance and insights. Thanks to Stephen Hodge and Erik Hazzard of the Florida Resources and Environmental Analysis Center (FREAC) of the Florida State University (FSU) for their keen eye for details, technical support, and GIS/programming skills.

Disclaimer

Any opinions, findings and conclusions, or recommendations expressed in this material are those of the author and do not necessarily reflect the view of Florida State University or The Nature Conservancy.

References

Bouchal LC (2010) Water chestnuts found in Lake Hopatcong, New Jersey H. New Jersey Herald. http://www.njherald.com/story/news/19Water-chestnuts

Crooks JA, Soule ME (1999) Lag times in population explosions of invasive species: causes and implications. In: Sandlund OT, Schei PJ, Viken A (eds) Invasive species and biodiversity management. Kluwer, Dordrecht

Fee J (2010) Give Me a Map Sandwich, SpatiallyAdjusted.com. http://www.spatiallyadjusted.com/2010/03/16/give-me-a-map-sandwich
Frye C (2010) The 'Map Sandwich,' Mapping Center. Environmental Science Research Institute (ESRI).http://cc.bingj.com/cache.aspx?q=map+sandwich&d=5027302320243823&mkt=en-US&setlang=en-US&w=56119670,610f1695
Hazzard E (2011) OpenLayers 2.10 beginner's guide. Packt, Publishing LTD, Birmingham
Lass LW, Prather TS, Glenn NF, Weber KT, Mundt JT, Pettingill J (2005) A review of remote sensing of invasive weeds and example of the early detection of spotted knapweed (*Centaurea maculosa*) and babysbreath (*Gypsophila paniculata*) with a hyperspectral sensor. Weed Sci 53:242–251
McMillan R (2000) Apache Power, Linux Magazine. http://web.archive.org/web/20050214074858/http://www.linux-mag.com/2000-04/behlendorf_02.html
North American Weed Management Association (2002) North American invasive plant mapping standards. Federal interagency committee for the management of noxious and exotic weeds
Owen SJ (1998) Department of conservation strategic plan for managing invasive weeds. Department of Conservation, Wellington
Randall J (2010) E-mail Interview, 10 Sep 2010
Tiemann M (2010) Open source software. In: Bates MJ, Maack MN (eds) Encyclopedia of library and information sciences, vol 1, 3rd edn. Taylor & Francis, London, pp 4031–4036
Tufte E (2010) Presenting data and information, Arlington. Lecture, 14 Apr 2010
Viger R, David O, O'Hara CG (2004) GEOLEM: a knowledge-handling infrastructure for integrating GIS and environmental modeling
Westbrooks RG (2004) New approaches for early detection and rapid response to invasive plants in the United States. Weed Technol 18:1468–1471

Websites

Apache (2010) The Apache Software Foundation. http://www.apache.org/
Django (2010) Django Software Foundation. http://www.djangoproject.com/
*i*MapInvasives (2010) The Nature Conservancy. http://www.iMapInvasives.org
KML (2010) Google. http://code.google.com/apis/kml/documentation/kml_tut.html
MooTools (2010) mad4milk. http://mootools.net
OpenLayers (2010) OpenLayers: free maps for the web, open source geospatial foundation. http://www.openlayers.org/
PostgreSQL (2010) PostgreSQL Global Development Group. http://www.postgresql.org/
Python (2010) Python Publishing Foundation. http://www.python.org/about
REST (2010) ArcGIS Server REST API, Environmental Science Research Institute (ESRI). http://resources.esri.com/help/9.3/arcgisserver/apis/rest/index.html
What's Invasive! (2010) Community data collection. http://whatsinvasive.com/
WIMS (2010) Weed information management system, The Nature Conservancy. http://www.imapinvasives.org/GIST/WIMS/index.html

Chapter 7
Towards a Dutch Mapping API

Edward Mac Gillavry, Thijs Brentjens, and Haico van der Vegt

Abstract Government departments in the Netherlands have come together to create a mapping Application Programming Interface (API) and a mapping services platform that will serve a core set of geographic data across government websites. The article presents the technical, legal and organizational considerations that helped shape the mapping API and the underlying platform. The article concludes with recommendations for creating mapping APIs and presents an outlook into the future of the Dutch mapping API.

7.1 Context

7.1.1 From Paper Maps to Online Services

Building the Dutch national spatial data infrastructure (NSDI), government departments in the Netherlands have come together to create a shared mapping platform that serves a core set of geographic data and mapping services online. Many national mapping and cadastral agencies (NMCA) in Europe are working towards a national spatial data infrastructure, providing their geographic information through online services. Notable examples can be found in the United Kingdom (Ordnance Survey), Switzerland (Swisstopo), Norway (Statkart) and France (IGN). While the NMCAs typically assume the pivotal role in these initiatives, in the Netherlands, four government organizations, the Cadastre, the Dutch Ministry of Economic Affairs, Agriculture and Innovation, the Ministry of Infrastructure and the Environment and TNO, an independent research organization, drive the initiative together. One of the advantages of this approach is the broader support and application of its

---

E. Mac Gillavry (✉)
Webmapper, Kenaustraat 12 zwart, 2011 MX Haarlem, The Netherlands
e-mail: edward@webmapper.net

results across the Dutch government. On the other hand, the challenge lies in its complexity, as all participants have to agree on a broad range of aspects, ranging from the technical architecture to the financial terms and conditions.

This common initiative, "Public Service on the Map" abbreviated as PDOK (from the Dutch "Publieke Dienstverlening op de Kaart"), is not only driven by the need to reduce government expenditure, but its objective is also to make national geographic data services widely accessible to society and to address requirements set forth in the European INSPIRE Directive. Another driver of the PDOK initiative is the development of a system of "Basisregistraties" (in English "Authentic" or "Key Registers") in The Netherlands. Finally, the increasing familiarity of the general public with maps through commercial mapping websites have paved the way for governments to inform the public through online maps.

7.1.2 Inspire

From a European perspective, the INSPIRE (Infrastructure for Spatial Information in Europe) directive from the European Commission is one of the main drivers of the PDOK initiative. The goal of INSPIRE is to build an infrastructure for the provision of harmonized spatial information across the members of the European Union (EU), supporting the definition of European and national environmental policies. However, geographic data sets from European member states currently differ in content, reference systems and semantics. Since Europe-wide issues such as environmental pollution or flooding do not stop at the national borders of its member states, harmonization facilitates the application of spatial data in solving these issues.[1]

Therefore, member states have to provide access to existing geographic data sets and services according to the so-called Implementing Rules. These rules define standards for data specifications, metadata, service architectures and terms and conditions to facilitate the creation, discovery provision and sharing of these spatial data sets. Finally, INSPIRE also defines what data themes have to be made available. As a result, not only is each member state required to set up its national spatial data infrastructure, but also many national organisations are involved as a provider of one or more data sets defined by INSPIRE. One of the important reasons for the PDOK initiative is to assist these national organisations in the Netherlands, building a common platform to create, discover, provide and share Dutch INSPIRE data.

---

[1] See http://inspire.jrc.ec.europa.eu/.

7.1.3 Authentic Registers

Until recently, the Dutch government stored its records about persons and their income, about roads, addresses and buildings at many different departments and in many systems. Apart from the problem of duplication, it also means all these data stores continuously has to be kept up-to-date. To address this situation, the Dutch government has agreed on a system of so-called Authentic Registers.[2] The collection and maintenance of every key record is the responsibility of only one government organisation and has to be re-used by all other government organisations. When your address changes, you only need to report this to your municipality, as the holder of the authentic register of addresses, instead of informing many other government organisations about your move.

PDOK builds on this system of Authentic Registers, creating a single point of entry and distribution of government information for its partners, instead of all partners creating their own solutions. The central PDOK portal can also be used to disseminate the contents of an Authentic Register to other government departments. In this way, PDOK is the hub for all Authentic Registers that have a geographic component, like the Authentic Registers of Topography, Addresses, Buildings, and Cadastre. One of the first Authentic Registers provided through PDOK is the "Basisregistratie Topografie" (BRT), in English the "Authentic Register Topography". This register forms the backbone of the all-purpose background map that is served from the shared mapping platform and is discussed in this article.

7.1.4 Commercial Mapping Websites

Since its launch in February 2005, Google Maps has had a dramatic impact on online mapping, particularly in the areas of usability and the ease of integration (Musser et al., 2007). Web visitors no longer have to sit around and wait for a full page refresh when zooming or panning around the map, as the new map images are loaded in the background. Web developers could easily start integrating the map functionality through a JavaScript-based mapping API that was released in June 2005. This novel approach has now become the de facto standard for delivering online mapping services and has been adopted by other online mapping providers such as Yahoo Maps, Bing Maps and MapQuest.

The richness of geographic data delivered through these commercial mapping APIs – aerial and satellite imagery, street-level photography, terrain maps – and the smooth user experience have not only raised the awareness and popularity of online mapping, but has also raised the expectations of the public with regard to geographic web applications.

---

[2] See http://www.rijksoverheid.nl/onderwerpen/basisregistraties/overzicht-basisregistraties.

7.1.5 Conclusion

Online mapping services are at the heart of our information society, delivering the information needed by governments, companies and the citizen in general. From the European to the municipal level, government bodies use online mapping services to deliver geographic information. It is therefore essential that these services are easy to implement and use. A mapping API may provide s such a solution. It typically simplifies the way mapping services can be used by Web visitors and integrated by Web developers.

The article continues in Chap. 2 with positioning the mapping platform and services as an adequate and relevant alternative for existing implementations of commercial mapping APIs on government websites, presenting the technical, legal and organizational considerations that helped shaping the shared mapping platform, touching on issues such as privacy, liability and accessibility.

The first application of the mapping platform, described in Chap. 3, is the Geographic Search and Portrayal Service abbreviated as GEOZET (from the Dutch "Geografische Zoek – en Toondienst") implemented on the government website Overheid.nl. The chapter highlights the various components, such as the all-purpose online base map, the geocoding service and the scripting API, addressing aspects such as interoperability, usability research, web cartography and web content accessibility guidelines.

The paper finally concludes with an evaluation of the project results delivered by the PDOK and GEOZET initiatives. It further aims to give a recommendation for the development of a Dutch mapping API based on these project results.

7.2 Why a Dutch Mapping Platform

Commercial mapping services and APIs do already exist, they are easy to use and to implement. Furthermore, they are also available for free. This raises the question why the Dutch government, within the framework of the PDOK programme, considers creating its own mapping API. The reasons for the government to create an alternative for commercial mapping APIs are described in the following paragraphs, giving the legal and commercial background for this alternative, but also explaining the usability and quality reasons behind it.

7.2.1 What Is an Online Mapping platform?

Before exploring the reasons behind the development of a Dutch mapping platform in the next paragraphs, it is important to understand its concept, as a mapping platform is more than just an API, an library of programming code, and a set of online maps. It typically consists of several components to implement geographic web applications. The following components can be identified:

- A set of **base maps**, for example street maps, terrain maps, and aerial or satellite imagery
- The mapping **Application Programming Interface** (API) itself, commonly a JavaScript or ActionScript library to create and control the base maps and to overlay thematic data
- A **geocoding service** (or more precisely: a gazetteer) to match addresses with their geographic locations
- A set of **documentation** to help web developers to implement the API, including user guides, tutorials, etc.
- A development **community**, where developers can share their questions, observations and experiences
- A **legal framework** of copyright statements, a service level agreement (SLA), the terms of service (ToS), privacy statements and an End User License Agreements(EULA) that describe the relationships, dos and don'ts between the data and service providers, the publishers and the end users.
- A **technical support desk** to assist developers in solving specific problems
- An **accounting system** to create invoices and to generate reports for usage statistics

7.2.2 Legal and Commercial Aspects

Other than commercial mapping services, that make money through business to business services and online advertising on their website, a government organisation's goal is to serve the public interest. The specific legal and commercial reasons for developing a Dutch mapping platform for government are described in the section below:

- Privacy: Commercial mapping APIs often collect user and usage statistics. Users for whom privacy is very important may not use the commercial mapping APIs for these reasons. Whereas it may not be a problem for commercial companies to exclude certain users from their services, for government bodies that serve the public this may be an issue though.
- No advertising: The free versions of commercial mapping APIs increasingly use advertising on the map directly. For government applications, this could raise problems. Offering mapping services by the (national) government is a better guarantee that maps are free of advertising.
- Service reliability: While commercial mapping APIs do provide service level agreements (SLAs) with their enterprise-level offerings, these typically exclude life-endangering situations just when governments step in to manage rescue and damage control efforts.
- Create once, use many: Many government organizations create their own background map for their internal production processes. This means that different versions of this map exist for the same purposes: creating a suitable base layer on top of which thematic information can be shown. By facilitating this base layer

from a central source, efficiency can be reached in terms of resources spent to build and manage such a base layer. Commercial APIs forces the use of all components together.
- Liability: Because the third parties providing commercial mapping APIs manage the map update-process, government cannot be held liable for issue concerning the currency of the data, but also the positional and attribute accuracy.

7.2.3 Usability and Quality

Besides the legal and commercial aspects, there are also a number of issues that deal with the usability and quality of the offered maps:

- Accuracy: Commercial mapping APIs use maps that may not have the suitable geometric or attribute accuracy for certain applications. In addition, the cartographic reprojection of geographic data available in the Dutch coordinate reference system onto the base maps of commercial APIs that use the Spherical Mercator reference system is not always accurate enough.
- Accessibility: By 2011, all Dutch websites of the national government, provinces, municipalities, and water boards have to comply with the official Dutch web accessibility guidelines. These guidelines ensure that websites have better accessibility, availability, usability and reliability for visitors, web browsers, search engines and mobile browsers. Furthermore, these guidelines provide cost reductions in website development and management. The Dutch accessibility guidelines are based on the Web Content Accessibility Guidelines 1.0 of the World Wide Web Consortium (W3C). Commercial mapping API's do not conform to all the guidelines for this.

We now know the background of a specific mapping platform for the Dutch government. In the next chapter a closer look will be given in how this mapping platform was created, with its different components as well as its first application in a government website.

7.3 Development of the Dutch Mapping Platform

7.3.1 PDOK: Global Platform Architecture

The Dutch mapping platform "Public Service on the Map" abbreviated as "PDOK" (from the Dutch "Publieke Dienstverlening op de Kaart") does not only have to deliver online map images, but also the raw geographic data and the meta data. To enable and improve the re-use of components and resources, standardization can be very effective. In most cases, standards of the Open Geospatial Consortium (OGC) and ISO have been implemented (Maus 2011):

7 Towards a Dutch Mapping API

Fig. 7.1 Global architecture of the mapping platform

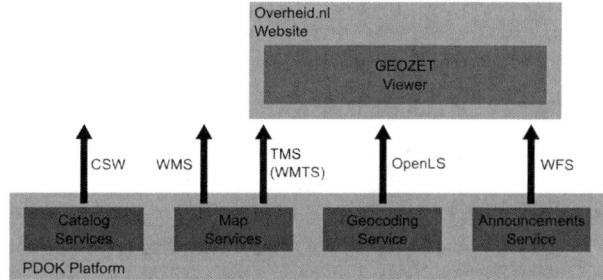

- A web portal to search and deliver meta data about mapping services and follows the Catalogue Service for the Web (CSW) standard of the OGC.
- A central access layer is a mechanism to control which mapping services are delivered to which government body
- The mapping services provide online maps following the Web Mapping Service (WMS), Web Feature Service (WFS) and Web Map Tile Service (WMTS) standards of the OGC. Before WMTS will be widely adopted, the map images, the so-called "tiles" are served according to the Tile Map Service (TMS) specification of the Open Source Geospatial Foundation (OSGeo).
- The geocoding service follows the Open Location Services Interface Standard of the OGC.
- Tooling for data extraction, transformation and loading (ETL) and data storage (Fig. 7.1).

The first application of the mapping platform is the Geographic Search and Portrayal Service abbreviated as "GEOZET" (from the Dutch "Geografische Zoek – en Toondienst") implemented on the government website Overheid.nl. The public will be able to search for local announcements from municipal and provincial governments in their area and see the results, for example building permit requests, public works, and changes to spatial plans, on a map. This has led to the additional requirement for the PDOK mapping platform to also provide an interactive map viewer that can also be re-used to facilitate implementations of geographic web applications on other government websites.

7.3.2 GEOZET: An Interactive and Accessible Map Service

In the context of a Dutch mapping API, the following components of the geographic search and portrayal service "GEOZET" are important:

- An interactive map viewer that conforms to the Dutch web accessibility guidelines.
- An all-purpose online base map, the BRT-Achtergrondkaart, is served from the PDOK mapping platform as small map images, so-called "tiles", conforming to the Tile Map Service (TMS) specification.

- A geocoding service that is served from the PDOK mapping platform for geographic lookups of full addresses and place names. The interactive map viewer uses the service to zoom to a specific location. The geographic search module uses the service for showing the nearest announcements.

This paragraph highlights the interactive map viewer. The all-purpose online base map and geocoding service will be discussed in subsequent paragraphs.

The interactive map viewer has been built using OpenLayers, an open source JavaScript mapping API. Other JavaScript libraries that have been incorporated are ExtJS and jQuery. Modifications have been made to improve the accessibility and to add specific functionality. The Dutch web accessibility guidelines range from generic principles (e.g., "information must be accessible, regardless the technology used") to very specific guidelines on HTML-elements (e.g., "always provide textual alternatives for images"). For the GEOZET mapping service, these guidelines apply to all aspects of the application: the geographic information, the tools to access the information, the geocoding service and the all-purpose base map.

Therefore, the mapping service uses well structured HTML, human readable URLs and separates content and presentation. It provides the information textually as an ordered list or graphically as an interactive map, depending on the capabilities of the technology or the Web visitor. The textual interface is used to guarantee accessibility: this interface does not depend on optional technology in a browser and thus works if only HTML is available. With HTML forms, Web visitors are able to search geographically through and browse through a result list ordered by distance to the supplied. Also, search engines, text browsers, and screen readers typically use the textual presentation (Figs. 7.2, 7.3).

Fig.7.2 Graphical view of the GEOZET mapping service: interactive map viewer

7 Towards a Dutch Mapping API

Fig. 7.3 Textual view of the GEOZET mapping service: ordered list of results

If optional technology such as JavaScript and Cascading Style Sheets (CSS) are enabled in the web browser, Web visitors can browse the results by browsing an interactive map. The map viewer provides controls to navigate the map, to display the announcements and to obtain further information on these announcements. Besides the main map and navigation controls, the viewer contains an overview map, an interactive legend to hide and show categories of announcements on the map. Although all controls are typically be operated with a pointing device like a computer mouse, Web visitors can also just use keyboard strokes. Because of its modular set up, the various tools of the map viewer can be enabled or disabled easily to accommodate implementations of the GEOZET interactive map viewer on other government websites.

7.3.3 BRT-Achtergrondkaart: An All-Purpose Online Base Map

The GEOZET mapping service can only be recognised as an adequate and relevant alternative for existing implementations of commercial mapping APIs on government websites if the interactive map viewer comes with an all-purpose base map. Besides the considerations presented in Chap. 2, the cartographic style of the all-purpose online base map also had to take into account the results from a usability survey of online mapping services and the national government visual identity.

The usability survey investigated the usage of online maps and customer preferences regarding online base maps. Both Google Maps (41%) and the online route planner of the Dutch automobile association ANWB (44%) are the two most popular mapping websites among the respondents. Almost 25% of respondents use online maps to search for location-based information. They noted that the following criteria influenced their preferences for a particular mapping website:

- Colour: The base map should have subtle, but clear colours that provide enough contrast and allows for colour-blindness.
- Distinction: The base map should have distinctive colours to differentiate road types (highway, secondary roads, local road) and land use (build-up area, forest, water). The lineage of the roads has to be easy to follow.
- Level of detail: The base map should strike the right balance in the amount of detail at each map scale.
- Legibility: The base map should have legible type faces and large enough font sizes.
- Tranquility: The style of the base map has to be tranquil

Map feature	Percentage of respondents (%)
Town names and street names	93
Roads	85
Railroads and railway stations	48
Water and waterways	37
Points of interest	30
Land use	28
Other public transport	26
Administrative boundaries	11

Respondents identified the following information to be the most important to be included on the base maps:

These survey results provided overall guidance for the cartographic style and information contents of the all-purpose online base map. Detailed instructions for the cartographic style were derived from the new national government visual identity. It does not only apply to letterheads and websites, but also uniforms, signage and government vehicles. The elements of the national government visual identity that apply to cartography are the Rijksoverheid Sans and Rijksoverheid Serif type faces, and the colour palette of 16 corporate colours, all inspired by the Dutch landscape and pictorial art. The consistent use of these visual identity elements creates a clear and recognisable brand image of the national government (Fig. 7.4).

7 Towards a Dutch Mapping API

Fig. 7.4 The BRT-Achtergrondkaart: the all-purpose online base map

Since the usage of Authentic Registers (discussed in Chap. 1) is not only mandatory for government bodies, but also free-of-charge, the "Authentic Register Topography" abbreviated as "BRT" from the Dutch "Basisregistratie Topografie" presented itself as the obvious source of mapping data for this all-purpose online base map. The TOP10NL topographic map data set at scale 1:10,000 was complimented with the TOP250Vector topographic map data set at scale 1:250,000 to render the small-scale map tiles. Also, there were various issues with the road infrastructure and railway stations in the TOP10NL map data set that prevented it from being used in the all-purpose base map. For example, many road names, particularly in built-up areas, and names of railway stations were not available in the TOP10NL topographic data sets, whereas this information proved to be very important for users (see above) Also, the placement of settlement names was still based on the structure of paper map sheets, resulting in many duplicate settlement names on the map. Finally, the slip roads of motorways could not be distinguished separately, so the visualisation of slip roads at the various scales could not be managed independently from other motorway segments to control the right level of detail.

As a temporary solution, the road infrastructure and railway stations were taken from OpenStreetMap, a map data set created from Volunteered Geographic Information (VGI) and licensed under a Creative Commons license. Settlement names were restricted to the names of municipalities that could be derived from the administrative borders data sets from the "Centraal Bureau voor de Statistiek" (CBS), in English the "National Bureau of Statistics".

The range of colours defined in the national government visual identity proved to be sufficient for defining the overall cartographic style, slightly influenced by the cartographic style of Google Maps as the most preferred online mapping service in the usability survey. The accessibility guidelines were also taken into account, particularly those regarding colour contrast between foreground and background information and colour consistency: each colour should only be used for one particular map object class. Finally, the guideline that not only colour should be used to convey a meaning was taken into account. For linear map features, line width or the line signature such as dashed lines were used to distinguish between map object classes, e.g., different road types. For areal map features, differences in grey scale or saturation were used to further distinguish between map object classes, e.g., different land use. The Rijksoverheid Sans type face was used for the lettering on the map. For the various point features on the map (i.e., railway stations, airports) the icons have been designed that are consistent with other icons within the national government visual identity, for example those for building signage.

The cartographic style of the map was formulated and formalised using the Styled Layer Descriptor (SLD) recommendation of the Open Geospatial Consortium (OGC). The all-purpose online base map is now available as a service conforming to the Tile Map Service (TMS) specification.

7.3.4 The Geocoding Service

In the GEOZET mapping service, Web visitors can search for full addresses, postal codes, towns, and municipalities. Since the Authentic Register for Buildings and Addresses that contains geographic coordinates for all postal addresses in the Netherlands is not yet available, the geocoding service uses the Address Coordinates of the Netherlands (ACN) data set from the Dutch Cadastre. The geocoding service conforms to the OpenLS interface specification of the OGC.

7.4 Conclusion and Recommendations

The PDOK mapping platform and the GEOZET mapping service implemented on the Overheid.nl website offer core functionality that has been identified as common components of most mapping APIs. While the interactive map viewer aims to be generic, there still remains some specific functionality that is particular for the GEOZET map service implemented on the Overheid.nl website. The map viewer components should therefore be made more generic in order to be implemented on other government web sites. From a technical point of view, the all-purpose online base map should be able to support the WMTS interface when this standard will be adopted more widely. Furthermore, Web visitors nowadays simply expect aerial imagery as a base map type, so in order to be an adequate and relevant alternative for commercial mapping APIs, aerial imagery is an essential service of the Dutch mapping API.

To facilitate the adoption and application of the Dutch mapping API on government websites beyond the implementation on the Overheid.nl website, efforts have to address both Web developers and the government policy makers.

Support among web developers is required in order to create a development community that helps to develop the current code base into a Dutch mapping API. Support among government policy makers is important in order to invest resources to sustain these efforts. It also provides a means to learn about the requirements of these communities that can be incorporated into future versions to improve the mapping API.

Already, the OpenLayers JavaScript-based mapping API has a strong and worldwide community. Efforts in this area should therefore be twofold: catering for the Dutch web development community, but also the wider OpenLayers community. While the focus for the Dutch web development community should be on providing particular support to quick start development through technical documentation, cookbooks and cheat sheets to understand the basics of OpenLayers and the details of the Dutch mapping API, the focus for the wider OpenLayers community should be on contributing to the OpenLayers documentation, the gallery of examples, but first and foremost the OpenLayers code base. Also, similar efforts have to be put in place for the textual interface, e.g., porting the Java code base to other languages and content management systems.

The Dutch government policy community should be addressed during road shows at particular events that draw many representatives of this community, e.g., trade shows and e-Government workshops. Fact sheets that highlight the business-level advantages of the Dutch mapping API together with case studies, best practices and road map information.

The life cycle management of the Dutch mapping API encompasses the smooth introduction of new versions of the API that not only address bug fixes, but also add new functionality. As new versions become available, existing customers have to be assisted in their migrations from older versions to the latest version.

Furthermore, life cycle management is also responsible for the day-to-day operations of the overall PDOK platform regarding the performance and service availability levels that have been committed to in Service Level Agreements between PDOK and the various government departments and organisations.

Life cycle management should provide first-line and second-line technical support to government departments not only for facilitating new deployments of the Dutch mapping API on new websites, but also for incident management. Finally, reporting on the performance and service availability levels and other Key Performance Indicators (KPIs) are part of its responsibilities.

References

Infrastructure for Spatial Information in the European Community. http://inspire.jrc.ec.europa.eu/. Accessed 14 Jan 2011

Maus J (2010) Technical design PDOK motor phase 1. http://www.geonovum.nl/sites/default/files/MOT_20100618_TechnicalDesign_v1.01.pdf. Accessed 14 Jan 2011

Musser J et al., (2007) Web 2.0 principles and best practices. O'Reilly Media, Sebastopol

Overzicht basisregistraties. http://www.rijksoverheid.nl/onderwerpen/basisregistraties/overzicht-basisregistraties. Accessed 14 Jan 2011

Chapter 8
LatYourLife: Applying Multiple API Services for Task Planning

Amin Abdalla

Abstract LatYourLife is a prototype application that uses web-mapping and mobile technologies in order to connect people's daily tasks with geography. It provides the user with a new perspective on task planning. Utilising three separate API's, the system helps to plan and schedule the day in relation to relevant geographic information. The main purpose is to illustrate, how information and online tools coming from different sources can be joined together to build a single system. After a description of how the service was built and what API's were utilised, findings which became evident from building the system are given.

8.1 Introduction

The aim of this article is to illustrate the added value of using more than one API as a means to build a web-mapping service. The LatYourLife-application (LYL) is used as an example of how such an integration can be achieved. The paper puts an emphasis on how the system is implemented. The reasons for choosing some of the API's and remarks on some of the problems that occurred are also addressed. After a short discussion about LBS in general and the motivation behind the LYL-application, the overall architecture and the API's utilised are introduced. This is followed by a detailed explanation of each function and how it is integrated. Finally, a conclusion of what was learned from working with the API's is drawn.

A. Abdalla (✉)
Research Group Geoinformation, Department of Geoinformation and Cartography,
Vienna University of Technology, Gusshausstr. 27-29, 1040 Wien, Austria
e-mail: abdalla@geoinfo.tuwien.ac.at

8.2 The LYL-Application

8.2.1 Motivation

LYL is an example of a LBS application that uses web-mapping and mobile technology to assist people in their daily life task-planning. A Location-based-service (LBS) is defined by Koeppel as: "any service or application that extends spatial information processing, or GIS capabilities, to end users via the Internet and/or wireless network" (Koeppel 2000). Jiang and Yao argue that "a true LBS application aims to provide personalised services to mobile users whose locations are in change" (Jiang and Yao 2007).

The range of topics covered by already existing LBS or prototyped services include restaurant, cash machine or real estate search queries (e.g., www.aloqa.com), as well as social services (e.g., Sohn et al., 2005; Miluzzo et al., 2008) or even location based gaming (Lonthoff and Ortner 2007).

According to Jiang and Yao (2007) the questions a LBS-user is concerned about are:

Where am I currently?
What and where are the nearest Locations of interest?
How to get there?

Or as summarised by Frentzos et al., in 2007, the current LBS solutions consist of three fundamental services: What-is-around, Routing and Find-the-Nearest.

Nevertheless, it was stated that current services "...are rather naïve, not exploiting current software capabilities and the recent advances in the research fields of spatial and spatio-temporal databases." Research fields like predictive modelling and the exploration of data collected from LBS (e.g., Karimi and Xiong 2003; Ashbrook and Starner 2003; Vu et al., 2009; Nanni and Pedreschi 2006) could significantly benefit from these advances. Frentzos et al., proposed three extensions of the current services in order to ensure the development of next-generation LBS. Amongst those the Get-together proposal, in which a meeting point for several users is calculated based on the future projection of the calling user's trajectory, is probably the most relevant to this application.

LYL connects tasks and events with geographical information and attempts to make innovative use of the geo-temporal data. Therefore the application not only deals with the three services listed above, but also seeks to create a When-to-act-service. This is achieved by integrating route time predictions as a basis for a pro-active alert feature.

8.2.1.1 User Requirements

Many studies have focused on the user requirements for LBS in terms of navigation in an urban area (e.g., Baus et al., 2002; Borntraeger et al., 2003) or the requirements for specific user groups (elderly people: Osman et al., 2003; blind people: Klante et al., 2004). A study particularly relevant for the LYL-application was conducted by Nivala et al., (2009) and analysed the potential users and their tasks during a hike. The resulting user requirements were grouped into three different phases of the hike: before, during and after. Thus, users need information to plan the hike in the before-phase, require services during the hike (e.g.: location of other hikers, navigation) and finally demand adequate functionalities after the hike (e.g., sharing experiences).

Using the analogy of hiking, planning a day could be divided into the same three phases, which call for similar services. In the before-phase, the functionalities should help the user in his process of decision making, by providing useful (geographical) information incorporated into the planning interface. The features in the second stage should assist users in their ambitions to achieve the aims mapped out for the day. In the last stage the user should be able to look back at past events and review information about it.

Depending upon the type of stage and its requirements, either a web- or mobile platform is better suited for implementing such a services. For example, a desktop-computer is not the best to remind the user of an event, because it may not be in or near the users way. A mobile device in comparison would be more appropriate, as it is mostly found in close proximity to the user and operates almost 24 h a day. Hence, the system consists of three components:

A Web Interface (used in the before- and after-phase);
A Spatial Database (important for the after-phase);
A Mobile Application (mainly for the during-phase).

The decision to include a web-based interface was made in order to separate the mapping and planning of the events from the mobile application.

Together, the parts build a structure enabling the user to conveniently plan and map upcoming tasks on a desktop computer and then navigate through the day having the information carried away on the mobile device.

8.2.2 System Architecture

The application architecture was designed to provide the user with a web interface as well as a mobile component which are tied together through a database and supplemented by API's.

The communication between the web interface and the database is achieved by using AJAX (Asynchronous Javascript and XML) requests to keep the interface responsive and avoid loading times when updating the event information. PHP (Hypertext Pre Processing) handles the interaction with the database. It is chosen due to the open nature of the scripting language. Since the acquisition and management of routing or weather data is a rather complicated undertaking, APIs are a convenient and elegant solution to acquire related information based on the

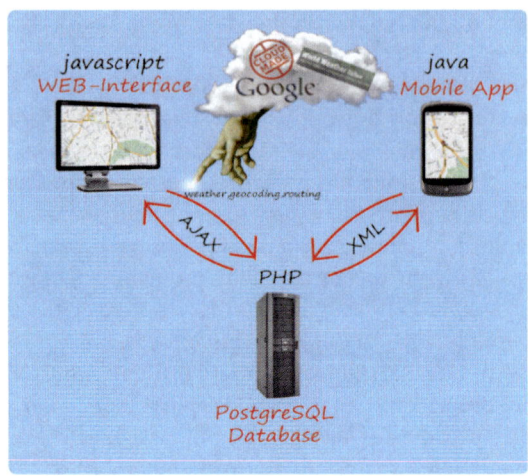

Fig. 8.1 An abstracted graphical illustration of the system structure

geographic location. As seen Fig. 8.1, there are three different API's utilised and provided by Google, Cloudmade and worldweather.com.

The Google Maps API is employed to provide the map and geocoding service and therefore plays a major role in the whole application. The Cloudmade API is serving geographic information from OpenStreetMap (OSM). Its main purpose is to act as an interface between the developer and the OSM data. Worldweather.com serves with worldwide weather information and forecasts.

8.2.3 The Database

The underlying database was built with PostgreSQL, an open-source object-relational Database Management System (www.postgresql.org). The open nature and the possible storage of geographic components in the DBMS were key factors for choosing the software. The data model designed to store the information was kept simple in order to avoid unnecessary complexity in further development of the application. An ER-diagram (Entity Relationship) was used to describe and design the model (see Fig. 8.2). For the sake of simplicity, it is assumed that an Event could only be related to one User/Attendee, but the design could be expanded, so that users are able to share events with others.

8.2.4 The Web Interface

The information about the upcoming events and tasks are stored in a PostgreSQL database, and PHP is used to create XML and JSON files from the database.

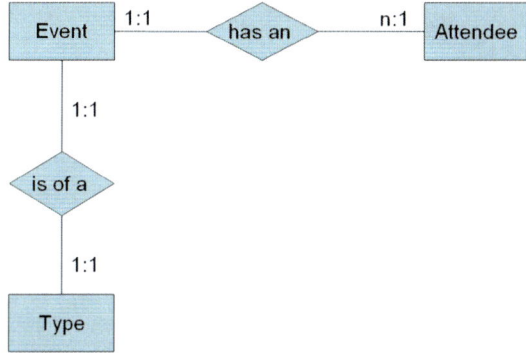

Fig. 8.2 A very simple, but satisfactory ER – model to store the basic information about the events

The calendar and buttons for the interface are realised using JQuery components. The "main-file" is where all the different data sources are brought together and where the bulk of javascript code is found to initialise and integrate the API's.

The Map is one of the main components in the interface and therefore occupies initially most of the space. For this project the Google Maps Javascript Api V.2 (Version 2) was used, although other Web Map services could be facilitated as well. The main reason for choosing it was the great amount of documentation and examples provided by the company. Throughout the process of building the application a newer version of the API was released (V.3) which allegedly was optimised for mobile usage and also incorporated some new functionalities. Unfortunately, along with that came a rather comprehensive change of the code syntax. This resulted in an unpleasant attempt to migrate to the newer version by rewriting the existing code. Finally, it was decided to stick with V.2 since the "upgrade" would have required considerable time and effort.

The core purpose of the interface is to plan and set events on the basis of a map as opposed to the traditional time-based calendar interface. The display is divided into three different parts: a Sidebar, a Map and a Calendar. The Sidebar shows general information such as weather, date, time and details of the next upcoming Event. The Calendar is initially hidden and can be opened by the user (see Figs. 8.3, 8.4, 8.5).

Events can be added onto the map and the calendar on the right allows the user to edit time and information about them. An input form stores, updates or deletes the events in the database (Fig. 8.6).

8.2.5 The Mobile Application

The mobile application was developed for the Android Platform (http://developer.android.com/index.html), an open source software stack created for mobile phones and other devices by Google. The platform was chosen due to its accessibility and coding in the java programming language. The mobile application is built out of

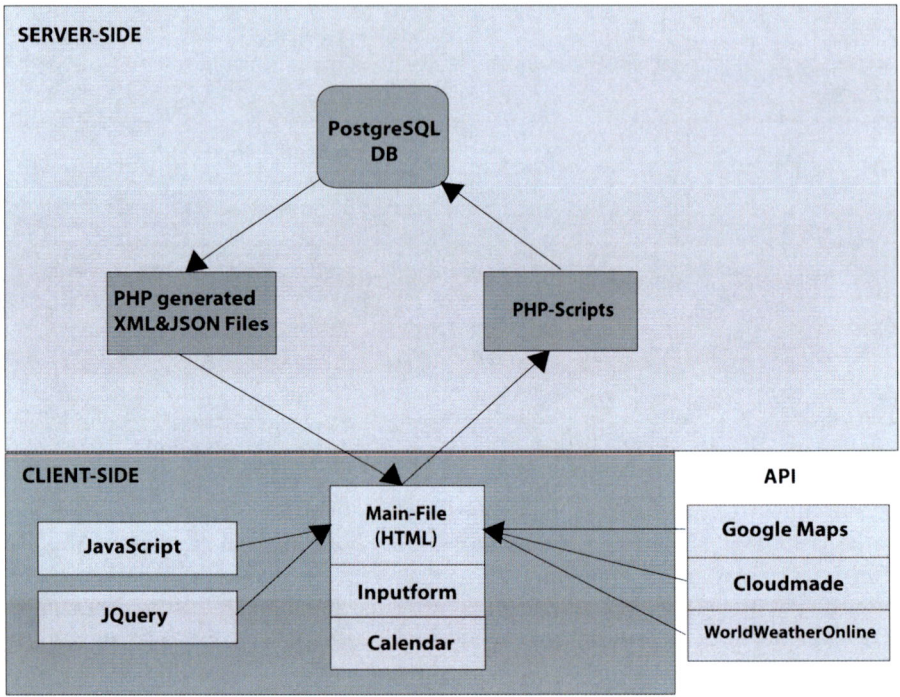

Fig. 8.3 Structure of the web component of the application

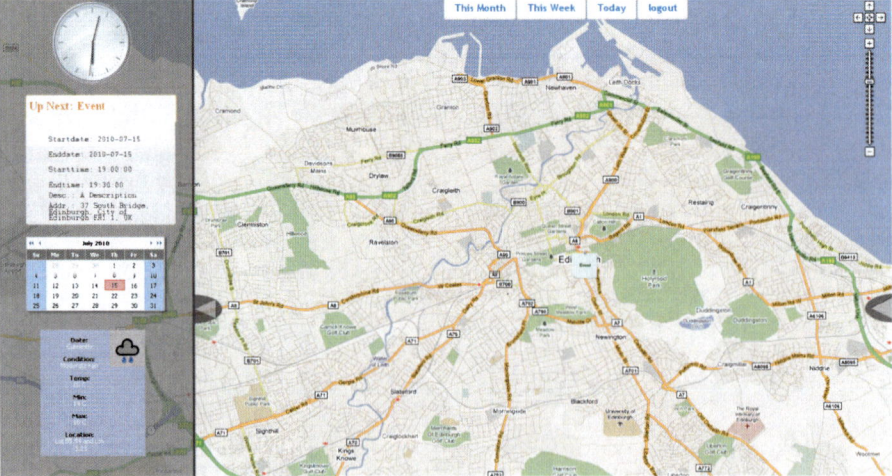

Fig. 8.4 The web-interface as it looks in its initial state. The sidebar on the *left* contains general information. The map in the *centre* shows the Events scheduled

8 LatYourLife: Applying Multiple API Services for Task Planning

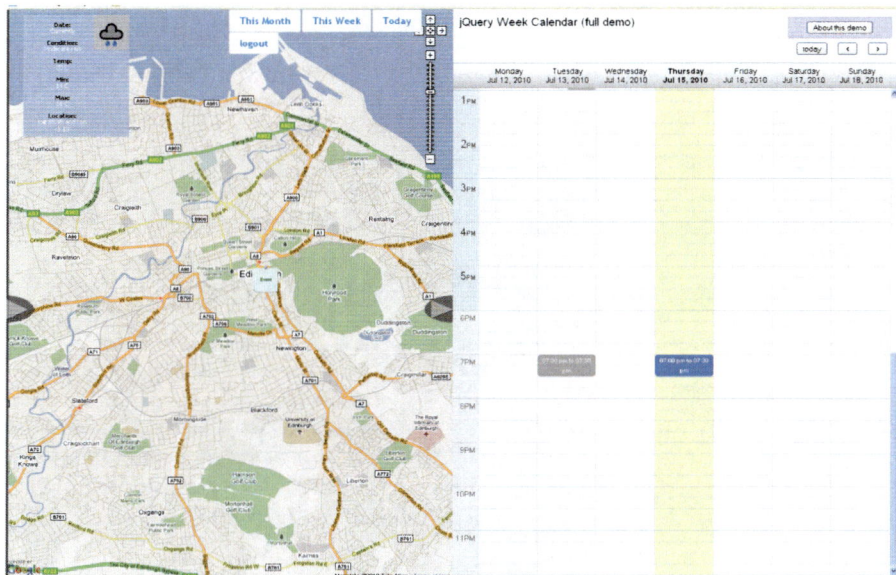

Fig. 8.5 The web interface with the calendar opened on the *right side*. Events are both visualised geographically and temporally

four different user interfaces and a "service" running in the background evaluating the user's position, downloading route information and setting the time for notifications. The application takes its information from the database, utilising php-generated xml-files, as well as from the mentioned API services.

Figure 8.7 shows three of the four views of the mobile application. From left to right the first view provides information about the next scheduled event, the middle view shows a list of all events scheduled for the current day and the last window contains a map with all today's events and the users current location mapped. The fourth view is simply an interface that enables the user to edit some of the event details.

Since the application is very much a prototype implementation, interface design and information display remains an area for improvement.

8.3 Implemented Functionalities

8.3.1 Geographic Planning Aid

The Geographic Planning Aid is the attempt to use geographic information to improve the task planning abilities of the user. The aim of it is to provide information

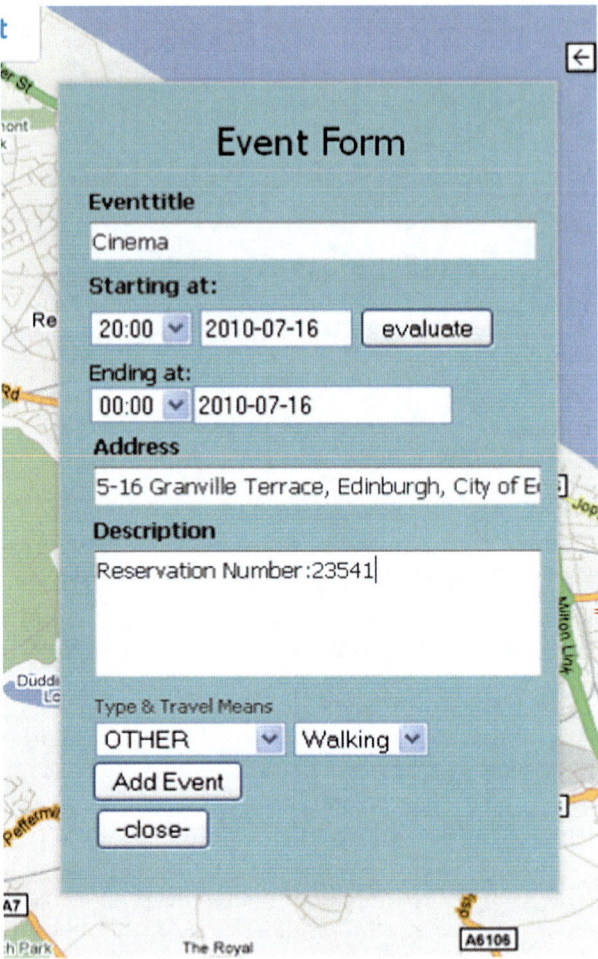

Fig. 8.6 The input form is used to store the event details once it was located on the map

about whether it is possible to get from one event to another at the scheduled time and location. In order to provide such a functionality, specific information has to be provided:

1. Location, Date and time of the event planned;
2. Transport means intended to be used;
3. Ending time and date of the Event before the planned event;
4. Routing time.

The first three need to be provided by the user. Once these details are provided, the system can fetch the time it takes to get from the last event to the next event, using the chosen means of transport. For this, the system utilises the API provided by Cloudmade.com. The API offers developers the ability to query the database and returns a file in JSON-format containing necessary information on how to navigate

Fig. 8.7 Three of the four different views from the mobile application interface

from A to B. Along with the detailed point-per-point navigation information comes the estimated time of travel. The application then takes the information, calculates the time available between the two events based on the ending and starting time of both and then compares it to the duration of travel acquired through the Cloudmade API. The reason for choosing this particular service instead of the Google Maps API was that it provided bicycle routes as opposed to just car and pedestrian routing. To request the routing information, a URL including query parameters, such as means of travel and start-/end-point of the trip needs to be defined. This returns a JSON-file containing the routing information.

In case the travel time exceeds the available time, the service informs the user. Figure 8.8 shows a snapshot of the functionality. We can see that the proposed route is superimposed onto the Google Map in the form of a polyline. Additionally a window is added to augment the line with information about duration and distance of travel. This information is again extracted from the XML-file acquired from the Cloudmade API. In the example case shown in Fig. 8.8, event 1 ends at 15:00 h, the next event is planned to be at 15:30 h and the mode of travel at the bottom of the input form is set to be "walking". Unfortunately, the estimated walking time between the two locations takes more than 30 min. Thus, the user has now three choices in order to be punctual:

1. Change in the mode of travel;
2. Change of the location;
3. Change of the starting time.

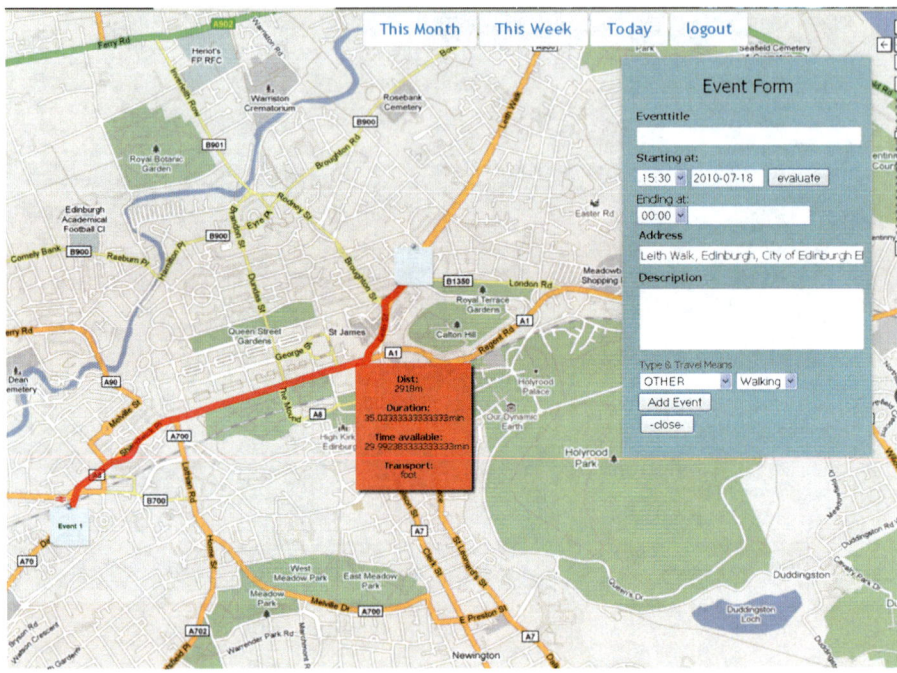

Fig. 8.8 The estimated travel time from event 1 to the next planned event exceeds the time available

In Fig. 8.9, we can see what happens if the user changes one of the three options. An AJAX-request downloads the new information in case mode of travel or location has been altered and updates the output projected on the map. The Google Maps API is hence utilised to detect a location change of the event and the new coordinate is determined and passed to a function which invokes the routing-file request (see Fig. 8.10). If time is the only component changed, there is no need to resend the request to the API.

8.3.2 Weather Information

Weather information could potentially change behaviour in terms of space, time or transportation. Rain might deter the user from using a bike, for example, or even cancel the event completely. Although weather information could potentially be acquired by the user from other web sources, the tight integration of the information into the application transforms it into an integral part of the planning process.

Ideally, the weather information would be added to each of the events according to time and venue, but, for the sake of simplicity, the web component shows

8 LatYourLife: Applying Multiple API Services for Task Planning 115

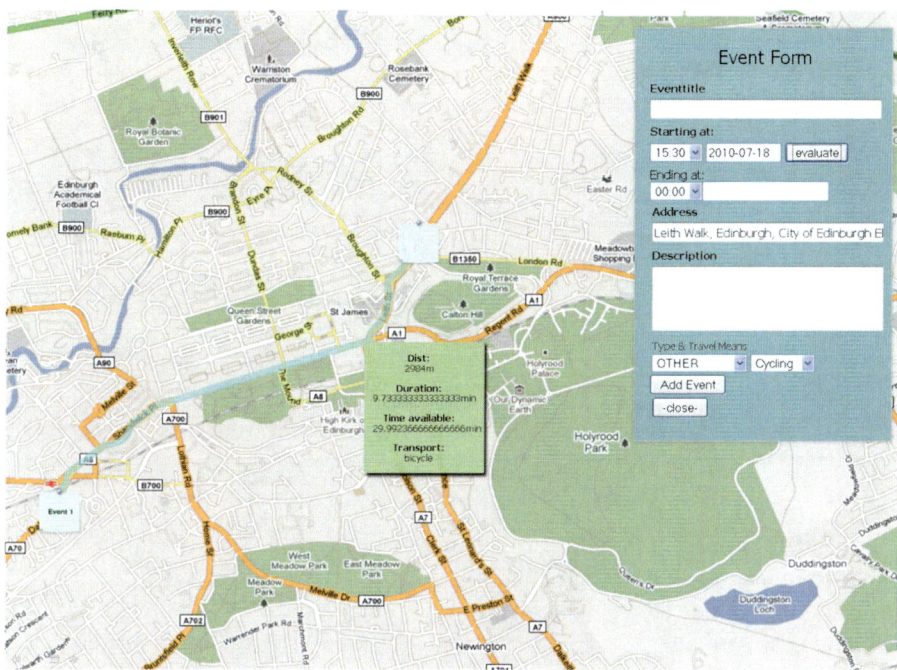

Fig. 8.9 The same situation as in Fig. 8.8 after changing the travel type from walking to cycling

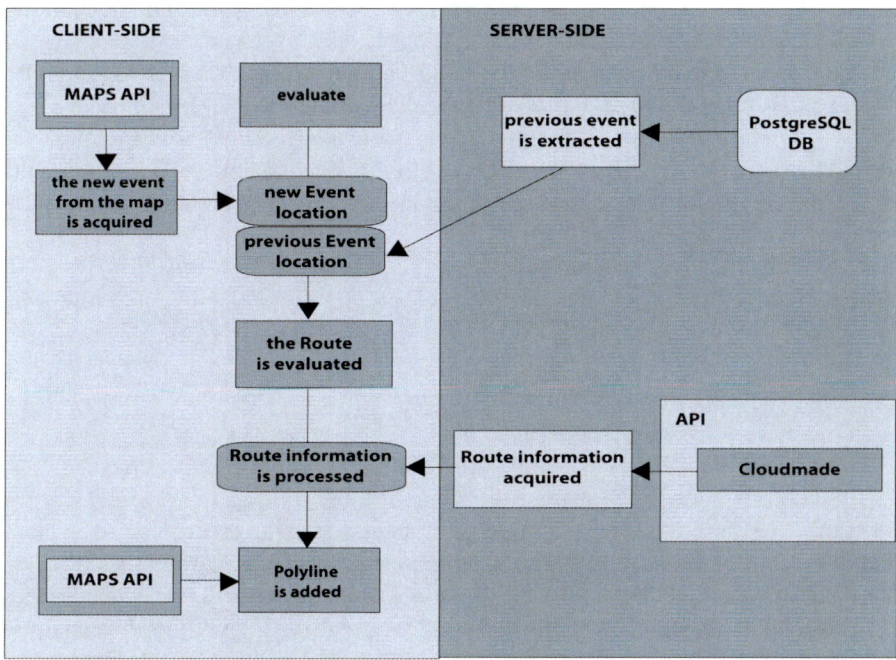

Fig. 8.10 A flow chart shows the process behind the meeting and planning aid

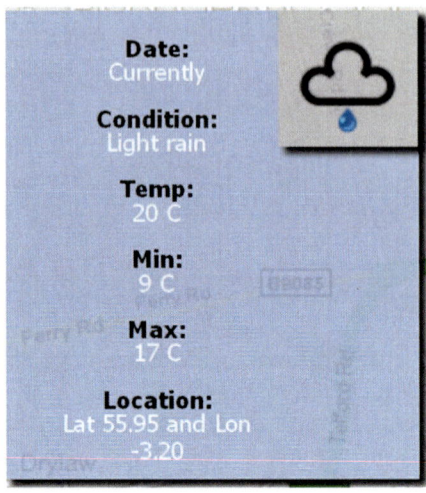

Fig. 8.11 The widget presents the data extracted from the JSON file provided through the Cloudmade API

weather details only for the centre of the map. The weather data is acquired by using an API provided by worldweather.com.

The API, similar to the Cloudmade API, produces a JSON-file when an URL containing a particular query parameters is invoked. The main parameters are a geographic coordinate and date for the weather forecast. In the application, a request is sent every time the user drags the map and centres it to another location. This movement is detected by a "listener" which is part of the Google Maps API. It then extracts the new centre-point and requests the weather information according to this point.

The JSON-file sent by the service contains various information such as humidity, degree celsius, condition and a forecast for the coming days. The number of days for the forecast can be specified in the query parameters. Along with the information comes a link to an image file visualising the weather conditions. These contents were used to build a widget presenting the information in an appropriate way, as seen in Fig. 8.11.

The implementation showed that weather information could easily be tied into such an application on the basis of geographical coordinates and temporal details, by using the right API services.

8.3.3 Routing and Location Viewing

The ability to get instant routing information on the mobile device was assessed to be the most useful feature in the application. One of the main advantages of using the Android platform for mobile development is its design which "...encourages the concept of reuse..." (Meier 2010) . The platform allows the developer to use other applications data or functionalities. Therefore, the Google Maps-App

Fig. 8.12 Directions are fetched and displayed by the Google Maps application on the mobile device

(http://www.google.com/mobile/maps/), which is very likely to be pre-installed on Android devices, was integrated to handle navigation and showing event locations on Google Street View. That approach is resembling the API utilisation in the web-interface, thus things which are too complex or simply exhausting to do manually are "outsourced" by integrating services from other applications (Fig. 8.12).

The application automatically recognises which event the user is supposed to be heading towards by comparing the current time to the appointments stored in the database. The locations from the users current position and the next event are then sent to the Google Maps application, along with a directions request so that the route information is automatically displayed on the map.

8.3.4 Punctuality Alert

This functionality is the second feature (after the Geographic Planning Aid) which attempts to make use of the geo-temporal data provided. The aim is to notify the user ahead of the next scheduled event, in order to ensure punctuality.

The notification time is set according to the user location, so that the alert would be triggered earlier if the user is located further away than it would be the case if it were nearby. The required information for implementing such an algorithm are:

- Current location of the user;
- Starting time of the next event;
- Travel time from the current position to the event position (implies knowledge of the travel mean).

The equation below formalises the procedure necessary to acquire the time for scheduling the alert. In order to provide the user with some time to prepare for the departure, an additional 10 min are subtracted. For this prototypic implementation, it is assumed to be sufficient, but ideally this amount will be chosen by the users themselves (Figs. 8.13, 8.14).

Fig. 8.13 A simple equation applied to calculate the alert time based on starting time and travel time

$$AT = ST_{ne} - TT_{ce} - 10$$

AT = Alerttime
ST_{ne} = Event Starttime
TT_{ce} = Traveltime from Current Position to the Event

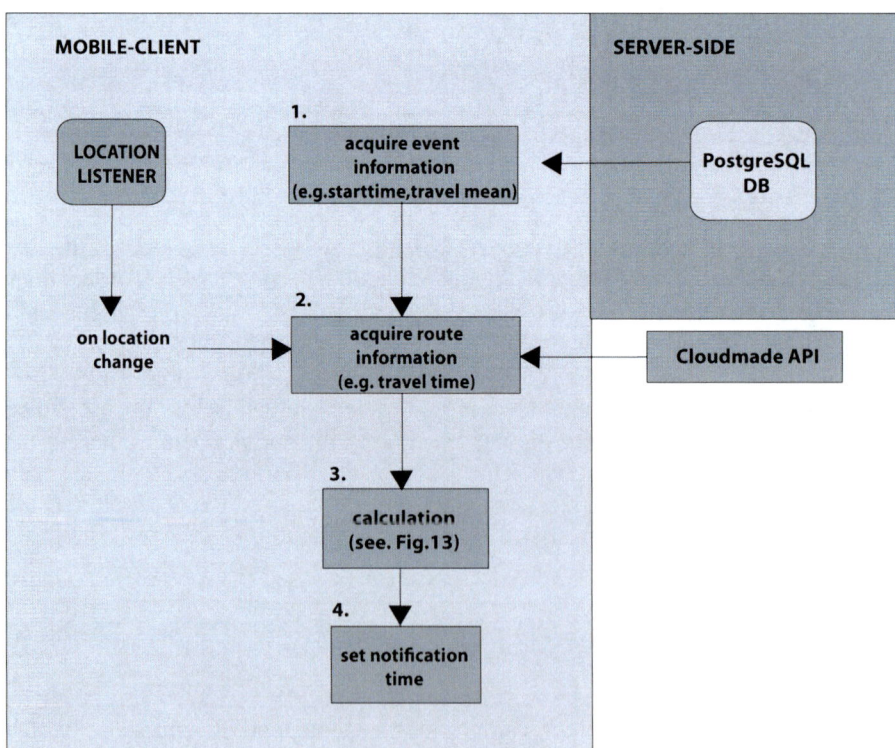

Fig. 8.14 Visualisation of the evaluation process applied for the Punctuality Alert

8 LatYourLife: Applying Multiple API Services for Task Planning

It shows that the process is done in four steps:

1. Fetch the necessary event details from the database;
2. Acquire the travel time through the API used to gain route information;

Apply the above quoted equation;
Set the Alert.

The location listener, is integrated to track the user movement and restarts the process from step 2 in case a location change of more than 50 m is detected. In this case the algorithm is rerun and step 4 will then update the alert time according to the new location and travel time.

The application in this research is restricted to walking, cycling and driving times, since public transport information is difficult to acquire and monitor. Ideally, a genuine public transport web API would need to be provided, that could be queried according to geographic location and time (Fig. 8.15).

Fig. 8.15 The user is notified in the notification bar (*upper left* frame) when it is time to depart

8.4 Findings

The article is an attempt to illustrate how we can use API's as a means to join different data together and build one consistent system. The LYL effectively showed that by using API's it is now possible for ordinary developers to integrate functionalities, that would be difficult if done manually. It was determined that for most functionalities, data is required that cannot easily be acquired or maintained by individuals, such as data on topological network. Even if the developer has access to such data, processing the data itself is a rather complex undertaking that would take time and effort. Thus, an API offering such functionalities is a very handy tool to overcome these issues and the developers does not have to concern themselves with the work done in the background.

Finally, we want to formulate some findings the process of building the application revealed:

8.4.1 Finding 1

API's are the web's form of "division of labour" and make it possible for individuals to perform tasks which they would not have been able to do before. So before you think about implementing a functionality for your application, look whether there are services providing it already, unless you have reasons to program it yourself.

In the case of the LYL-application we used the Google Maps API V.2, but soon discovered that it does not offer all the information we require for the system, such as bicycle routes. Also, public transport is not supported. Unfortunately, there was no solution for the public transport information for the regions we were interested in, but at least there was a solution for the bicycle problem by utilising another API. So, the Google API was used to visualise and query the routing information from another data source. This illustrates the capabilities we gain by supplementing API's with others. In our case, combing the Google Map API with the Cloudmade API.

8.4.2 Finding 2

API's are tools which provide a limited set of functionalities. Joining several API's together is an easy and efficient way to expand their capabilities.

While the LYL application was being written, a newer version of the Google Maps API was released that promised better performance and introduced some new functionalities. This was accompanied by major change in code syntax, that lead to a considerable amount of time spent in an attempt to transform the old code into the new syntax. Soon, it became apparent that this was a rather exhausting undertaking and that the effort would exceed the benefit. Fortunately, the Google Maps V.2 was still maintained by Google so the decision was taken to stay with the version.

8.4.3 Finding 3

Spending time studying the API and provided functionalities can save trouble. Make sure there are no newer versions and assess whether the migration to the next version is worth the effort.

8.5 Conclusion

The main goal of the LYL-application was to serve the user with useful tools for daily life task planning. For this purpose, several functionalities were implemented that afforded routing and weather information. It became evident that the service would not have been possible the utilisation of several API's since data such as routing information or weather is not provided by a single API. It also illustrated the power of combining various API's together such that lacking information of one API is compensated by the integration of another source. Besides that, it highlighted some of the things we have to be aware of when working with API's, such as versions.

An important point that was not mentioned, but not to be underestimated, are legal issues. Although API's seem to be a nice and easy way to gain data and resources, developers need to be well aware of what and how much they are allowed to do with the data.

References

Ashbrook D, Starner T (2003) Using GPS to learn significant locations and predict movement across multiple users. PersUbiquitous Comput 7:275–286

Baus J, Krüger A, Wahlster W (2002) Resource-adaptive Mobile Navigation System. In: Proceedings of the 7th international conference on intelligent user interfaces, San Francisco, pp 15–22

Bornträger C, Cheverst K, Davies N, Dix A, Friday A, Seitz J (2003) Experiments with multi-modal interfaces in a context-aware city guide. In: Chittaro L (ed) Proceedings of the 5th international symposium of human-computer interaction with mobile devices and services, Udine, pp 116–129

Frentzos E, Gratsias K, Theodoridis Y (2007) Towards the next generation of location-based services. In: Proceedings of the 7th international conference on Web and wireless geographical information systems, Cardiff, pp 202–215

Jiang B, Yao X (2007) Location based services and GIS in perspective. Location based services and telecartography, Lecture Notes in Geoinformation and Cartography, Springer Berlin Heidelberg, pp 27–45

Karimi HA, Xiong L (2003) A predictive location model for location-based services. In: Proceedings of the 11th ACM international symposium on advances in geographic information systems, New Orleans, pp 126–133

Klante P, Krösche J, Boll S (2004) Accessights: a multimodal location-aware mobile tourist information system. In: Proceedings of the 9th international conference on computers helping people with special needs, Paris, pp 287–294

Koeppel I (2000) What are location services?- From a GIS Perspective. Tech Rep, Sun Microsystems

Lonthoff J, Ortner E (2007) Mobile location-based gaming as driver for Location-Based Services (LBS) – exemplified by mobile hunters. Informatica 31:183–190

Meier R (2010) Professional android 2 application development, Wrox programmer to programmer. Wiley, UK

Miluzzo E, Lane ND, Fodor K, Peterson R, Lu H, Musolesi M, Eisenman SB, Zheng X, Campbell AT (2008) Sensing meets mobile social networks: the design, implementation and evaluation of the CenceMe application. In: Proceedings of the 6th ACM conference on embedded network sensor systems 2007, Raleigh, pp 337–350

Nanni M, Pedreschi D (2006) Time-focused clustering of trajectories of moving objects. J Intell Inf Syst 27(3):267–289

Nivala, AM, Sarjakoski T, Laakso K, Itäranta J, Kettunen P (2009) User requirements for Location-Based Services to support hiking activities. In: Proceedings of the 5th international conference on location based services and teleCartography (II), Salzburg, pp 167–184

Osman Z, Maguire M, Tarkiainen M (2003) Older users' requirements for location based services and mobile phones human-computer. In: Proceedings of the 5th international symposium of human-computer interaction with mobile devices and services, Udine, pp 352–357

Sohn T, Li KA, Lee G, Smith I, Scott J, Griswold WG (2005) Place-Its: a study of location-based reminders on mobile phones. In: Proceedings of the 7th international conference of ubiquitous computing, Tokyo, pp 232–250

Vu THN, Ryu KH, Park N (2009) A method for predicting future location of mobile user for location-based services system. Comput Ind Eng 57(1):91–105

Chapter 9
Guidelines for Implementing ArcGIS API for Flex Developers

Georgianna Strode

Abstract The ArcGIS Flex API from the Environmental Science Research Institute (ESRI) can produce rich, high-performing, engaging web mapping applications. It is usable out-of-the-box, with no programming needed. At the same time, there are significant programming resources available for those interested in customizing their maps. This discrepancy can be puzzling to those considering the option of using Flex because it is unclear how much programming is actually needed.

This incongruity carries over to the information and resources available for Flex developers. For example, there is ample information available for beginners who are not interested in programming, and there are also generous resources for advanced programmers. But at the time of this writing there is a gap in the information available to users who are in the process of determining how much programming they will need. This chapter attempts to fill this knowledge gap by providing general information on the Sample Flex Viewer (SFV) and a practical ten-step guide for beginning Flex developers.

9.1 Introduction

The term "Rich Internet Application" (RIA) was coined by Macromedia to describe a client technology with high interactivity, native multimedia support, the ability to increase the capabilities offered by traditional web applications, and the ability to overcome browser incompatibilities. The RIA combines the advantages of the desktop with the advantages of the web. RIAs can run the same on all major browsers and operating systems, they can combine raster and vector data for

---

G. Strode (✉)
Florida Resources and Environmental Analysis Center (FREAC), Florida State University (FSU), Tallahassee, FL, USA
e-mail: GStrode@admin.fsu.edu

presentation, and they make good use of client resources (Linlin and Hu 2009). The user experience is the same regardless of which browser is used. These characteristics explain why RIAs are dominant in online gaming.

RIAs have asynchronous client and server communication, resulting in web pages being refreshed less often. Avoiding continuous page refreshment increases the performance because only data that has been changed needs to be updated. Making good use of network resources reduces bandwidth needs (Guo et al., 2009).

RIAs combine the advantages of the desktop and the web. From the desktop, RIAs have: an interactive interface, fast response that updates partial pages, and real-time interaction. From the web, RIAs have: cross-platform compatibility, one-time loading with multiple uses, efficient network data transmissions, the ability to run multiple applications at the same time, and rapid development and deployment (Guo et al., 2009).

There are many frameworks for creating RIAs. Some have colorful names such as: Cappucino, Curl, Appcelerator Titanium, Lively Kernel, Sproutcore, Tersus, Panda3D, and others. Several of the more well-known frameworks are listed below:

- Openlaszlo – an open source product that generates Dynamic Hypertext Markup Language (DHTML) or Flash™.
- Adobe's Flex – a mature product that requires a Flash™ plug-in (installed on over 90% of computers).
- Google Web Toolkit (GWT) – an open source AJAX development kit, used for making Google Maps.
- Sun Microsystems's Java applets – small applications written in Java that run inside a browser plug-in. This is the most mature RIA technology but is not widely adopted (Machado et al., 2007).
- Asynchronous Javascript and XML (AJAX) – a group of technologies that can communicate with a server in the background without interfering with the current web page.

Critics of the term RIA point out that the term lacks clear definition. How does one rate the richness of an application? There is no clear distinction between what can be called an RIA and what cannot. RIAs range from applications with minimum richness to those that encompass many RIA capabilities. The definition can include applications with all characteristics, or no characteristics, and things in-between (Schmelzer 2009).

9.2 Framework Background

9.2.1 Adobe Products

Flex is an open source framework from Adobe. Flex applications are Flash™ movies (SWF files) that run inside HTML pages. This design makes it easier to deploy and access applications from the web.

Flex applications use MXML and ActionScript languages. MXML is similar to XML in the sense that tags lay out the page, and ActionScript compares to Javascript in that they both are programming languages that perform similar functions. Flex projects consist of MXML tags that configure the page and ActionScript code that drive the events of the page. The ActionScript language uses a class library which contains classes for many features, including manager and data classes. ActionScript syntax is an object-oriented language similar to C# and Java. MXML is converted to ActionScript which is converted to SWF files.

Adobe has released a Software Development Kit (SDK) for building applications in Flex. Flash Builder 4™ is the integrated development environment (IDE) for building applications using Flex SDK. Flash Builder™ has a debugging option that requires Flash Player 10™.

9.2.2 ArcGIS API Background

The ArcGIS API for Flex is a code library maintained by ESRI that allows the Flex SDK to work with ArcGIS services. The API was first released a number of years ago, but never captured much of the GIS programming audience because it was difficult to use, required extensive programming knowledge, and did not deploy rapidly. To address these drawbacks, ESRI created the Sample Flex Viewer (SFV) to simplify application development. The SFV acts as a "template" where developers can create GIS RIAs quickly without extensive programming. Because of the SFV, Flex is gaining momentum as a platform for mapping applications. While the original Flex API required extensive programming knowledge, the SFV places mapping technology in the hands of GIS analysts.

The SFV is offered in two versions: Compiled and Un-compiled. The Compiled Version is customizable by editing XML files while the Un-compiled Version contains the MXML source code. Figure 9.1 shows the relationship between the API and the two versions of the Sample Flex Viewer.

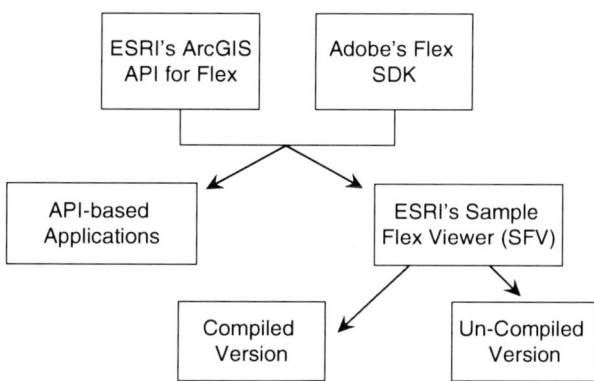

Fig. 9.1 Framework relationships

9.2.3 Complete Software Environment

At the time of this writing, the current product versions are: ArcGIS Viewer for Flex version 2.3.1, Adobe Flash Builder 4™, and ArcGIS 10. A *crossdomain.xml* may be needed. This is a Flash™ security requirement to allow internet access. Without this file, a browser plug-in can only access resources on the same web server that runs the plug-in.

Both versions of the SFV are available from ESRI at no charge. However, compiling source code requires Flash Builder™ from Adobe, Inc. The SFV can use mapping data from multiple sources. Some possibilities are: use published data from other sources (including ESRI), self-publish data using ArcGIS Server, self-publish data using an open source solution in Web Map Service (WMS) format, or a combination of these. If data is to be published using ArcGIS Server, then ArcMap 10 must be purchased from ESRI.

9.3 The Sample Flex Viewer: Compiled or Un-compiled?

When ready begin, developers are faced with the choice of which of the two versions of the SFV to use. There can be confusion between the SFV options because the names are similar and the short descriptions cannot fully explain what each option delivers. The paragraphs below give more detailed information of what developers can expect from each option.

ArcGIS Viewer for Flex Compiled Version – The webpage reads "Download Viewer – compiled and ready to go" (ESRI 2011). This option has only compiled code (SWF) and XML files. These files can be placed inside the internet resource folder (e.g., /Inetpub/wwwroot) and the SFV will be ready for use. Developers can customize their application by editing the XML files without programming. Only customization changes using XML files are available as the source code is not supplied with this option. Developers are limited to making edits to the individual XML files associated with each widget and the main Widget Configuration File (WCF), *config.xml*.

ArcGIS Viewer for Flex Un-compiled Version – The webpage reads "Download source code for Viewer – for Flex developers" (ESRI 2011). This option has the source code (MXML) and XML files. These files should be loaded into an Adobe Flash Builder™ project where they can be edited and compiled. Programmers have full access to the source code and are free to customize as they wish. After the project is compiled, the SWF and XML files can be placed in an internet-ready folder for access over the web. Note that even simple tasks such as changing a font style require editing the source code. Choosing the ArcGIS Viewer for Flex Compiled Version is preferable if the budget allows the purchase of Flash Builder™ because very simple modifications often must be made in the source code.

9 Guidelines for Implementing ArcGIS API for Flex Developers

Table 9.1 Comparison of product capabilities and mapping requirements

	No purchase option	Purchase option
Adobe's Flash Builder™	No purchase necessary if not planning to edit or compile source code. Customization is limited to editing XML files using a text editor	Must purchase Flash Builder™ to edit and compile source code
ESRI's ArcGIS Server	No purchase necessary if not planning to publish data using an ESRI format. Can consume free map services from ESRI for basemaps. Can self-publish using open source options (such as MapServer). Other options for self-publication without purchase of ArcGIS Server could be available in the future	Must purchase ArcGIS Server to publish data in an ESRI format (such as REST)

API-Based Applications – The original API is available for developers who do not want to use the Sample Flex Viewer (SFV). Building an application from scratch requires the API library to be downloaded independently. This option requires a skilled developer. This option is beyond the scope of this chapter.

A convincing argument for use of the Flex Viewer comes from ESRI's Derek Law in "5ive Reasons to Use the ArcGIS Viewer for Flex:"

1. The viewer can be used immediately out of the box
2. The viewer can be configured; no programming is needed
3. The viewer has an extensive widget library
4. The viewer framework is extensible with the ArcGIS API for Flex
5. The viewer is free and fully supported by ESRI (Law 2011)

The ArcGIS Viewer for Flex Compiled Version can be configured out-of-the-box in 6–8 min by downloading and unzipping the application and placing it in a web resources directory. It is very easy to make minor configuration changes, such as change to the title or logo by simply editing XML files. There are 20 widgets as part of the core software, plus an online gallery of widgets contributed by ESRI staff and the user community at large. When there is a new release of ArcGIS, there is also a new release of the viewer software. The software and widgets are available at no cost (Law 2011).

Table 9.1 details the capabilities that developers can expect with purchase of Adobe and ESRI products. Developers should balance their mapping requirements with their financial budget to determine if a product purchase is necessary.

9.3.1 Helpful Resources

Regardless of which of the Sample Flex Viewer is used, identifying and bookmarking useful resources at the start of a project saves time. Web pages for different versions of a product can look alike, and performing a web search is likely to return links to older product versions. Be sure to bookmark resources for the current version.

The correct forum to use (*ArcGIS Viewer for Flex* or *ArcGIS Flex API*) does not necessarily depend upon which SFV was downloaded but rather whether the solution involves editing XML files or editing MXML files. For example, suppose a developer downloads the SFV Un-compiled Version and has always used the *ArcGIS Flex API* forum for programming questions. Now the developer wants to change the background color of the web page. While this question might seem like a programming question, it is not. The question should be posted to the *ArcGIS Viewer for Flex* forum because this question can be answered by editing XML files and does not require editing of source code. It can be confusing to know which forum to post to because it can be difficult to determine whether the problem can be solved with XML or MXML. It is recommended for SFV Un-compiled Version users to bookmark both forums in case of a situation like the one described. If a question is posted to the wrong forum, the staff will courteously move the post to the correct forum.

There is a list of resources at the end of this chapter. It is recommended to bookmark at a minimum: forum(s), API, and REST/WMS information for data sources (to verify a field's spelling and type), and to bookmark the ArcGIS Server and REST cache for use when the data is updated.

9.4 ArcGIS Viewer for Flex Compiled Version

9.4.1 Widget Configuration File: config.xml

The *config.xml* file is responsible for much of the appearance, functionality, and data of the interactive map. There is much that can be accomplished by utilizing the capabilities in this one file. This section is a brief overview of what the *config.xml* offers.

9.4.1.1 User Interface (UI) Elements

The User Interface Elements are those that are considered basic to a map, for example, navigation, overview map, and table of contents. These elements are already programmed and are easily added to the map via the *config.xml* file. The sample code below shows how elements can be placed at a specific screen location using *left, top, bottom,* and *right* attributes. The *MapSwitcher* element allows users to switch between basemaps. The *Navigation* widget provides nice zooming and navigation functions.

9 Guidelines for Implementing ArcGIS API for Flex Developers

```
<!-- UI elements -->
<widget left="15" top="95"
config="widgets/Navigation/NavigationWidget.xml"
url="widgets/Navigation/NavigationWidget.swf"/>

<widget right="20" top="55"
config="widgets/MapSwitcher/MapSwitcherWidget.xml"
url="widgets/MapSwitcher/MapSwitcherWidget.swf"/>

<widget right="1" bottom="1"
config="widgets/OverviewMap/OverviewMapWidget.xml"
url="widgets/OverviewMap/OverviewMapWidget.swf"/>
```

9.4.1.2 Initial Extent

The Initial Extent command allows the map to conveniently display the chosen geographic area. The code sample below shows how easy it is to set an initial extent.

```
<map initialextent="-14083000 3139000 - 10879000 5458000">
```

ESRI has published a handy application named *Create Extent – Map Corners Method* for easily determining a map extent (the URL is a published at the end of this chapter). Using this tool, developers navigate the sample map to the geographic area of interest. The map extent is updated to display Web Mercator coordinates, which can be used for the Initial Extent. ESRI has conveniently included a one-click method for copying the coordinates to the computer's clipboard memory to save typing.

9.4.1.3 Basemaps

A map can have several basemaps, such as a street map or aerial, and users can toggle between them. The sample code below shows how to add multiple ESRI-published basemaps.

```
<basemaps>
    <mapservice label="Streets" type="tiled" visible="true"
    alpha="1"
    url="http://server.arcgisonline.com/ArcGIS/rest/services/Wo
    rld_Street_Map/MapServer"/>

    <mapservice label="Aerial" type="tiled" visible="false"
    alpha="1"
    url="http://server.arcgisonline.com/ArcGIS/rest/services/Wo
    rld_Imagery/MapServer"/>

    <mapservice label="Topo" type="tiled" visible="false"
    alpha="1"
    url="http://server.arcgisonline.com/ArcGIS/rest/services/Wo
    rld_Topo_Map/MapServer"/>
</basemaps>
```

Fig. 9.2 A sample map using the default Navigation, Basemap, and Overview Map Widgets

The ESRI website has links to hundreds of basemaps covering terrain, topology, aerials, hybrid aerials, shaded relief, and even land parcels. Scales range from worldwide to country- and city-wide. Publishers include OpenStreetMap, Bing, government agencies, and others. ESRI-published basemaps are available from a URL published at the end of this chapter, but other non-ESRI basemaps are available from the ESRI Map Gallery by searching on 'basemaps.'

Figure 9.2 shows a sample map made using the previous code samples. The *Navigation* bar allows users to zoom and pan. The *Overview* map in the lower right corner shows a "birds-eye" view. The *Basemaps* are labeled by "Streets," "Aerial," and "Topo," and users can easily select a basemap.

9.4.1.4 Operational Layers

Operational layer are data layers of the map. They can be self-published or published by other entities. These layers are displayed in the Table of Contents and can be toggled on and off by the map user. The following code displays a combination of ESRI-published layers in the Table of Contents.

```
<operationallayers>
    <layer label="Boundaries and Places" type="tiled"
    visible="false"
    url="http://server.arcgisonline.com/ArcGIS/rest/services/Re
    ference/World_Boundaries_and_Places_Alternate/MapServer"/>

    <layer label="Fires" type="feature" visible="false"
    alpha="1"
    info="widgets/InfoTemplates/SimpleInfoWinWidget.swf"
    infoconfig="widgets/InfoTemplates/IWT_Fires.xml"
    url="http://sampleserver3.arcgisonline.com/ArcGIS/rest/serv
    ices/Fire/Sheep/FeatureServer/0"/>
</operationallayers>
```

9.4.2 Directory Structure

The compiled ArcGIS Viewer for Flex contains executable code for the Sample Flex Viewer. The SFV uses a modular programming approach by allowing "widgets" to be added to the application as needed. A widget is a program that fulfills a single purpose, such as showing a legend, zooming to a street address, or other clearly-defined task. A mapping application can have as many widgets as needed. The base code contains at least 20 widgets, and others are available for download. The Widget Configuration File (WCF), *config.xml*, controls much of the appearance and functionality of the interactive map.

The main directory structure looks like:

```
\Inetpub\wwwroot\MyMap
    \assets
    \widgets
    config.xml
    << other files >>
```

There are two main folders: *assets* and *widgets*. The *assets* folder contains images that are used throughout the map. The *widgets* folder contains the widgets available with the original download. Each widget has its own folder, as seen in the example below.

```
\Inetpub\wwwroot\MyMap\widgets
    \Bookmark
    \Coordinate
    \Draw
    \Edit
    << other widgets >>
```

Each individual widget folder contains two file types: SWF and XML. SWF files are the compiled executable code and cannot be edited. SWF files can be obtained by downloading from ESRI (the initial download) or they can be downloaded from other places. A limited amount of customization is possible by editing the XML files. The code example below shows that each widget folder contains SWF and XML files.

```
\Inetpub\wwwroot\MyMap\widgets\Bookmark
    BookmarkWidget.swf
    BookmarkWidget.xml
```

9.4.3 Customizing a Widget

Editing XML files is an easy way to customize. For example, in conjunction with *Bookmark.swf*, the sample *Bookmark.xml* file below would show users a list of place names, allow users to click on a place (e.g., San Francisco), and zoom the map to that location. Adding an additional bookmark for a new area, such as New York, is easily accomplished by editing *Bookmark.xml* to include a new entry with the appropriate coordinates. Other XML files can usually be edited easily by following the existing formats.

\Inetput\wwwroot\MyMap\widgets\Bookmark.xml

```
<?xml version=1.0 ?>
<configuration>
    <bookmarks>
        <bookmark name="San Francisco">-13638000 4541000 -
            1362000 4551000</bookmark>
        <bookmark name="Los Angeles">-13211400 3993400 -
            13119200 4056100</bookmark>
        <bookmark name="Paris">214700 6218400 308200
            6281700</bookmark>
    </bookmarks>
</configuration>
```

9.4.4 Adding a New Widget

To add a new widget, download the SWF and XML code and place them in a separate folder inside the *widget* folder. To connect the widget to the map, edit the Widget Configuration File, *config.xml*, following the existing format.

```
<widget label="MyNewWidget"
    icon="assets/images/new_icon.png"
    config="widgets/MyNewWidget/MyNewWidget.xml"
    url="widgets/MyNewWid/MyNewWidget.swf" />
```

9.4.5 Changing Icons

Each widget comes with a default icon, but customized graphics can enhance the branding of existing web pages. Replacing graphics is a simple process:

1. Create the new graphic (JPG, GIF, or PNG)
2. Place the new graphic in the /assets/images folder
3. Edit *config.xml* to call the icon as in the example below:

```
<widget label="Bookmarks"
    icon="assets/images/my_new_bookmark_icon.png"
    config="widgets/Bookmark/BookmarkWidget.xml"
    url="widgets/Bookmark/BookmarkWidget.swf" />
```

9.5 ArcGIS Viewer for Flex Un-compiled Version

The Un-compiled ArcGIS Viewer for Flex contains the source code, MXML files, as well as XML files. Much of the information provided earlier for SFV Compiled Version is the same for the Un-compiled Version. This section provides additional information specific to this option.

9.5.1 Compiling Code

Compiling is a simple step of clicking a button, resulting in executable SWF files in each widget's folder inside the *bin-debug* directory. But sometimes there can be confusion when programming changes are made but the results are not visible. Usually this is because a widget is not compiling.

If programming changes are not visible, the first thing to look for is an error in the MXML code. The Flash Builder™ program is useful for notification of programming errors. If there is no error, the second thing to check for is to see if the widget is registered in the *Flex Modules* section: right click on the project name, select *Properties*, then select *Flex Modules*. Be sure that the name of the widget is listed. Another way to determine if a widget is on the *Flex Modules* compilation list is to look for a small blue icon in the widget name. Widgets listed in the *Flex Modules* list should have a blue icon on the folder name. Programmers from the Flex environment are likely to know this information, but GIS programmers often learn this lesson the hard way.

If there are further doubts about what is being compiled, the *clean* command will erase all existing SWF files. The *build* command will rebuild the project and compile all modules in the *Flex Module* list.

9.5.2 Referencing Layers and Fields

Below are two examples of XML files for two different widgets that use the same mapservice that contains *joined* tables. The examples show two ways to reference layers: by name and by number. Additionally, the examples show two different ways to reference *joined* fields: by short name and by fully qualifying the field name.

Sample code from *Identify.xml*:

```
1.    <layer>
2.    <name>Preliminary Points</name>
3.    <fields>id,name,description,abstract</fields>
```

Sample code from *Search.xml*:

```
4.    <layer>
5.    <name>Preliminary Points</name>
6.    <url>http://myserver/ArcGIS/rest/services/MyMap/
         myservice/3</url>
7.    <fields all="false"
8.    <field name="basedata.sde.prelim.id</field>
9.    <field name="basedata.sde.prelim.name</field>
10.   <field name="basedata.sde.related_table.description</field>
11.   <field name="basedata.sde.related_table.abstract</field>
```

Lines 2 and 6 both reference the layer. In line 2, the layer is implicitly referenced by name, but line 6 shows referencing the layer by number. Line 3 lists the fields using the short name and does not require the full field name. In contrast, lines 8–11 use the fully-qualified field names.

Both widgets, the *Identify* and the *Search*, are similar in that they need to find layers and fields, and that functionality is not the reason for the differences in XML code. The reason that these examples are different can be traced to the original programming of the MXML files. Because the widget collection is contributed by the community at large, it is natural that widgets may have different programming approaches that are reflected in variations in the XML files.

9.6 Ten-Step Guide for Flex Developers

The Ten-Step Guide for Flex Developers was created as a guideline to those new to using either version of the Sample Flex Viewer.

1. **Write down the map functionality to be included.** While this step may seem obvious, it is important to focus on the expected mapping outcome and not become sidetracked. There are many impressive widgets that might initially appear to fully address a mapping issue, but further inspection can show otherwise. Clearly identifying the map functionality in writing increases a developer's focus.
2. **Make or find your data sources**. You can use yours or someone else's data or a combination of both. These can be WMS and/or REST services.
3. **Download the source code Sample Flex Viewer** (if budget permits). While the source code is free of charge, compiling the source code requires the

purchase of Adobe Flash Builder™. It is important to note that even minor changes, such as changes to font size or color, require edits to the source code because this information is embedded in the MXML files. If the budget allows for the purchase of Flash Builder™, this is the best choice.
4. **Bookmark references**. Having resources handy saves time and confusion. References should include URLs for: both Viewers (Compiled and Un-compiled), forums for both Viewers, Adobe Flex support, and the widgets collection. For data sources (whether yours or someone else's), bookmark the URLs for the WMS and/or REST services. If you are publishing your own data, bookmark URLs for refreshing the ArcGIS Server and for clearing the REST cache.
5. **Watch ESRI videos**. There are many excellent videos giving instruction on customization using XML files. These videos are introductory in nature and are useful in gaining an overview of the Compiled Viewer.
6. **Investigate widget collection**. There is much community support for Flex and new widgets are continually being added. Allow enough time to investigate the pre-existing widgets because finding a useful widget saves a tremendous amount of time. Exploring the work of others gives insight into possibilities and limitations. The best approach is to look through the widgets, download a few, and start trying them. Note that it is possible to ask a widget's developer to include additional functionality in the next widget release.
7. **Make the map**. The map navigation code and over 20 basic widgets are already included. Depending on the functionality identified in the first step, adding additional widgets could significantly contribute to map completion. It is possible that the map could be complete at this point and there is no need for programming. If so, go to step 10.
8. **Create or customize widgets, if needed**. Widgets can be edited or new ones can be created. There are ESRI videos showing how to make new widgets. Another option is to find a widget with similar functionality, make a copy of it, and begin editing. Remember to maintain the *Flex Module* registration list so that the new widget will be compiled.
9. **Use bookmarked references** as needed.
10. **Document the steps for data update**. For example, if a module has hard-coded database layer names or numbers, this will need to be adjusted every time the data is updated. This will usually occur in XML files but it is possible that MXML files can have hard-coded layer and field names. Having these instructions in writing will save time later.

9.7 Conclusion

This chapter is intended to serve as a guide for beginning Flex developers who may not have decided which version of the Sample Flex Viewer, Compiled or Un-compiled, is best for them. There is a gap in information available for beginning developers who are not sure how much programming, if any, will be needed to complete their projects. This chapter attempts to provide general information on what to expect from both versions of the Viewer as well as a practical ten-step guide for initial project development. Hopefully, this information will be of use and can save developers time.

Acknowledgements The author would like to thank Edwin Cake III and Michael Peterson for their thoughtful reviews and comments.

References

Environmental Science Research Institute (2011) ArcGIS for Flex Viewer. http://help.arcgis.com/en/webapps/flexviewer/
Guo L, Jianhua G, Jun S, Xiangwang W (2009) Study on GIS architecture based on SOA and RIA. National Sci-Tech Major Special Item of China No. 2009ZX10004-720
Law, D (2011) 5ive reasons to use the ArcGIS Viewer for Flex. ArcUser Online. Environmental Science Research Institute (ESRI). http://www.esri.com/news/arcuser/0311/five-reasons-to-use-the-arcgis-viewer-for-flex.html
Linlin W, Hu D (2009) Research and realization of RIA WebGIS based on Flex. In: 2nd international workshop on intelligent systems and applications. IEEE, Wuhan. http://ieeexplore.ieee.org/xpls/abs_all.jsp?arnumber = 5073036
Machado LC, Orlando BF, João AR (2007) RIA technologies comparative study applied to GIS. Seminário de Informática
Schmelzer R (2009) The dissolution of the Rich Internet Application (RIA) Market. ZapThink, 3 Sept 2009. http://www.zapthink.com/2009/09/03/the-dissolution-of-the-rich-internet-application-ria-market

Resources

Environmental Systems Research Institute (ESRI)
ArcGIS API for Flex Samples (not using the Sample Flex Viewer). http://help.arcgis.com/en/webapi/flex/samples/index.html
ArcGIS Server Blog (Flex Announcements). http://blogs.esri.com/Dev/blogs/arcgisserver/archive/tags/Flex/default.aspx
ArcGIS API for Flex Video Gallery. http://resources.arcgis.com/gallery/video/arcgis-api-for-flex
ArcGIS Viewer for Flex (Compiled Version). http://help.arcgis.com/en/webapps/flexviewer/
ArcGIS Viewer for Flex Forum (Compiled Version). http://forums.arcgis.com/forums/111-ArcGIS-Viewer-for-Flex
ArcGIS Viewer for Flex Code Gallery (Widgets). http://help.arcgis.com/en/webapps/flexviewer/gallery.html
ArcGIS API for Flex (Un-compiled Version). http://help.arcgis.com/en/webapi/flex/index.html
ArcGIS API for Flex Forum (Un-compiled Version). http://forums.arcgis.com/forums/18-ArcGIS-API-for-Flex
ArcGIS API for Flex Code Gallery (Widgets). http://help.arcgis.com/en/webapi/flex/gallery.html
ArcGIS Services Directory (Sample REST services). http://server.arcgisonline.com/ArcGIS/rest/services
ArcGIS Server REST API. http://help.arcgis.com/en/arcgisserver/10.0/apis/rest/index.html

ArcGIS Resource Centers. http://resources.arcgis.com
ArcGIS.com. http://www.arcgis.com
Create Extent MXML – Map Corners Method. http://trainingcloud.arcgis.com/baaf/apps/CopyExtentToClipboard/GetExtent_MapCornersMethod.html
ESRI Basemaps. http://services.arcgisonline.com/ArcGIS/rest/services
ESRI Developer Network (EDN). http://www.esri.com/edn
Adobe, Inc.
Flex in a Week Video Training. http://www.adobe.com/devnet/flex/videotraining.html
Flex Developer Center. http://www.adobe.com/devnet/flex.html
Flash Builder 4™. http://www.adobe.com/products/flashbuilder/

Part III
Symbolization

Chapter 10
Web Services for Thematic Maps

Otakar Cerba and Jachym Cepicky

Abstract The current world of information technologies (including geo-information technologies /GIT/ as well as internet maps) heads towards the new generation of Web (as a collection of interlinked documents accessed via the Internet). The next development stages of Web (Web 2.0, Web 3.0, Semantic Web etc.) are associated with terms – user-centred applications, information sharing, collaborative work and interoperability. Just the last word is very important because it represents the necessary condition to functioning of complete system and its components like blogs, wikis, mushups, web applications and services. Geo-information technologies work also with one important term associated with Web – Spatial Data Infrastructures (SDI).

The chapter "Web Services for Thematic Maps" is divided into two main parts. The first part (sub-chapters 10.2-10.5) is focused on common questions of thematic cartography and application of web services in cartography. There are also mentioned problems associated with web services in cartography like spatial data heterogeneity, standardisation or lack of semantics. Because this theme is very large, we have selected web services standardized by Open Geospatial Consortium (OGC) and thematic maps as the important representatives of both domains. This part describes general service-oriented architecture and four main OGC web services, including their advantages, disadvantages and "cartographic" abilities.

The second section (10.6, 10.7) of this text is centred on a design of application of Web Processing Service (WPS) to thematic cartography. We describe current implementations of WPS (e.g., project Humboldt or open-source solution PyWPS). The main part is focused on the concrete application of WPS to generating of different types of thematic maps in the format SVG (Scalable Vector Graphics). The general principle of the generating of SVG maps through XSLT (Extensible

---

O. Cerba (✉)
Section of Geomatics, Department of Mathematics, Faculty of Applied Sciences, University of West Bohemia in Pilsen, Univerzitni 22, Plzen 306 14, Czech Republic
e-mail: ota.cerba@seznam.cz

Stylesheet Language Transformation) as desktop solution was used in the project VisualHealth and it was published in various papers in the past years (Chile 2009, Austria 2009, Czech Republic 2009, Bulgaria 2008, Moscow 2007). This text describes a combination of existing desktop solution and WPS. The aim is the using XSLT processor, pre-defined XSLT styles intended for transformation of spatial data to SVG format, and RELAX NG (REgular LAnguage for XML Next Generation) scheme (the method of description of a maps and their attributes) as web application.

10.1 Introduction

Thematic maps are present in all types of media. They are a part of today's TV news, newspapers and the Internet. Many books use thematic maps to explain spatial questions. They are also used in transport, economy, demography, statistics, tourism, agriculture, education, or promotion campaigns. Many experts, such as geographers, spatial planners, architectures etc., work with thematic maps and use them as a source of information or type of presentation of their results. Thematic maps are a routine part of our daily life in a modern information society.

Thematic maps are an ideal cartographic product for many different types of users. For example, people who are not very knowledgeable in working with maps, and the large amount of information in topographic maps, could discourage them from reading maps, or not allow them to obtain key information. Another group of users may consist of advertising applications, with simplified map drawings to get a lot of space for other graphical elements. A third group may be users with different types of physical or mental disabilities, who, like the first group welcome particularly maps that are easy to read. Another application is for those people who need to get information from map quickly like users of car navigation systems or orienteering runners.

This chapter is focused above all on the interconnection of thematic maps and web services. Web services represent one of the most forward-looking technologies in the current world of information processing. Web services as a collection of interlinked documents accessed via the Internet, represent the new generation of the Web (Web 2.0, Web 3.0, Semantic Web etc.). They represent the described interface making possible the full exploitation of software products from external sources and their sharing and combination. Web services enable an automated processing of large amounts of data, rapid response to changes in data, real-time applications, the use of external resources, elimination of redundancies in data and software resources (re-use of existing applications). Use of web services leads to increasing efficiency in the use of financial, personnel and time resources.

Web services means the chance for cartographers to find and use the best software products through a standardized way without the necessity of installing and buying unneeded commercial software solutions. Also, the results of cartographic activities connected with web services will be published through Internet. Last, but not least, it is important to mention the role of web services in the process of building Spatial Data Infrastructures (SDI), because discovery services, view services, download services or transformation services are the essential part of SDI.

This chapter is divided into two main parts. The first part is focused on common overview of thematic maps, service-oriented architecture and web services, including the potential of web services in cartography. Problems are introduced with application of web services in cartography like spatial data heterogeneity, standardization or lack of semantics. Because the "web service" theme is very large, we want to describe, first of all, the web services standardized by Open Geospatial Consortium (OGC) as the main authority in Geoinformation technologies (GIT).

The second section Sect. 10.6 of this chapter is centered on a design of application of Web Processing Service (WPS) as a tool for creating thematic maps. We describe current implementations of WPS (e.g., project Humboldt or open-source solution PyWPS). The main part is focused on an interconnection of WPS and system for the generating of different types of thematic maps. This process is based on markup languages – the general principle for generating SVG (Scalable Vector Graphics) maps through XSLT (Extensible Stylesheet Language Transformation) styles that was used in the project VisualHealth (http://zdravi.geogr.muni.cz; only a part of this portal is in English) that has been described in other papers (Cerba 2007, 2008, 2009a, b). This paper describes a combination of existing desktop solutions and WPS.

Both parts are summarized in last two sections Sects.10.7 and 10.8.

10.2 Thematic Maps: A Short Overview

Thematic maps (also called special-purpose maps, single-topic maps, statistical maps) are defined as "maps designed to demonstrate particular features or concepts. In conventional use this term excludes topographic maps." (Meynen 1973). Thematic maps (some examples in Fig. 10.2) are ranked among the most interesting, used, and important cartographic products for visualizing spatial data. There are some arguments for the ever-growing importance of thematic maps (based on Cerba 2009b):

- The rising popularity of cartography, which relates to the development of new products such as Internet maps, mobile maps, Earth browsers or LBS (Location-Based Services).
- Thematic maps prefer model approach, where the relevant information (topic of maps and its spatial location) are highlighted, and marginal information are simplified or removed entirely.
- Another goal of the thematic maps is connected with the dual view of the map. The user can see the global situation, but also can study and compare details.
- Thematic maps are more intelligible and clear than very complicated spatial or statistical data, text documents, tables or charts. They represent very simple, but

very sophisticated tool that makes it possible to transform the complicated language of numbers to a simple graphical language.
- Via maps, users are able to connect information and location (concrete place). Such a relation can evoke an higher level of consciousness or weight of information.
- Simple interconnection to new attractive technologies and tools, such as digital globes or multimedia solutions.

10.3 Web Services in Cartography (Conceptual Level)

The Web is very important medium for the distribution of cartographic products distribution. In the book "Cartography: visualization of spatial data" (Kraak and Ormeling 2003) the authors answer the question – "Why is the WWW (World Wide Web) an interesting medium for maps?" The answer is that information on the web is virtually platform-independent. Also, many users can be reached at minimal costs and it is easy to update the maps frequently, although this last argument is only valid when the data provider is geared for it. Furthermore, the WWW allows for a dynamic and interactive dissemination of geospatial data. This results in new mapping techniques as well as new possibilities for uses have that not been seen before with traditional printed maps and most on-screen maps. While preparing maps for the WWW, one has to remind oneself that its access is currently geographically skewed. Some parts of the world still have poor access, such as Africa or Latin America and parts of Asia. Also, one has to remember that the WWW is a fast medium used by often impatient people.

Web services enable an exploitation of software applications available from web servers. Users can make use of them through clients supporting a concrete standard interface of web services (Fig. 10.1). Exploitation of web services have many benefits. For example users need not own any robust complete software solution, but only choose necessary services.

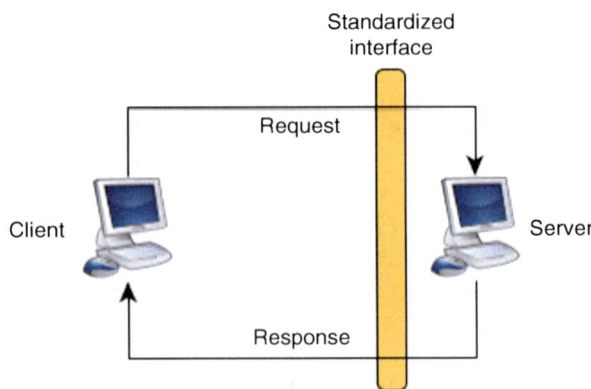

Fig. 10.1 Fundamental principle of web services

The current implementation of web services is limited due to many problems connected with

- Description of services (including metadata);
- Heterogeneity of source spatial data and lack of data semantics;
- Chaining, combination and interconnection of web services;
- Adherence, support and level of implementation of standards.

10.4 OGC Web Services

In geoinformation (GI) sciences including cartography, the most used web services are standardized by the Open Geospatial Consortium (OGC). Therefore, in this chapter use the term "web services" to mean "OGC Web Services". OGC Web Services form a subset of general web services, as defined by World Wide Web Consortium (W3C). Furthermore, W3C web services are using SOAP (Simple Object Access Protocol) envelope for communication between server and client, where OGC Web Services are using their custom XML (Extensible Markup Language) request/response format without any further envelope. At this time, OGC Web services standards are synchronized with "standard" W3C web services format, but many standards are still using a simple request/response format.

A short overview of OGC web services suitable for web cartography is provided here (see also Table 10.1). In particular, the benefits of web services for cartography are emphasized.

Web Map Service (WMS) produces maps of spatially referenced data dynamically from geographic information. WMS-produced maps are generally rendered in a pictorial format such as PNG (Portable Network Graphics), GIF (Graphics Interchange Format) or JPEG (Joint Photographic Experts Group), or occasionally as vector-based graphical elements in SVG (Scalable Vector Graphics) or WebCGM (Web Computer Graphics Metafile) formats (de la Beaujardiere 2004). Web Map Service is also standardized by International Organisation for Standardisation – ISO/DIS 19128.

As an extension to allow user-defined symbologies, a WMS may provide the capability of accepting SLD (Styled Layer Descriptor), an XML-based description language for extending OGC Web Services such as Web Map Services (WMS) and Web Coverage Services (WCS) with a user-defined symbolization. Only simple point symbols/markers, lines, polygons, text and raster images are integrated in the

Table 10.1 Overview of OGC web services

OGC standard	Processed information
WMS	Images (georeferenced)
WCS	Spatial data (above all coverages)
WFS	Spatial data (vector data)
WPS	Responses and request to external servers
KML	Spatial data

SLD language, which is why WMS instances are often used for topographic map services, but not for thematic map services (Iosifescu-Enescu et al., 2007).

A WMS without SLD would only have the possibility to visualize the map using some pre-defined styles from the server. This can be done more properly using SLD as it allows classifications also from the client (Dietze and Zipf 2007).

Another OGC standard – Symbology Encoding (SE) – together with the Styled Layer Descriptor Profile for the Web Map Service Implementation Specification is the direct follow-up of Styled Layer Descriptor Implementation (SLD) (Iosifescu-Enescu et al., 2007). The SE specification only contains the encoding part of SLD without any interfaces so that it could also be used outside of web services, e.g., in desktop applications. New possibilities with the SE specification compared to SLD include more advanced linear signatures, labeling, better scaling of symbols, and others. The need for the hereby proposed extension of the SE specification was caused through the currently limited possibilities for thematic mapping. Current possibilities in this direction are restricted to choropleth maps and simple point maps. But, the current specification is not adequate enough if pie charts or line charts should be drawn on point maps. It would also be desirable to simplify the classification process for dynamic choropleth maps (Dietze and Zipf 2007).

The Web Coverage Service (WCS) supports electronic retrieval of geospatial data as "coverages" – that is, digital geospatial information representing space-varying phenomena. A WCS provides access to potentially detailed and rich sets of geospatial information in forms that are useful for client-side rendering, multi-valued coverages, and input into scientific models and other clients. Unlike the WMS, which portrays spatial data by returning static maps (rendered as pictures by the server), the Web Coverage Service provides available data together with their detailed descriptions; defines a rich syntax for requests against these data; and returns data with its original semantics (instead of pictures) which may be interpreted, extrapolated, etc. – and not just portrayed. Unlike WFS (Web Feature Service), which returns discrete geospatial features, the Web Coverage Service returns coverages representing space-varying phenomena that relate a spatio-temporal domain to a possibly multidimensional range of properties (Whiteside and Evans 2008).

Web Processing Service (WPS) defines a standardized interface that facilitates the publishing of geospatial processes, and the discovery of and binding to those processes by clients. A WPS can be configured to offer any sort of GIS (Geographic Information System) functionality to clients across a network, including access to pre-programmed calculations and/or computation models that operate on spatially referenced data. This interface specification provides mechanisms to identify the spatially referenced data required by the calculation, initiate the calculation, and manage the output from the calculation so that the client can access it. This Web Processing Service is targeted at processing both vector and raster data. The WPS specification is designed to allow a service provider to expose a web accessible process, such as polygon intersection, in a way that allows clients to input data and execute the process with no specialized knowledge of the underlying physical process interface or API (Application Programming Interface). The WPS interface standardizes the way processes and their inputs/outputs are described, how a client can request the execution of a process, and how the output from a process is handled. Because WPS offers a generic interface, it can be used to wrap other existing and planned OGC services that focus on providing geospatial processing services (Schut 2007).

Because of this chapter is focused above all on OGC WPS, we would like to show some examples of WPS applications on geospatial domain. In the Humboldt project, WPS is used in the forest scenario in supervised image classification. the resulting data – classified images – are returned back to the client as a PNG, that is directly displayed in the mapping application. The process can be configured in a way that it gets list of to-be-classified land-cover categories (as GML – Geography Markup Language) and their corresponding color list that indicate how they should be displayed to the user. Another usage in Humboldt project is that WPS (more concretely PyWPS) is used as the interface between Humboldt Conceptual Schema Transformer (CST) and Internet application. CST is called from within WPS Process, which gets an input file (GML), it's corresponding schema file, schema file for the result, and an ontology mapping file (OML – Ontology Mapping Language). Humboldt CST is called from the (Py)WPS process, and at the end, result of the transformation is returned back as a GML file.

KML (former abbreviation of Keyhole Markup Language) is relatively new standard adopted by OGC. It is an vector file format, introduced originally by Keyhole, Inc, later purchased by Google. KML is relatively simple format, that is not able (compared to e.g., GML) to hold complex attributes to particular features. On the other hand, it has many cartographic applications. The producer of KML can style the features either by using in-file description or by using an external style file URL (Uniform Resource Locator). KML is not provided by any OGC service, but it has found its way in to many server and client applications. Through Google Earth the format has become more popular.

The complete list of all OGC Web Services standards (e.g., Grid Coverage Service or Web Coverage Processing Service and other) with detailed description is offered by the OGC at http://www.opengeospatial.org/standards.

10.5 Thematic Maps and OGC Web Services

The section describes three cases illustrating the possibilities of generating of thematic maps through OGC web services.

References to web services are very poor in current cartographic publications and books. Publications on thematic cartography do not incorporate the international standards like OGC Web Services and use proprietary services (e.g., ArcIMS – Slocum 2009) or general descriptions (e.g., server-side mapping – Dent et al., 2009; map generator – Slocum 2009) or non-standardized terms (e.g., Web Mapping Server – Slocum 2009). Theoretical aspects for using of web services in thematic cartography are describe in more detail by Cammack (2005) and (2007).

The publication (Sae-Tang and Ertz 2007) describes the different types of thematic maps and possibilities of WMS and SLD to generate these types.

- Choropleth map: One rule with a filter (class boundaries) per class, each rule having its polygon symbolizer with the fill color to apply.
- Proportional symbols: A point symbolizer with a built-in graphic mark like circle, a fill color, and its size controlled by an attribute data.
- Bivariate symbols: A mix of the filter and color of a choropleth map, and the point symbolizer of proportional symbols. Both size of the symbol and fill color are controlled by the two attributes.
- Overlayed symbols: Two rule with a filter for the rendering order (the fact that the smallest symbol has to be in front of the greatest). And, two point symbolizers per rule, each with its size controlled by an attribute date.

To be able to publish thematic maps through web services, it is necessary to add new items to existing standards. In the publication (Dietze and Zipf 2007), the Thematic Symbology Encoding (TSE) specification is introduced. This specification was defined as an extension (W3C XML Schema) of SE specification. The TSE schema contains:

- ThematicSymbolizer (advanced cartographic thematic elements of areal symbols);
- DiagramSymbolizer (diagram symbols – pie charts, bar charts and line charts, including their detailed description, e.g., size, labels, placement);
- ChoroplethSymbolizer (different types of classification of areal symbols).

The similar approach appeared in papers (Iosifescu-Enescu et al., 2007, 2010; Iosifescu et al., 2009). The Map and Diagram Service is based on and enhances the WMS and SLD standards. The Map and Diagram Service Specification, extend the WMS/SLD standards with cartographic features like various diagram types (e.g., pie diagrams, bar diagrams), definition of complex point symbols, data distributions, transparency levels for individual layers, patterns, user-defined symbols, patterns and gradients based on other open standards (SVG), advanced texture mapping and definitions of label placement rules (Iosifescu-Enescu et al., 2007).

The Map and Diagram Service Specifications extend the offer of operations – getDiagram, getLegendGraphic, getLayerDescription, and getStyle (detailed description in Iosifescu et al., (2009). The Map and Diagram Service specifications have been implemented in the open source software QGIS mapserver (Iosifescu et al., 2009).

The third similar solution called SLD-T (in the form of W3C XML Schema as the extension of SLD) is introduced in (Sae-Tang and Ertz 2007). The SLD-T contains:

- CategoryThematicSymbolizer (classification types);
- SimpleThematicSymbolizer (proportional symbols);
- MultiThematicSymbolizer (overlayed and juxtaposed symbols);
- ChartThematicSymbolizer (chart symbols).

10.6 Proposal of WPS Application

All solutions mentioned in this point are applicable just for simple thematic maps. It is very difficult to extend them and add other types of methods of cartographic interpretation. In this section we want to describe the solution that uses the power of web services as implemented by the OGC Web Processing Service. The main goal of this system is re-using of any existing solution through WPS interface.

The system (Fig. 10.4) is based on principles of generating thematic maps through XSLT (Extensible Stylesheet Language – Transformation) transformations published in Cerba (2009a) or Cerba (2008). The desktop solution applied in project VisualHealth (Fig. 10.2) or in creating maps for a printed publication – Atlas of International Relationships (Waisova et al., 2007; Cerba 2007, 2008).

In Fig. 10.3, there are shown four basic elements used to generate thematic maps by using XSLT and related technologies. The whole system is composed of four fundamental parts:

1. Maps description – particular thematic maps are described in so-called control file. This file, based on XML (Extensible Markup Language), specifies the fundamental attributes of maps like source data, cartographic interpretation methods, used colors, etc. The control file is connected with the schema file (written in the compact syntax of RELAX NG/REgular Language for XML Next Generation/format), which defines elements, attributes, their limits and restrictions. The control and schema files and advantages of formalized description of maps are described in Cerba (2009b).
2. Source spatial data – original data must be kept in some format based on XML. We have tested the GML format and JML (JUMP Markup Language, native format of JUMP GIS). In the future we would like to implement KML (Keyhole Markup Language), LandXML and other markup-languages format describing spatial data.
3. Transformation styles coded in XSLT 2.0. Currently, three interconnected style files were created – Atlas (the basic framework, including graphical user interface), Map (elements of particular maps) and Methods (generating of thematic maps) (see detail description in Cerba (2009b)).
4. Transformation processor – we decided to use the open-source version of Saxon 9.1.0.1 J. We also tested processors integrated into web browsers (Cerba 2009a).

The other benefit of XSLT transformation in cartography arises from the previous list. All technologies used are based on or connected with markup languages. This fact brings the homogeneity of this approach and many other advantages described in Ray (2003), Marchal (2000), and Harold (2001).

The implementation of WPS and other web services should limit the main problem of desktop application – lack of interoperability and user comfort. The WPS will serve as a container of XSLT processor. Figure 10.4 shows the first version of a proposed architecture based on WPS.

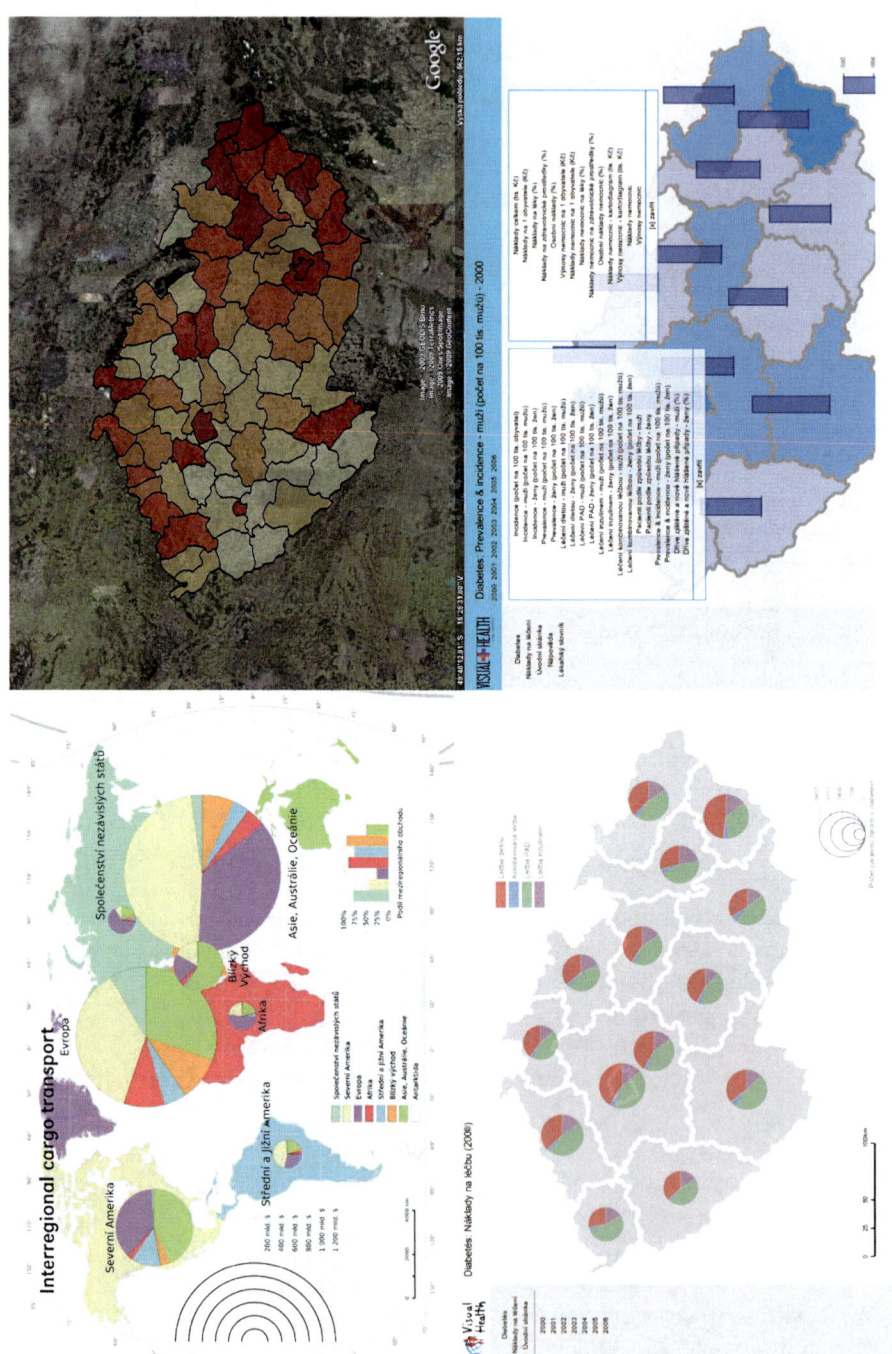

Fig. 10.2 Examples of maps defined through XSLT templates

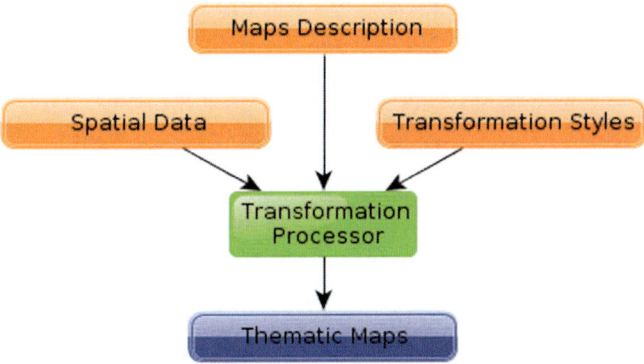

Fig. 10.3 Basic scheme of generating thematic maps by using XSLT

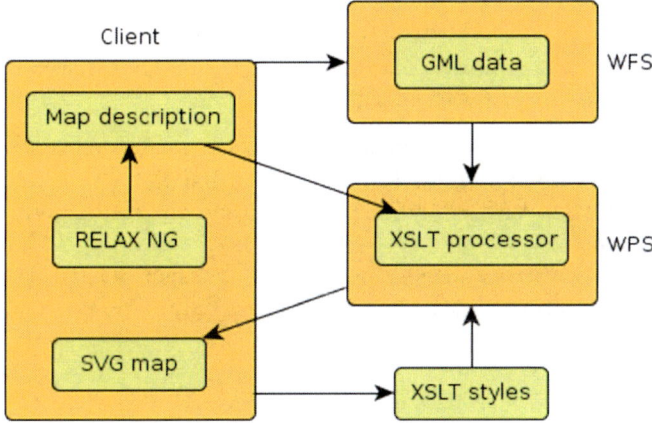

Fig. 10.4 Architecture of using of XSLT transformation as the WPS

The main part of the application is hidden on the server-side and is configured as a WPS process. The process get two inputs: GML data to be processed and XSLT template file with processing rules. The WPS process returns one output back to the client, namely a new file in SVG format, what is produced via XSLT transformation.

Such service can be now used by any client application, which does implement OGC WPS standard, without the need to have any additional software installed on the client-side. Because of this type of service is kind of "format transformation" as defined in INSPIRE Network architecture, it is more than just useful to store this transformation service into metadata of data providing service (WFS). If done in a way as suggested, it should be possible, with future tools, that applications will be able to display the raw vector data in a desired style, thanks to automatic transformation to target the SVG format.

The emphasis of standardization and the interconnection of frequently used and widely accepted standards is the most important benefit of the proposed application. In the final solution, the following standards will be used that are managed by three distinguished organizations – World Wide Web Consortium (XML, SVG, CSS – Cascading Style Sheets, XPath, Xlink), Open Geospatial Consortium (GML, WPS) and International Organisation for Standardisation (RELAX NG). Just the re-using and combining of existing standards (not developing new formats, languages, schemas etc.) is necessary to reach the higher level of interoperability in geoinformation sciences, including web cartography.

10.7 Future Steps and Ideas

In the previous section the proposal was made to use WPS for the creation of thematic maps. For this end, it is necessary to accomplish the following steps to support the better exploitation of existing resources (standards, schemes) and better adaption for users, especially those without experience:

- Extension of current desktop solution to a system based on XSLT style will require added new source formats (e.g., KML or LandXML), new output format (e.g., KML, VML – Vector Markup Language) and new cartographic methods to the XSLT templates.
- Improving output maps with graphical user interface (GUI) of target maps (zooming, interactive viewing of data values, help), addition of multimedia items (hypertext, animations, video and audio files) to existing maps (details in Cerba (2010)).
- Optimization of current desktop solution through modularization of XSLT templates, building of complete schema file of control file, transformation of original schemes to other languages (e.g., W3C XML Schema).
- Changes in the control file – Graphical User Interface (GUI) enabling the comfort editing and modification of control file describing the maps. The GUI should be interconnected with developed schema files and vocabularies to eliminate errors and mistakes in the process of map description.
- Integration of existing tools (schemes) such as Symbology Encoding Implementation Specification (Muller 2006), DiaML – Diagram Markup Language (Schnabel 2006) or CartoOWL (Karam et al., 2010).
- Transition to expert system (visualisation based on user requirements, data characteristics and cartographic rules), involving of artificial intelligence, methods of (semi)automatic analyses of spatial data.
- Integration of social media and collaborative (community) techniques of data capturing and map creating.

10.8 Conclusion

This chapter shows the possibilities what the world of web services, especially those standardized by Open Geospatial Consortium, offer to thematic cartography. There are many examples how to create thematic maps via Web Map Service (WMS). But, the possibilities of such solutions are limited and do not fully exploit the potential of thematic cartography. Therefore, a solution based on Web Processing Service is describe. This solution enables the re-use of existing systems for spatial data processing and the generation of non-trivial thematic maps and similar products. The goals of the proposed system are

- Extensibility (new components or external server solution could be added);
- Interoperability (system is based on international and well-documented standards);
- Exploitating the power of cartography (many similar systems offer only very simple spatial data visualization and not sophisticated thematic maps – Fig. 10.2).

The essential part of creating maps, including thematic maps, is the transformation of source spatial data. The result of this transformation is a graphical form of the spatial data being represented by a map and/or a similar cartographic products (such as the so-called enhanced reality, or a spatial data view). This process is not the result of sudden enlightenment of cartographers, or an application of learned knowledge (in the meaning of a strict procedure) or using a computer program or web application. Always, it is a combination of knowledge, experience and creativity (intuition, talent, "cartographic intelligence"...) of cartographers. Web services presented in this chapter can automate some parts of this process and allow cartographers to focus mainly on creative activities.

From a cartographic point of view, the OGC standards offer limited graphical representation possibilities in comparison with state of the art cartographic products (Sykora et al., 2007). But, on the other hand, web services and their influence in cartography must not be underestimated. All of the above mentioned projects and activities (and also this document) try to emphasize the importance of applying web services in cartography (including construction of thematic maps) and benefits resulting from the use of web services. Cartography has to reflect the new methods and technologies of geoinformation sciences. It enables us to keep the standing of cartography as a bridge between complicated spatial data and intelligible information and knowledge.

It is possible to find interconnections of web services and cartography in the activities of International Cartographic Association (ICA). In ICA Research Agenda (Virrantaus and Fairbairn 2007) there are emphasized terms like interoperability, accessibility, web technologies, data harmonization, semantics, metadata, visualization models, etc. The web services and other technologies, methods and principles mentioned in this chapter are relating to this document. Described applications support mainly these aspect of ICA Research Agenda focused on web technologies, and standards.

Finally, it is necessary to ask the question – "Could cartographers be replaced by web services or other automatic tools?". The answer is certainly "no". Cartographers work not only with spatial data or exact data models. They also must apply their knowledge and experience. The development of maps depends on many other factors and information, which are not processable by any automatic tool including web services. A participation of automatic tools will increase within the context of the rapid arise of technologies such as ontologies, expert systems, fuzzy systems and artificial intelligence, but the role of cartographers will be still very important. This can change. The difference between cartographers and map-makers will deepen. Common maps will be created by map-makers by using geoinformation technologies. Cartographers will assist in making cartographic products to be compliant with cartographic rules, standards, methods and procedures with regard to the user requirements. Furthermore, cartographers will irreplaceable for the final completion of maps prepared automatically.

References

de la Beaujardiere J (ed) (2004) OGC web map service interface. Version 1.3.0. Open Geospatial Consortium Inc., 2007. http://portal.opengeospatial.org/files/?artifact_id=4756

Cammack RG (2005) Distributed cartography: a look at web-mapping service and how it changes the mapping process. In: Auto carto proceedings 2005, Las Vegas, 21–23 Mar 2005. http://www.cartogis.org/autocartoarchive/autocarto2005-pdfs/cammack.pdf

Cammack RG (2007) Cartographic approaches to web mapping services. In: Cartwright W, Peterson MP, Gartner G (eds) Multimedia cartography, 2nd edn. Springer, Berlin/Heidelberg, ISBN: 978-3-540-36650-8

Cerba O (2007) XML technologies for cartographers. In: XXIII international cartographic conference, International Cartographic Association, Moscow, 4–10 Aug 2007

Cerba O (2008) Cartographic scales & XSLT. In: Second international conference on cartography and GIS, Borovets, Sofia, 2008. ISBN 978-954-724-036-0

Cerba O (2009a) XSLT templates for thematic maps. In: ICA symposium on cartography for central and Eastern Europe proceedings, Vienna University of Technology, Vienna, Feb 2009, pp 867–878

Cerba O (2009b) Thematic maps on the web. In: 24th international cartographic conference, International Cartographic Association, Santiago, 2009

Cerba O (2010) Multimediální tematické mapy. In: XXII. Sjezd České geografické společnosti, Ostrava (Czech Republic)

Dent BD, Hodler TW, Torguson J (2009) Cartography: thematic map design. 6. vyd. McGraw-Hill Higher Education, New York, p 336, ISBN 978-0-07-294382-5

Dietze L, Zipf A, (2007) Extending OGC styled layer descriptor (SLD) for thematic cartography. Towards the ubiquitous use of advanced mapping functions through standardized visualization rules. In: 4th international symposium on LBS and telecartography, 2007. http://tolu.giub.uni-bonn.de/karto/Thematic-SLD.LBS-Telecarto2007.pdf

Harold ER (2001) XML bible, 2nd edn. Hungry Minds, New York, p 1206

Iosifescu-Enescu I, Hugentobler M, Hurni L (2007) Cartographic extensions to OGC web map services specifications for a map and diagram service. In: XXIII international cartographic conference, International Cartographic Association, Moscow, 2007. http://icaci.org/documents/ICC_proceedings/ICC2007/documents/doc/THEME%2012/Oral%204/Cartographic%20Extensions%20to%20OGC%20Web%20Map%20Services.doc

Iosifescu I, Hugentobler M, Hurni, L (2009) Cartographic web services and cartographic rules – a new approach for web cartography. In: XXIV international cartographic conference, Santiago, 2009. http://sany-ip.eu/filemanager/active?fid=288

Iosifescu-Enescu I, Hugentobler M, Hurni L (2010) Web cartography with open standards – a solution to cartographic challenges of environmental management. Environ Modell Softw 25(9):988–999

Karam R et al., (2010) Cartographic integration on mobile devices from several providers' LBS by means of map symbol ontology. In: WebMGS 2010, Como, 2010

Kraak MJ, Ormeling FJ (2003) Cartography: visualization of spatial data, 2nd edn. Addison Wesley Longman, Harlow/Essex

Marchal B (2000) XML by example. Oue, 2000, p 505

Meynen E, International Cartographic Association (eds) (1973) Multilingual dictionary of technical terms in cartography. Franz Steiner Verlag, Wiesbaden

Muller M (ed) (2006) Symbology encoding implementation specification. Version: 1.1.0 (revision 4). Open Geospatial Consortium Inc., 2006. http://portal.opengeospatial.org/files/?artifact_id=16700

Ray ET (2003) Learning XML. O'Reilly & Associates, Sebastopol, p 2003

Sae-Tang A, Ertz O (2007) Towards web services dedicated to thematic mapping. OSGeo Journal, vol 3. http://www.osgeo.org/files/journal/v3/enus/final_pdfs/ertz.pdf

Schnabel O (2006) Diagram markup language for maps. Version 0.9 (2006-12-01). Institute of Cartography, ETH Zurich, Switzerland, 2002–2006. http://www.carto.net/schnabel/mapsymbolbrewer/schemas/diaml.xsd

Schut P (ed) (2007) OpenGIS® web processing service. Version 1.0.0. Open Geospatial Consortium Inc., 2007. http://portal.opengeospatial.org/files/?artifact_id=24151

Slocum TA (2009) Thematic cartography and geovisualization. 3. vyd. LinkPearson/Prentice Hall, Upper Saddle River, p 561, ISBN 978-0-13-229834-6

Sykora P, Schnabel O, Iosifescu I (2007) Extended cartographic interfaces for open distributed processing. Cartographica 42(3):209–218

Virrantaus K, Fairbairn D (2007) ICA research agenda on cartography and GI science. In: XXIII international cartographic conference, International Cartographic Association, Moscow, 4–10 Aug 2007

Waisova S et al., (2007) Atlas mezinarodnich vztahu. Vydavatelstvi a nakladatelstvi Ales Cenek, s.r.o., Plzen

Whiteside A, Evans JD (eds) (2008) Web coverage service (WCS) implementation standard. Version: 1.1.2. Open Geospatial Consortium Inc., 2007. http://portal.opengeospatial.org/files/?artifact_id=27297

Chapter 11
A Technical Survey on Decluttering of Icons in Online Map-Based Mashups

Haosheng Huang and Georg Gartner

Abstract Recent years have witnessed rapid advances in online map-based mashups with Application Programming Interfaces (APIs) and web services. Map-based mashups often display different kinds of information (e.g., POIs, represented as icons) on base maps, such as Google Maps and Bing Maps. The visualization of large number of icons in a map on web browsers or mobile devices often results in the icon cluttering problem with icons touching and overlapping each other. This problem decreases map legibility, and thus prevents users from effectively processing the information presented in the map. It also leads to a dramatic degradation of performance, and a high transmission load. All these problems will greatly decrease the usability of a mashup application.

This paper surveys and assesses approaches from different disciplines (i.e., computer science and cartography) for avoiding icon cluttering in online map-based mashups. We focus on two issues: filtering of irrelevant POIs, and icon placement and aggregation. Different techniques from information filtering research are analyzed and compared for reducing the number of icons to be displayed in a map. For the latter issue, approaches of aggregating and placing icons from map generalization research are discussed and assessed for their applicability in online mashups. Some related APIs and typical mashup examples are also discussed and compared. This paper concludes that in order to provide more cartographically pleasing maps in mashups, techniques from computer science and cartography should be seamlessly integrated.

H. Huang (✉)
Institute of Geoinformation and Cartography, Vienna University of Technology,
A-1040 Vienna, Austria
e-mail: haosheng.huang@tuwien.ac.at

11.1 Introduction

Compared to Web 1.0, Web 2.0 provides users with more user-friendly interface and software. Users can also contribute their own data (user-generated content) and control the data (Wikipedia 2011a). Some well-known Web 2.0 sites are Facebook, Twitter, Wikipedia, and OpenStreetMap. Mashup is a new web development method in the era of Web 2.0. The term mashup implies easy, fast integration, frequently using open Application Programming Interfaces (APIs) to combine data, presentation or functionality from two or more sources to create new services (Wikipedia 2011b). According to the survey made by ProgrammableWeb (2011), mapping is the most popular category in mashup applications.

Map-based mashups often overlap different information (such as Points of Interest (POIs), represented as icons) on base maps, e.g., Google Maps, Bing Maps, and OpenStreetMap. The visualization of a great many icons in a map often results in a dramatic degradation of performance, and a high transmission load. More importantly, it may lead to the icon cluttering problem, with many icons touching and overlapping each other (Burigat and Chittaro 2008). This problem decreases map legibility, and thus prevents users from effectively processing the information presented in a map.

The goal of this paper is to survey and assess approaches from different disciplines for avoiding icon cluttering in online map-based mashups. After carefully analyzing the problems in Sect. 11.2, we focus on two key issues: filtering of irrelevant POIs, and icon placement and aggregation. Related filtering methods from the field of computer science are then analyzed and compared for reducing the number of icons to be displayed in a map (Sect. 11.3). Section 11.4 surveys and analyzes approaches for aggregating and placing icons from research in cartographic generalization. In Sect. 11.5, we discuss the implementation (existing APIs) of the described methods and some typical mashup websites. Finally, we summarize the work in Sect. 11.6.

11.2 Icon Cluttering in Online Map-Based Mashups

A mashup can be simply understood as a new service, which combines functionality or content from different existing sources. In recent years, many APIs and web services have been made available to not only programmers but also to end users. Users with some basic programming knowledge can easily create their own mashups. There are also some tools and editors, such as Yahoo! Pipes, providing graphical interface for building mashups. All users have to do is drag and drop content from one source to another. They are very useful to end users and require little technical understanding. As a result, more and more mashups are created and published on the web.

Map-based mashup is the most popular type of mashup. This might be due to the fact that much of information is spatially-related, and map is a logical interface for visualization. Furthermore, mapping APIs, for example, Google Maps API, provide completed base maps, and can be easily integrated with some other data APIs. Map-based mashups often display different kinds of information (e.g., POIs) on base maps. These POIs are usually visualized as icons (e.g., push pins, markers in Google Maps) in a base map, with some multimedia information – mostly text, images, and videos – attached to the icons (Haklay et al., 2008).

Showing a small number of icons or markers in a map (e.g., a Google map) is fairly simple. Currently, with the impetus of Web 2.0 applications, such as Facebook, Flickr, and Foursquare, huge amounts of user-generated content (UGC) or Volunteered Geographic Information (VGI) are being continually created. More and more mashups need to display a large number of icons (POIs) in a single map.

Therefore, some key challenges and problems appear. Firstly, the performance of the mashup will be dramatically decreased. Users' browsers may freeze or become unresponsive for a period of time. The memory occupied will also sharply increase. For example, GMarker is used in Google Maps API v2 for showing icons. A test by Gabriel Svennerberg on a PC with a 3.60 GHz Pentium 4 HT processor and 2 Gb RAM running Windows XP shows that visualizing 500 markers in a map on Internet Explorer 8.0 takes 3,329 ms (Svennerberg 2009). Additionally, visualizing a great many of icons in a map also brings a high transmission load, which may be impractical for mobile applications with slower connection speeds.

Furthermore, visualizing large number of icons in a map often results in cluttering problems, especially at a small map scale (Burigat and Chittaro 2008). Icons will touch and overlap each other. They may also mask other important map features such as roads and place names. It becomes worse when the map changes from a larger scale to a smaller scale. This cluttering problem decreases map legibility, and thus prevents users effectively processing the amount of information presented in the map. For example, Fig. 11.1 shows restaurants near the first district of Vienna as icons in a Google map. In this map, icons overlap each other and mask most of the other map features. End users cannot really get some useful information from it.

Generally, the above problems can be alleviated by two approaches: filtering of irrelevant POIs, and icon placement and aggregation (Burigat and Chitttaro 2008). Filtering of irrelevant POIs is used to reduce the number of icons to be displayed in a map. The later issue displaces and aggregates icons, and shows them in a cartographically pleasing way (Kovanen and Sarjakoski 2010).

In the literature, many papers have addressed the cluttering problems from a cartographic perspective, and focused on icon placement and aggregation. Different map generalization methods were designed to displace and aggregate icons (Burigat and Chitttaro 2008; Kovanen and Sarjakoski 2010). However, the cluttering problems can also be tackled from a computer science perspective. For example, by applying information filtering methods, irrelevant POIs can be filtered, and thus the number of icons to be displayed in a map is reduced.

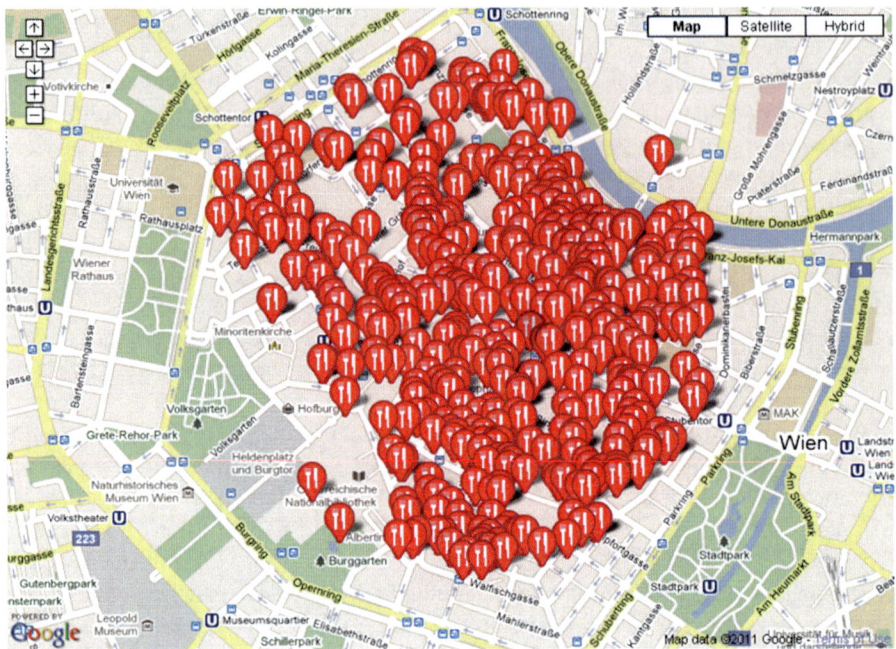

Fig. 11.1 An example of icon cluttering in a map-based mashup. Restaurants near the first district of Vienna are displayed

In order to provide uncluttered maps, both filtering of irrelevant POIs, and icon placement and aggregation should be seamlessly combined. In the following, we survey and assess techniques from the fields of computer science and cartography to address the above problems.

11.3 Filtering of Irrelevant POIs

Information filtering (IF) is a useful technique for delivery of relevant information. An IF system removes redundant or unwanted information from a source using (semi)automated or computerized methods prior to presentation to a human (Wikipedia 2011c). It can help alleviate the problem of "information overload" (Wikipedia 2011d), and thus support users' decision-making. Recommender systems (RSs) are active IF systems that attempt to provide the user with relevant information (for example, relevant to his/her interests, needs and context). For map-based mashups, IF/RSs techniques can help filter irrelevant POIs, and therefore reduce the number of icons to be displayed in a map.

Varieties of techniques have been proposed for RSs/IF. Previous work proposed different classifications of RSs (Hanani et al., 2001; Burke 2007; Ricci et al., 2011).

Among them, Ricci et al., (2011) provided an up-to-date classification of RSs: content-based, collaborative filtering, demographic, knowledge-based, community-based, and hybrid RSs. In the following, we analyze and assess these six techniques for reducing the number of icons to be displayed in a map.

A mashup providing users with relevant restaurants (POIs) in maps will be used as the illustrated example. It obtains POI data by using Qype API, and visualizes these POIs in Google Maps with Google Maps API (v2).

11.3.1 Content-Based Filtering

Content-based RSs recommend items (e.g., POIs, restaurants) similar to those the user has liked in the past. These systems build a model or profile of user's preferences based on the features (description) of the objects rated/chosen/liked by that user (Lops et al., 2011). The profile is a structured representation of user's preferences. The recommendation process matches up the attributes of the user profile with the attributes of an object (object profile). The result is a relevant judgment that represents the user's level of interest in that object. Objects with higher relevant judgment values are often recommended to the end user. The performance of content-based RSs mainly depends on how accurate the profile reflects the user's preferences. For a state-of-the-art survey, please refer to Lops et al., (2011).

In a simple form, the user himself can also explicitly provide a user profile. For example, in our restaurant finder mashup, a user can explicitly choose some types of restaurants (e.g., Japanese restaurants) to be displayed in a map. In Qype.com, every restaurant (POI) has been labeled with different tags, which can be viewed as attributes (description) of the restaurant. Therefore, irrelevant POIs (restaurants) can be filtered according to the user's choices. Figure 11.2 illustrates an example that only shows Japanese restaurants.

As the increasing popularity of social networking websites, such as Facebook and MySpace, more and more users store (explicitly or implicitly) their personal preferences and interests on these social websites. Therefore, a user profile can also be obtained from his/her social networking accounts, which may exempt the user from active involvement (i.e., explicitly choosing from a list). There are some open APIs available for this purpose, such as OpenSocial API by Google, and Facebook Platform by Facebook. Morandell (2010) provided an example. He used the OpenSocial API to inquire user's preferences that are stored in her/his MySpace account, and then used the preferences to filter irrelevant POIs.

Content-based RSs have several limitations (Adomavicius and Tuzhilin 2005; Lops et al., 2011), such as limited content analysis in building profiles (both users and objects), and overspecialization. Sometimes, as little description about objects is available, it is hard to build profiles that accurately reflect the characteristics of objects and preferences of users. Overspecialization means content-based RSs have no inherent method for finding something unexpected (Lops et al., 2011).

Fig. 11.2 An example of content-based filtering. Only Japanese restaurants are displayed

For example, a person with no experience with Thai cuisine would never receive a recommendation for Thai restaurants even if the best restaurant in town is a Thai restaurant.

11.3.2 Collaborative Filtering

Collaborative filtering (CF) recommends a user the items (POIs) that other users with similar tastes liked in the past. It is the most popular and widely implemented technique in RSs. Amazon-like recommendation ("people who bought ... also bought ...") is a well-known CF application.

Algorithms for CF can be grouped into two general classes: model-based and memory-based (or heuristic-based). Model-based CF uses the collection of ratings to learn a model, which is then used to make rating predictions. Some probabilistic models (such as Bayesian network and cluster model) are often employed for model learning. Heuristic-based CF can be divided into user-based approach and item-based approach. Given an unknown rating (of an item by the current user) to be estimated, heuristic-based CF first measures similarities between the current user and other users (user-based), or, between the item and other items (item-based). Then the unknown rating is predicted by averaging (weighted) the known ratings of

Fig. 11.3 An example of popularity-based recommendation in the restaurant finder mashup. Only the 30 top rated restaurants are displayed

the item by similar users (user-based), or the known ratings of similar items by the current user (item-based). User-based CF can be viewed as a heuristic implementation of the "Word of Mouth" phenomenon (Wang et al., 2006). For an up-to-date survey, please refer to Desrosiers and Karypis (2011).

Popularity-based recommendation (e.g., "the most viewed" at YouTube, "the most popular tags" at Flickr) is a non-personalized CF. These "most popular (viewed, discussed)..." like recommendations have been shown to be very useful for end users. Figure 11.3 depicts our restaurant finder mashup implementing the popularity-based technique to reduce the icons.

The biggest advantage of CF is that it requires no previous knowledge about the content of the data, and thus can be applied to any type of data, regardless of content. However, pure CF has some disadvantages (Desrosiers and Karypis 2011). Two of them are cold-start problem (new user problem and new item problem), and data sparsity (too few common ratings). In order to make accurate recommendations, the system must first learn the user's preferences from the ratings that the user gave. For a new user, as he/she has few or no ratings available in the system, it is hard to make recommendations to him/her. It is also impossible to recommend a new item to users when using pure CF. The problem of data sparsity is caused by the fact that users typically rate only a small proportion of the available items. When the rating data are sparse, two users or items are unlikely to have common ratings, and consequently, heuristic-based CF will predict ratings using a

Fig. 11.4 An example of demographic recommendation in the restaurant finder mashup. Only restaurants tagged with "disabled access" are shown

very limited number of neighbors (Desrosiers and Karypis 2011). Therefore, the recommendation performance may suffer from data sparsity.

11.3.3 Demographic Recommendation

This type of system recommends items based on the demographic profile of a user. The assumption is that different recommendations should be made for users with different demographics. Many websites adopt simple and effective personalization solutions based on demographics. For example, users are dispatched to particular websites based on their language or country (Ricci et al., 2011).

Figure 11.4 shows an example of using demographic information to filter irrelevant POIs in a map: making recommendations for a disabled person. Only restaurants tagged with "disabled access" are shown.

11.3.4 Knowledge-Based Recommendation

Knowledge-based RSs recommend items based on predefined knowledge bases that contain explicit rules about how certain item features meet user needs and preferences and, ultimately, how the item is useful for the user (Felfernig et al., 2011). Compared to CF and content-based filtering, knowledge-based RSs have no cold-start problems since the user's requirements are directly elicited within a recommendation session through a series of dialogs. However, they suffer from "the knowledge acquisition bottleneck in the sense that knowledge engineers must work hard to convert the knowledge possessed by domain experts into formal, executable representations" (Felfernig et al., 2011, pp. 187–188).

A knowledge base is typically defined by two sets of variables (V_C, V_{PROD}) and three different sets of constraints (C_R, C_F, C_{PROD}) (Felfernig et al., 2011). Customer Properties V_C describe possible requirements of customers (users), i.e., instantiations of customer properties, which may be explicitly provided by users via a series of dialogs. Product Properties V_{PROD} describe the properties of a given product assortment. Constraints C_R systematically restrict the possible instantiations of customer properties. Filter Conditions C_F define the relationship (rule) between potential customer requirements and the given product assortment. Products C_{PROD} store all the products, and represent them by using the properties defined by V_{PROD}. Among them, Filter Conditions C_F plays a key role. An example of C_F rule can be

$$C_F = \{CF1 : With\_cash = not \rightarrow Credit\_cards\_accepted = Yes\}$$

It can be explained as "users without cash should receive recommendations (restaurants) which accept credit cards". Figure 11.5 shows an example using this rule.

11.3.5 *Community-Based Recommendation*

Community-based RSs recommend items based on the preferences of a user's friends (Ricci et al., 2011). Evidence suggests that people tend to rely more on recommendations from their friends than on recommendations from similar but anonymous individuals (Sinha and Swearingen 2001). These types of RSs acquire information about the social relations of the user and the preferences of his/her friends. Community-based recommendation can be viewed as a special CF, which only uses the ratings provided by the user's friends. These RSs follow the rise of social networking applications. Research in this area is still in its early phase.

11.3.6 *Hybrid Recommendation*

The above techniques have some advantages and disadvantages. Table 11.1 compares them according to their information sources (input), advantages, and disadvantages.

Fig. 11.5 An example of knowledge-based recommendation in the restaurant finder mashup

As mentioned above, each RS technique has advantages and disadvantages. Hybrid RSs combine two or more of the above techniques. A hybrid system combining techniques A and B tries to use the advantages of A to fix the disadvantages of B (Ricci et al., 2011). For instances, pure CF suffers from the cold-start problem (new item and new user), i.e., they cannot recommend items that have no ratings, and they cannot make recommendations to users who have not given ratings. These can be alleviate by applying a knowledge-based technique at the beginning. Burke (2007), and Adomavicius and Tuzhilin (2005) provided some surveys on hybrid RSs.

To summarize, different techniques lead to results with different qualities, and require different inputs (i.e., information sources). When choosing a suitable filtering method for a mashup application, it is important to consider each of the aspects in Table 11.1.

11.4 Icon Placement and Aggregation

The techniques described in Sect. 11.3 can be used to filter irrelevant POIs, and thus reduce the number of icons to be displayed in a map. Reducing the number of POIs can greatly alleviate the problems of performance degradation, high transmission

Table 11.1 Comparison of different techniques (Adapted from Bruke (2002))

Technique	Information sources (input)	Advantages	Disadvantages
Content-based filtering (CBF)	1. The features associated with items	a. Domain knowledge not needed	A. Quality dependent on large historical dataset
	2. The ratings a user has given to items	b. Implicit feedback sufficient	B. Overspecialization C. New user problem
explicit_CBF: explicitly providing profile	1. The features associated with items 2. The profile explicitly provided by user	a, c. No historical dataset required	B, D. User must input their profile
Collaborative filtering (CF)	1. Ratings for items from different users	a, b, d. Can identify cross-genre niches	A, C, E. New item problem F. Data sparsity
CF: popularity-based	1. Ratings for items from different users	a, b, d	E, G. Non-personalized
Demographic recommendation	1. Demographic information about a user 2. The features associated with items 3. Knowledge about features of item and demographic information	c	H. Must gather demographic information I. Knowledge engineering required
Knowledge-based	1. User's need provided by the user via a series of dialogs 2. The features associated with items 3. Knowledge about how these items meet a user's need	c, e. Can explain the reason for recommending a specific item	I, J. Must gather user's need
Community-based	1. Ratings for items from the user's friends	a, d	K. Privacy concerns

load, and "information overload". However, it may not solve all the overlapping problems (refer to Fig. 11.2 for an example).

There is research from cartographic generalization focusing on automatic symbol placement. The aim is to place symbols on maps while avoiding the overlap with other symbols and underlying map features. Many techniques have been developed for automatic symbol placement (Mackaness and Fisher 1987; Harrie et al., 2004; Burigat and Chitttaro 2008; Kovanen and Sarjakoski 2010). In the following, we focus on displacement and aggregation.

Displacement: In order to solve the problem of competition of limited map space between symbols or map features that overlap or lie too close to each other, the displacement operation is often applied. It shifts symbols or map features to other places to prevent coalescence. According to Foerster and Stoter (2008), displacement is the most important operation when considering how often the operation is applied and how dominant a role it plays. However, the displacement problem is Non-deterministic Polynomial-time hard (NP-hard) (Marks and Shieber 1991). Many methods have been proposed for the displacement problem, e.g., Ruas (1998), Harrie (1999), Mackaness and Purves (2001), and Lonergan and Jones (2001). Most of them work in an iterative manner. Each iteration is composed of three phases: detection of conflict (overlap), calculation of a new location, and evaluation of the result (Mackaness and Purves 2001). Multiple iterations are needed when new conflicts are created due to the previous iteration or when some constraint is still violated (Kovanen and Sarjakoski 2010). In worst cases, the conflict problem cannot be solved at all.

In general, these displacement techniques proposed in the literature are also suitable for map-based mashups. However, two important issues have to be kept in mind. Firstly, the purpose of placing icons in a map is to show the existence of specific POIs at specific places. A POI icon usually has a predefined fixed location. As a result, the displacement movement of POI icons should be as small as possible. Secondly, as the map features of the underlying map such as roads and streets cannot be detected or changed, displacement operation sometimes cannot solve the problem of poor legibility.

Aggregation (cartographic): The basic idea of aggregation operation is to identify clusters of mutually overlapping icons, and replace them with special aggregator icons (Burigat and Chitttaro 2008). The purpose of aggregator icons includes pointing out the presence of these icons as well as providing users with a means to access information about each of them. Burigat and Chittaro (2008) first created a conflict graph to store overlap relationship between icons, and then developed a maximum aggregation algorithm and a fast aggregation algorithm to identify a set of aggregator icons without conflicts. They added some mouse events to the aggregator icons. With the events, clicking an aggregator icon opens a pop-up window that lists the aggregated POIs and allows the user to obtain more information about each of them.

In addition to the visual overlapping aspect, aggregation can also be based on some semantic aspects, for example, categories and features of POIs. Therefore, more meaningful and semantic-enhanced cartographic aggregation can be provided.

11.5 Implementation, APIs and Mashup Examples

In this section, we discuss and analyze the implementation through existing APIs of the described methods in the previous sections. Two typical mashups that address the problem of icon cluttering are also introduced.

11.5.1 Implementation of Irrelevant POIs Filtering

As discussed in Sect. 11.3, different filtering techniques can be used to reduce the number of POIs to be displayed in a map. All of them have some advantages and disadvantages, and may lead to results with different qualities. In terms of implementation, the needed technical skills are also different.

Technical speaking, explicit_CBF and simple demographic recommendation can be viewed as filtering by some querying parameters (such as category, location and language). There are some data APIs enabling developers to specify querying parameters. For example, Qype API can be used to query POIs near a certain position, search POIs in a category, and find POIs by their name or a keyword. YouTube data API defines different query parameters that can be used for filtering and ordering results, such as category, format, caption, language, and location. Yahoo! Local Search Web Service allows users to search the Internet for POIs near a specific location, as well as search by categories. These APIs and web services can be easily used to implement explicit_CBF and demographic recommendations. For knowledge-based filtering, the Filter Conditions C_F (i.e., rules) plays a key role. When these rules are identified, the above APIs and web services can also be employed to implement simple knowledge-based recommendation.

There are also some APIs and web services implementing popularity-based CF. For example, Qype API provides "order" to sort results by "distance" or "rating." YouTube data API enables developers to order results by specifying an "orderby" parameter, such as relevance, published (chronological), viewCount, and rating. For Yahoo! Local Search Web Service, the "sort" parameter is used to order the results by the chosen criteria (e.g., relevance, title, distance, or rating).

Compared to the above four kinds of filtering methods, standard content-based filtering and standard collaborative filtering require more advanced programming skill to implement. Currently, there are few open APIs available on the web. To implement them, developers need to code the algorithms. To get implementation ideas, please have a look at some state-of-the-art surveys, such as Lops et al., (2011) and Desrosiers and Karypis (2011).

The community-based recommendation is a special collaborative filtering. As discussed in Sect. 11.3.1, there are some open APIs for obtaining information from user's social networking accounts (Facebook, MySpace), such as OpenSocial API by Google, and Facebook Platform by Facebook. These APIs provide a basis for community-based recommendations by obtaining preferences from friends. Simple community-based recommendations may just display all the friends' favorite POIs, while complex community-based recommendations may also have to adopt the method of standard collaborative filtering. As a result, different programming skills may be needed.

To summarize, different methods lead to different qualities of filtering, and require different technical skills to implement. When choosing a suitable filtering method for your application, it is critical to consider what you can get from the data sources. Candidate methods can be identified by checking whether the required inputs are available. When different filtering methods are feasible with the current data sources, the best method can be determined by analyzing the ease of implementation, advantages, and disadvantages of each method.

11.5.2 Existing APIs on Icon Displacement and Aggregation

There are few open APIs available for icon displacement. To implement, developers should design their own algorithms. Implementation can be focused on the three components of icon displacement (i.e., detection of conflict, calculation of a new location and evaluation of the result).

For icon aggregation in map-based mashups, there are some open APIs available. In the following, we discuss and compare some existing APIs designed for Google Maps API (v2 and v3).

1. **MarkerManager**: Provided by Google for addressing the problem of slow rendering of maps and visual cluttering when adding a large number of markers to a Google map (GMaps Utility Library 2011). When using MarkerManager, you have to define at which zoom-levels a marker will be visible. MarkerManager keeps tracking of all added markers, and shows them according to the defined zoom-levels. Recently, an updated version of the MarkerManager API is developed to support managing markers in Google Maps API v3 (MarkerManager v3 2011).
2. **Clusterer**: Designed by Jef Poskanzer from ACME labs (ACME 2011). Two techniques are implemented in it: Only the markers currently visible actually are created; if too many markers would be visible, then they are grouped together into cluster markers. With these, adding thousands of markers to a map can also maintain decent performance.
3. **ClusterMarker**: Developed by Pearman (2011). It detects any group(s) of two or more markers whose icons visually overlap when displayed. Each group of overlapping markers is then replaced with a single cluster marker. The cluster marker, when clicked, simply centers and zooms the map in on the markers whose icons overlap. Currently, it only works for Google Maps API v2.
4. **MarkerClusterer**: Developed by Xiaoxi Wu and is part of the Google Maps Open Source Utility Library (Wu 2011). It groups markers into different clusters and displays the number of markers in each cluster with a label, creating new clusters as the zoom level of the map changes. The clustering algorithm is simple: for each new marker it sees, it either puts it inside a pre-existing cluster, or creates a new cluster if the marker does not lie within the bounds of any current cluster.

Gabriel Svennerberg compared the above APIs on major browsers (Internet Explorer, Firefox, Google Chrome, Safari and Opera), and showed that Clusterer is the fastest technique when only considering how long it takes before the markers are passed to the map (Svennerberg 2009). However, when considering the actual time until all markers are visible on the map, MarkerClusterer is the fastest one, followed by ClusterMarker.

In the following, we analyze the functions provided by the above APIs, and their applicability in map-based mashups from the perspective of cartography. The MarkerManager only implements a Level of Detail (LoD) method of showing large number of icons. It can be easily incorporated with some techniques from cartographic generalization to provide comprehensive solutions for avoiding icon cluttering in map-based mashups. The last three APIs implement the cartographic aggregation function (Sect. 11.4) that groups markers into different clusters to avoid icon cluttering. However, the clustering algorithms in them are simply based on the visual overlapping of icons. In order to provide meaningful and semantic-enhanced clustering, some other features of the icons (POIs) can also be considered, such as clustering overlapping icons according to their semantic categories.

11.5.3 Mashup Examples

The following two map examples show, how uncluttered maps are provided in mashups. The example in Fig. 11.6 combines information filtering techniques and icon aggregation techniques. POIs can be filtered by categories and keywords. It develops its own aggregation algorithm. Overlapping icons are aggregated into a translucent rectangle. When moving the mouse to the top of a rectangle, the number

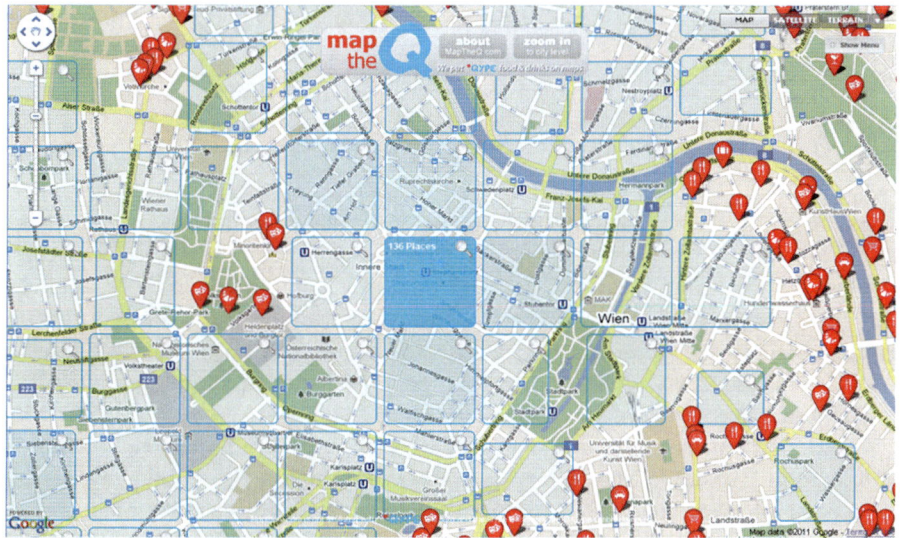

Fig. 11.6 An example of decluttering icons in mashup taken from http://www.maptheq.com

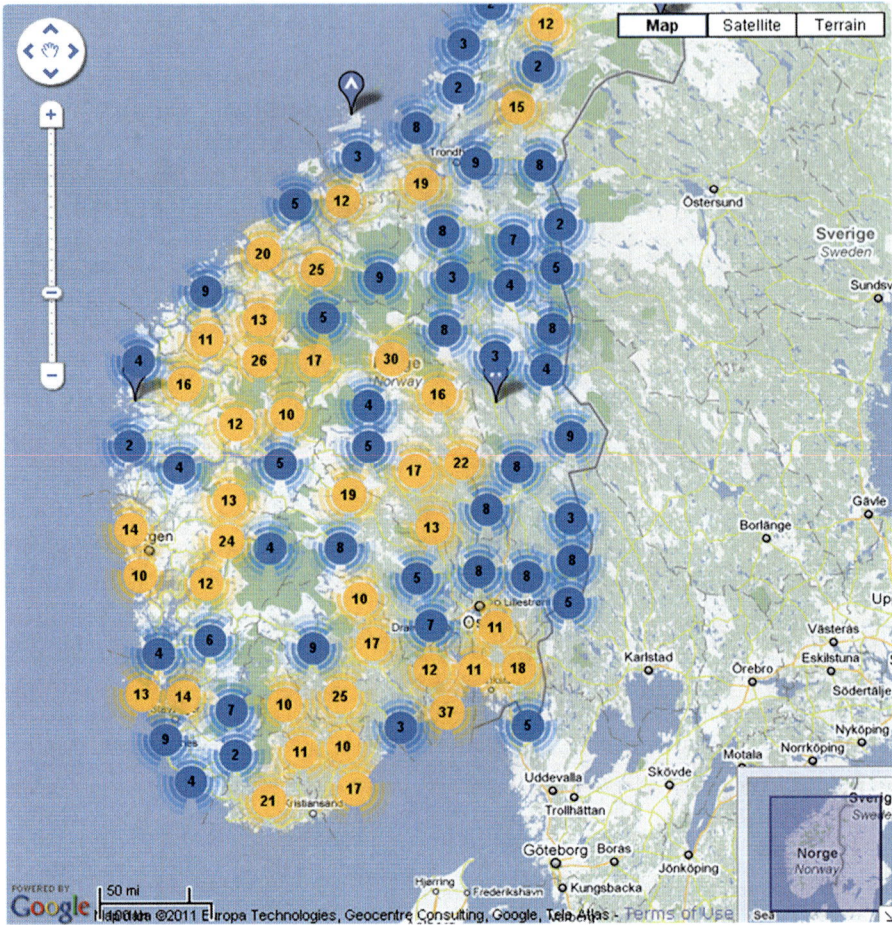

Fig. 11.7 An example of decluttering icons in mashup taken from http://www.norwegen-reise.com. The MarkerClusterer API is employed to aggregate icons in maps

of icons in the rectangle is showed. Clicking the rectangle will show and zoom the map to the area of the rectangle.

The mashup in Fig. 11.7 also combines information filtering and icon aggregation. POIs can be filtered by categories. The MarkerClusterer API (see Sect. 11.5.1) is employed to aggregate icons in maps. The number of markers in a cluster is displayed on the cluster marker. Clicking the cluster marker shows the markers in this cluster.

It is important to note that, the above two examples only implement some very simple information filtering techniques. More information filtering techniques can also be implemented to filter irrelevant POIs, which will greatly alleviate the problems of information overload, and high transmission load. In addition, the clustering/aggregating algorithms in them are simply based on the visual aspect.

More advanced techniques (e.g., semantic-enhanced aggregation and displacement) from cartographic generalization should be employed to address the icon cluttering problem.

11.6 Conclusions

Recent years have seen an increased interest in online map-based mashups. Visualizing large number of icons in a map often results in dramatic degradation of performance, high transmission load, information overload, and icon cluttering problems. In order to alleviate the above problems and provide a cartographically pleasing map, improving the methods of visualizing many icons in a map is becoming more and more crucial.

In this paper, we surveyed and assessed different techniques from computer science and cartography. Related techniques from information filtering were analyzed and compared for reducing the number of icons to be displayed in a map. Additionally, approaches for aggregating and placing icons from map generalization research were surveyed and analyzed for providing uncluttered maps.

Implementation using existing APIs of the described methods and some typical mashup examples were discussed and analyzed. The analysis showed that decluttering icons in map-based mashups is still in the early stage of development. Solutions from computer science and cartography should be seamlessly integrated to provide more cartographically pleasing maps in online map-based mashups.

References

ACME (2011) JavaScript utilities. http://www.acme.com/javascript/#Clusterer. Accessed June 2011

Adomavicius G, Tuzhilin A (2005) Towards the next generation of recommender systems: a survey of the state-of-the-art and possible extensions. IEEE Trans Knowl Data Eng 17(6):734–749

Burigat S, Chittaro L (2008) Decluttering of icons based on aggregation in mobile maps. In: Meng L, Zipf A, Winter S (eds) Map-based mobile services – design interaction and usability. Springer, Berlin, pp 13–32

Burke R (2002) Hybrid recommender systems: survey and experiments. User Model User-Adap 12(4):331–370

Burke R (2007) Hybrid web recommender systems. In: The adaptive web. Springer, Berlin, pp 377–408

Desrosiers C, Karypis G (2011) A comprehensive survey of neighborhood-based recommendation methods. In: Ricci F, Rokach L, Shapira B, Kantor P (eds) Recommender systems handbook. Springer, Boston, pp 107–144

Felfernig A, Friedrich G, Jannach D, Zanker M (2011) Developing constraint-based recommenders. In: Ricci F, Rokach L, Shapira B, Kantor P (eds) Recommender systems handbook. Springer, Boston, pp 187–215

Foerster T, Stoter JE (2008) Generalisation operators for practice: a survey at national mapping agencies. In: Proceedings of the 11th ICA workshop on generalisation and multiple representation, Montpellier, 20–21 June

GMaps Utility Library (2011) http://code.google.com/apis/maps/documentation/javascript/v2/overlays.html#Marker_Manager. Accessed June 2011

Haklay M, Singleton A, Parker C (2008) Web mapping 2.0: the neogeography of the geoweb. Geogr Compass 2(6):2011–2039

Hanani U, Shapira B, Shoval P (2001) Information filtering: overview of issues, research and systems. User Model User-Adap 11(3):203–259

Harrie L (1999) The constraint method for solving spatial conflicts in cartographic generalisation. Cartogr Geogr Inf 26(1):55–69

Harrie L, Stigmar H, Koivula T, Lehto L (2004) An algorithm for icon placement on a real-time map. In: Fisher P (ed) Developments in spatial data handling, Proceedings of the 11th international symposium on spatial data handling. Springer, Leicester, pp 493–507

Kovanen J, Sarjakoski LT (2010) Displacement and grouping of points of interest for multi-scaled mobile maps. In: Proceedings of LBS 2010, Guangzhou, 20–22 Sept 2010

Lonergan M, Jones CB (2001) An iterative displacement method for conflict resolution in map generalization. Algorithmica 30:287–301

Lops P, de Gemmis M, Semeraro G (2011) Content-based recommender systems: state of the art and trends. In: Ricci F, Rokach L, Shapira B, Kantor P (eds) Recommender systems handbook. Springer, Boston, pp 73–105

Mackaness WA, Fisher PF (1987) Automatic recognition and resolution of spatial conflicts in cartographic symbolisation. In: Proceedings of AutoCarto 8, 29.03-03.04, Baltimore, pp 709–718

Mackaness WA, Purves RS (2001) Automated displacement for large numbers of discrete map objects. Algorithmica 30:302–311

MarkerManager v3 (2011) http://google-maps-utility-library-v3.googlecode.com/svn/tags/markermanager/1.0/docs/reference.html. Accessed June 2011

Marks J, Shieber S (1991) The computational complexity of cartographic label placement. Technical Report TR-05-91, Center for Research in Computing Technology, Harvard University

Morandell C (2010) Möglichkeiten der nutzerspezifischen Gestaltung von Location Based Services mit Daten aus Social Networks. Master thesis of Vienna University of Technology

Pearman M (2011) Google Maps API Projects. http://googlemapsapi.martinpearman.co.uk/readarticle.php?article_id = 4. Accessed June 2011

ProgrammableWeb (2011) http://www.programmableweb.com/mashups#topt-2. Accessed June 2011

Ricci F, Rokach L, Shapira B (2011) Introduction to recommender systems handbook. In: Ricci F, Rokach L, Shapira B, Kantor P (eds) Recommender systems handbook. Springer, Boston, pp 1–35

Ruas A (1998) A method for building displacement in automated map generalisation. Int J Geogr Inf Sci 12(8):789–803

Sinha R, Swearingen K (2001) Comparing recommendations made by online systems and friends. In: DELOS workshop: personalisation and recommender systems in digital libraries

Svennerberg G (2009) Handling large amounts of markers in Google maps. http://www.svennerberg.com/2009/01/handling-large-amounts-of-markers-in-google-maps/. Accessed June 2011

Wang J, Vries A, Reinders M (2006) Unifying user-based and item-based collaborative filtering approaches by similarity fusion. In: Proceedings of the 29th ACM SIGIR conference on information retrieval. Seattle, Washington, USA, pp 501–508

Wikipedia (2011a) http://en.wikipedia.org/wiki/Web_2.0. Accessed June 2011

Wikipedia (2011b) http://en.wikipedia.org/wiki/Mashup_%28web_application_hybrid%29. Accessed June 2011
Wikipedia (2011c) http://en.wikipedia.org/wiki/Information_filtering_system. Accessed June 2011
Wikipedia (2011d) http://en.wikipedia.org/wiki/Information_overload. Accessed June 2011
Wu X (2011) MarkerClusterer: a solution to the too many markers problem. http://googlegeodevelopers.blogspot.com/2009/04/markerclusterer-solution-to-too-many.html. Accessed June 2011

Chapter 12
Web Map Design for a Multipublishing Environment Based on Open APIs

Pyry Kettunen, L. Tiina Sarjakoski, Salu Ylirisku, and Tapani Sarjakoski

Abstract The aim of the study described in the paper is to carry out research on the utilization of web-based multipublishing for the purpose of outdoor activities. The idea behind a multipublishing service is that the service is able to deliver different kinds of outdoor maps through a number of channels and at varying scales from a single data core. The paper focuses on the technical solutions, design principles and usability testing of a web map that was created to serve as one of the publishing channels. The other channels of the implemented multipublishing service are map applications for a mobile phone and a multi-touch screen, and printed graphic maps. The environment is based upon a web server that provides raster and vector geodata for the channels through open standard web services. Both the server and the web map client application are built on free and open source geospatial software, to which modifications were made to achieve the design goals of the multipublishing environment. The web map user interface (UI) aims at providing a complete tool to interact with the maps being published and it shares similar design with the other publishing channels. We applied the "minimalist" and "direct manipulation" design paradigms for UI design to limit the user's cognitive overload.

12.1 Introduction

Network-delivered interactive and dynamic digital maps with high cartographic quality and user-friendly interfaces have been a dream of cartographers for years (Moellering 1984; MacEachren 1994; MacEachren and Kraak 1997; Cartwright 1997; Slocum et al., 2001). It has proven difficult, though, to design and implement maps that satisfy all of these aspects. The problem has been divided into a number

---

P. Kettunen (✉)
Department of Geoinformatics and Cartography, Finnish Geodetic Institute, P.O. Box 15, 02431 Masala, Finland
e-mail: Pyry.Kettunen@fgi.fi

of sub-problems which range from web technology to usability issues. Typically, the development of the sub-problems has followed the progress of the dedicated research fields. For instance, advances in web technology have had a major influence on solving the problems of the delivery of digital maps, and usability research has guided the development of the map UIs (Haklay and Zafiri 2008; Roth and Harrower 2008).

Digital cartography utilizes and integrates advanced information technologies, but it also has some specific problems to solve, most prominently that of cartographic generalization (Sarjakoski 2007). A stable integration of the technologies has become a reality only during the first decade of the twenty-first century and digital web maps are still often considered to have insufficient performance and poor usability (Nivala et al., 2008; Coltekin et al., 2009). The holistic design of web maps is necessary in order to provide a satisfactory user experience and to fulfill the expectations and needs of the users, while taking into account the usability, cartographic quality and an efficient computing architecture. Such a design is of noteworthy complexity and can significantly benefit from open application programming interfaces (APIs) that reduce the need for implementing the underlying intricate processes.

12.1.1 Background of the Study

The research described in the paper is part of two projects: "Multipublishing in supporting outdoor leisure activities" (MenoMaps), and MenoMaps II (Map services for outdoor leisure activities supported by social networks). The aim of the MenoMaps (2008–2010) project has been to carry out research on utilizing web-based multipublishing for the purpose of outdoor activities, and to develop a user interface (UI) that is easy to use, useful and challenging and which provides a pleasurable use experience (Sarjakoski et al., 2009; Flink 2009). The main outcome of the first project was a map-based multipublishing service prototype.

The idea of a multipublishing service is that the service is able to deliver different kinds of outdoor maps through a number of channels and at varying scales from a single data core. The data core utilizes integrated data sources, such as high-resolution digital terrain models based on airborne laser scanning. Figure 12.1 shows the implemented channels of the MenoMaps service: the map applications for an iPhone (Kovanen et al., 2009) and a multi-touch screen (Sarjakoski et al., 2010), printed graphic maps (Oksanen et al., 2011) and a web map, the channel which is the focus of this paper. The map-based multipublishing service prototype will be further developed in the follow-up project, MenoMaps II (2010–2013) and it will be exhibited in the Nuuksio Nature Centre in southern Finland in the beginning of 2013.

12.1.2 Structure of the Paper

In the following section, we review previous research in the field. The viewpoints for the research works come from three perspectives: the technological point of

Fig. 12.1 The channels of the MenoMaps multipublishing service: the map applications for a mobile phone (iPhone) and a multi-touch screen, printed graphic maps and a web map (Flink et al., 2011). The data for all the channels is derived from the same data core

view, the design point of view and the usability point of view. After Sect. 12.2, we present a case study, in which a multipublishing service prototype for outdoor activities was implemented based on geospatial Free and Open Source Software (FOSS), for example OpenLayers application programming interface (API). Finally, we summarize the experiences and success of the prototype in terms of architecture, design and usability.

12.2 Previous Studies

The following points of view have been considered relating to studies on web-based map applications: technological perspectives, UI design and usability.

12.2.1 Technological Solutions for Web Maps

Web-based APIs have recently become common platforms for building web maps. A web map API is a collection of programming utilities which can be used for the implementation of a web map without the need to program the elementary functionalities from scratch. Web map portals such as Google Maps, Yahoo Maps and Bing Maps provide APIs for using their maps on web pages. On the other hand, stand-alone, data-independent client APIs such as OpenLayers API and ArcGIS Web Mapping APIs are also available. Current APIs allow users to add a default map application to a web page with very basic programming skills, but they also provide a developer with a number of tools for customized implementation. The APIs are generally based upon Asynchronous JavaScript and XML (Ajax) web technologies that run on ordinary web browsers without additional plugins.

The Open Web Service (OWS) specifications of Open Geospatial Consortium (OGC) define a set of network interface standards for transferring geospatial data between clients and servers. Web maps typically use the Web Map Service (WMS) for requesting and receiving georeferenced raster data that are static map images, and the Web Feature Service (WFS) for the retrieval and remote manipulation of vector geodata (e.g., González et al., 2009; Henrie 2009). The OWSs allow developers to collect multiple independent data sources from distinct servers into a single web map application as well as distribute geodata from a single server to multiple independent clients. In this way, the OWSs enable the creation of federated and heterogeneous GIS server architectures that can grow into wide data infrastructures (Webster 1988). Versatile and flexible web-based system architectures are possible, like the multi-channel publishing of maps for diverse devices (Lehto et al., 2001).

The OWSs have been actively implemented in Free and Open Source Software (FOSS) GIS applications to fit into open GIS architectures. An open architecture means that both the system components and the binding architecture are accessible to a developer on the server and client side (Dunfey et al., 2006). The developer can fully control an open system since he can read the source code to understand how the system operates and is also capable of modifying the code when needed. FOSS software provides, often easily, the much needed interoperability in GIS when several data formats and coordinate systems must be used in concert. Dunfey (2006) achieved promising results in building a proof-of-concept open architecture for a SVG map viewer on top of FOSS components, confirming their flexibility and functionality. Henrie (2009) reports similar results in building a web map for historic town plans based upon FOSS GIS software. He points out that it was easy to implement with these tools, and that the open architecture allowed good performance.

The rendering of vector data into bitmap graphics, which can be done either on the client or server side, is, computationally, one of the most demanding parts of web map architecture. The server-side rendering usually outperforms the client-side rendering because of the superior computing power of the server. However, client-side rendering is necessary when the vector attributes should be passed to the client application. In both cases, the styling rules for rendering vectors are needed to specify the appearance of the output.

Styled Layer Descriptor (SLD) is an open OGC standard for defining vector styles. An XML-derived SLD enables graphically rich, dynamic display styles for georeferenced vector features such as arbitrarily dashed and round-ended lines. The styles can be piled up to form complex styles and filters can be applied to restrict the effect of a style on specific zoom levels or targeted features. Many FOSS GIS applications apply SLD either on the side of the client or the server (Zipf 2005).

12.2.2 Web Map UI Design and Usability

MacEachren and Kraak (1997) discuss exploratory cartographic visualization (ECV) within the context of the use space cube of maps (MacEachren 1994). They state that map display systems need to be designed following the intended use of every single map, whether that use is exploration, analysis, synthesis or the presentation of geographic data. Several web map UIs have been built in order to study the UI design for ECV. One of the earliest such UIs was in a software system called Descartes, which provided versatile dynamic manipulation tools for exploring spatial data using maps (Andrienko and Andrienko 1999). The UI of Descartes applied multiple linked views (MLVs) which are central to the graphical UI (GUI) design: the contents of the multiple graphical windows in the UI are actively tied together so that changing the view in one window updates the other windows accordingly (Roberts 2005).

The MLVs fit well for creating UIs for ECV because they provide parallel visualizations of the same data and, thus, prepare the way for new, interesting cognitive linkages (Roberts 2008). Recently, the study of the ECV has continued through web map UIs like London Profiler (Gibin et al., 2008) and Nunaliit Cybercartographic Atlas Framework (Pulsifer et al., 2008). Both of these interactive, dynamic digital maps have been designed with an amateur in mind, taking care to avoid the undesired complexity of usage which is common in the professional GISs. In addition to combine the data, these systems also aim at displaying cartographically polished geographical data that can be easily explored by a content-oriented user.

The UI design must consider usability issues in order to bring the user experience of the interactive and dynamic digital maps to a pleasing level (Nivala 2007). Our UI research applies design methods called "minimalism" and "direct manipulation". Minimalist design means that every component included within a UI must have a good reason to be included and spare components are left out (Carroll 1985). Direct manipulation design places the functionalities of a UI on the objects so that the objects can be manipulated just where they are (Hutchins et al., 1985). For a successful final product, a usability evaluation of the already implemented software is also necessary in order to detect the remaining problems in usage situations (Nivala et al., 2008).

Usability evaluation methods distribute on a continuum between controlled experimentation using quantitative measures such as eye-movement tracking, and usability testing using qualitative measures such as focus groups (Roth and Harrower 2008). Comparative usability studies have been made in order to create design guidelines for interactive and dynamic digital maps. Nivala et al., (2008) conducted a usability evaluation of web mapping sites using user tests and expert evaluations. They found that web mapping sites should be improved and harmonized in order to better guide the user. Coltekin et al., (2009) compared two separately designed web map UIs using eye-tracking combined with performance metrics, system usability scale (SUS) and participant interviews. The findings revealed that the users found the design with the map in the focus and with floating menus more attractive than traditional map UI design with a separate, hierarchical layer-based menu. Users also felt that the former maps performed better.

12.3 Case Study: MenoMaps Web Map

A multipublishing service prototype was designed and built as part of the MenoMaps project for outdoor leisure activities in Nuuksio National Park in the southern Finland. The MenoMaps web map was implemented as one of the channels for personal exploration and annotation of the maps within the multipublishing environment. The intended use of the web map is for exploring the national park and planning activities, for example before a hike, as well as memorizing and sharing the activities after a hike. In this section, we present the MenoMaps web map from the viewpoints of the architecture, UI design and usability evaluation.

12.3.1 System Architecture

12.3.1.1 Overall Architecture of the Multipublishing Environment

The multipublishing service prototype of the MenoMaps project was implemented according to the open architecture model so that both the architecture and the components can be accessed (Dunfey et al., 2006). The FOSS software was used for the implementation because we wanted to test the capabilities of such a solution. We wanted to create a stand-alone system that would be flexible for a developer. We also wanted to use our own data independently from the web map portals such as Google Maps and Bing Maps. The architecture (Fig. 12.2) consists of three functional levels that inter-operate on the same or distinct computers:

1. The user interface level displays the maps on the client side through the following channels: the web map on a web browser, a mobile application on an iPhone (Kovanen et al., 2009), a multi-touch screen map application in a public space, (Sarjakoski et al., 2010) user-made map printouts and professional printed maps (Oksanen et al., 2011) all utilize the same data core for the map contents.
2. The data access level supplies the geodata requested by the clients and stored on the servers. The data access level also performs data transformations when

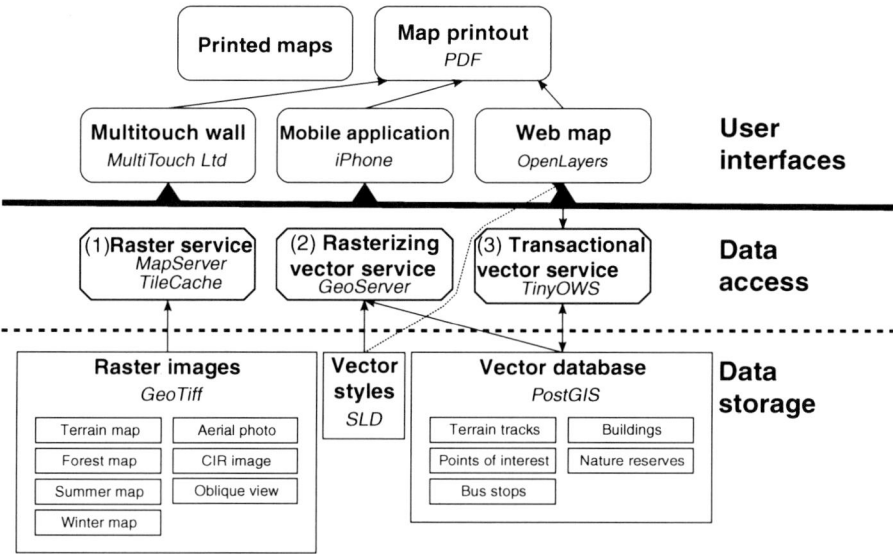

Fig. 12.2 Overall architecture of the MenoMaps multipublishing service prototype. The utilized technologies are *italicized*. The *arrows* denote the data flow

needed. The MapServer is used for supplying the raster data together with the TileCache caching service. The GeoServer and TinyOWS supply the vector data as rasterized images and geospatial vectors, respectively.

3. The data storage level contains and delivers the source data on the server. The raster data is stored as static, georeferenced and tiled GeoTiff image files on the hard disks. The vector data is stored in an open multi-relational PostgreSQL database that is spatially enabled as a PostGIS database.

12.3.1.2 Web Map Architecture

The web map client application was implemented using Asynchronous JavaScript and XML (Ajax) web technologies. HyperText Markup Language (HTML) was used for creating the general web site layout, the appearance of which was defined using Cascading Style Sheets (CSS). The web map interactivity is produced with JavaScript programming language, which enables the dynamic modification of HTML through the Document Object Model (DOM). Web map functionality is realized using OpenLayers API, which is a free and open source web map UI programming library in JavaScript that takes care of the OWS data requests, data rendering and map interaction.

The web map client displays the geodata that is requested from our three data access services (Fig. 12.2). The WMS interfaces of the services are used for retrieving raster data (1, 2) and a WFS for vector data (3):

1. The raster service provides the web map with static map images that are used as background layers in the map view. The TileCache delivers tiled Portable Network Graphic (PNG) files directly from a disk cache when the requested tiles are available. Otherwise, the TileCache requests the tiles from the neighbouring MapServer, which generates PNG tiles from the GeoTiff data store.
2. The rasterizing vector service provides the web map with static PNG images that are rendered as overlaid layers in the map view. The vector data is rasterized by the GeoServer using styling rules from the SLD files. The rasterization is needed for some vector layers because processing large amounts of vectors is inefficient on the client-side. Another reason for using server-side rasterization is that the GeoServer is capable of styling features that OpenLayers cannot render, for example, multiple overlaid styles and grouping of point symbols (Fig. 12.3).
3. The transactional vector service provides the web map with vectors that are rendered as overlaid layers in the map view and carries their data attributes along to be utilized on the client. The vectors are provided in the Geographic Markup Language (GML) format, which is transformed into SVG by the OpenLayers API and rendered into the map view by a SVG renderer of a web browser. In the case of Internet Explorer, which does not support SVG, OpenLayers uses Vector Markup Language (VML). Vector appearance is drawn on the client by the OpenLayers API according to styles defined in SLD files, which are separately collected from the server.

In addition to querying and displaying raster and vector geodata, the MenoMaps web map allows users to modify certain vector database tables through the transactional operations of the WFS. The INSERT, UPDATE and DELETE operations of the WFS are handled by the TinyOWS service and passed to the PostGIS database as Structured Query Language (SQL) queries.

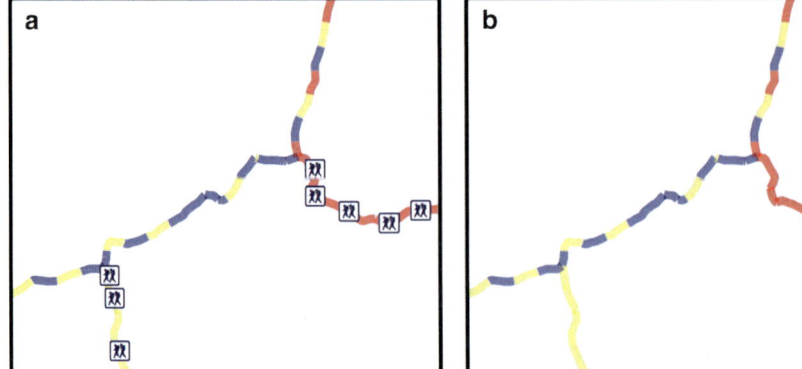

Fig. 12.3 Multiple styles are used for displaying three overlapping routes (**a**) without and (**b**) with grouping of point symbols. Piled line styles are used on single vectors

12.3.2 Web Map User Interface Design

The web map UI is directed at casual users interested in outdoor life who possess basic web browsing skills. Elderly users are considered as a special user group because they are an important subgroup of outdoor people with special needs. The aim of the UI design was to create an easy-to-use map UI which supports map exploration and which can easily be learned. We viewed the "minimalist" and "direct manipulation" design paradigms, as well as "expand-in-context" design pattern, as appropriate for achieving these goals. The web map UI is comprised of three main usage environments: a home page for configuring the UI, a map view for interacting with the map and a tool menu for controlling the map functionalities.

12.3.2.1 Home Page

The home page is initially opened when the user arrives at the MenoMaps web map page. The page lets the user easily configure the display parameters for the map usage (Fig. 12.4). The user chooses the parameters from a floating window on top of

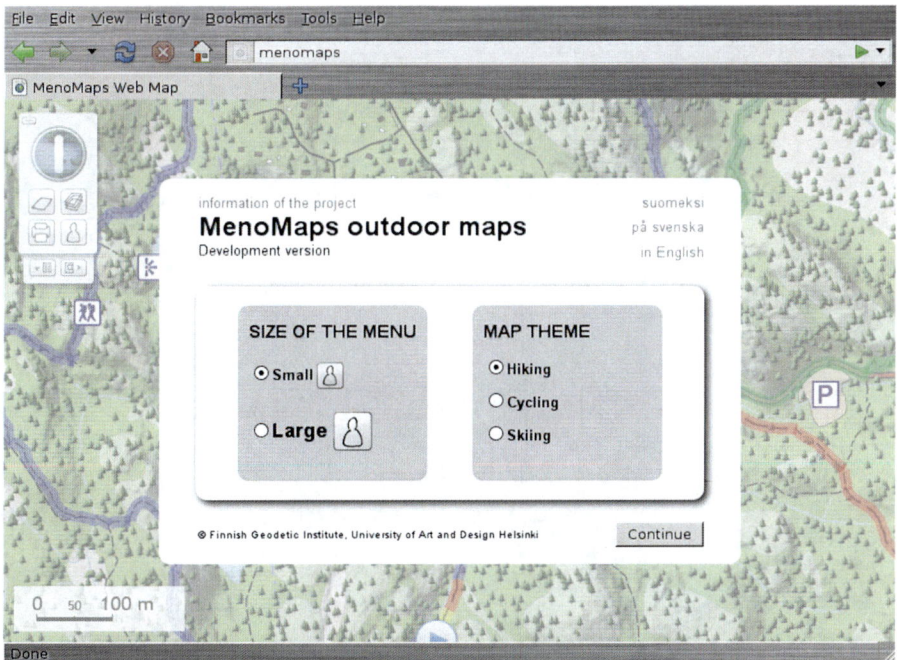

Fig. 12.4 The home page is initially opened when the user arrives at the MenoMaps web map page. The default choices of display parameters are selected (Hiking theme)

the faded map view so that the effects of the changes can be simultaneously observed on the map below. Figure 12.4 shows the default choices.

The main choices of the home page are "size of the menu" and "map theme". The "size of the menu" lets the user change the sizes of the icons and fonts in the UI, a selection that is particularly targeted at elderly users who may be far-sighted or have difficulties in pointing accurately with a mouse. The "map theme" provides a simple means to change the contents of the map view according to the three popular outdoor activities of hiking, cycling and skiing. When the user selects the map theme, this activates pre-defined selections of background map and overlaid map layers (Figs. 12.5, 12.6, 12.7). The contents of the overlaid layers are adapted according to the season and the interests in each outdoor activity. For example, swimming places are not shown in the skiing theme.

The user can select the language of the UI at the web map's home page, a selection that applies to textual contents in the UI. An information page about the project is available. This page opens via a separate tab or window in the browser. The home page can be closed and the map view entered by using the "continue" button or by clicking on the faded map view.

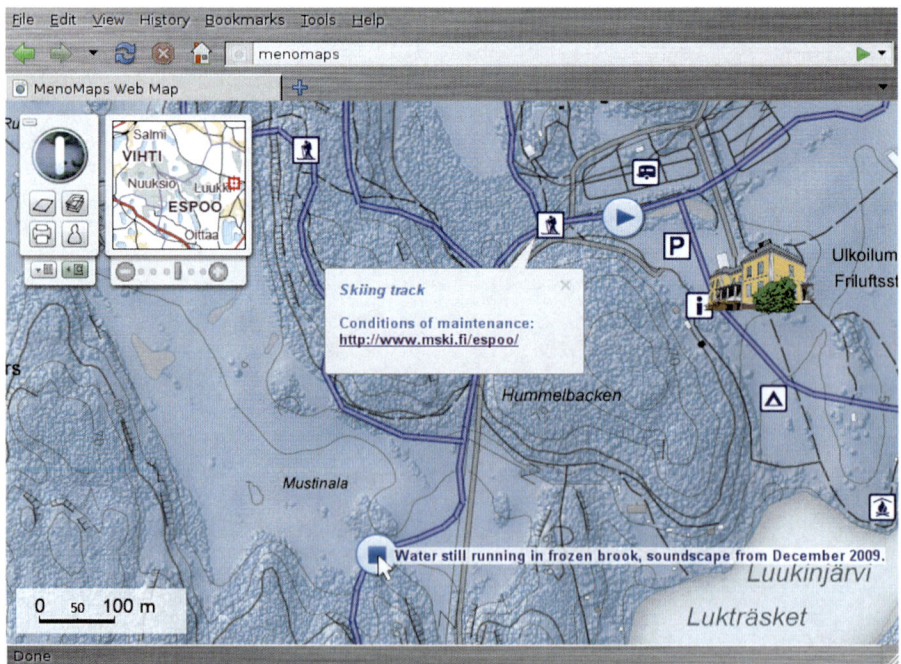

Fig. 12.5 The map view is based upon "minimalist design". The primary functionalities are embedded within the map (Skiing theme)

12 Web Map Design for a Multipublishing Environment Based on Open APIs 187

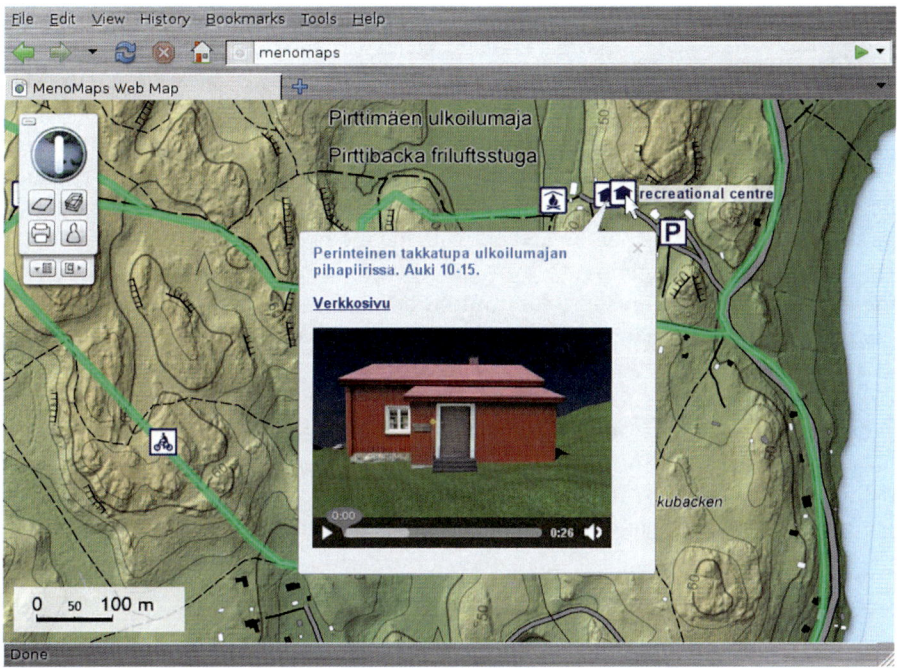

Fig. 12.6 Multimedia objects such as video clips are shown in the pop-ups (Cycling theme)

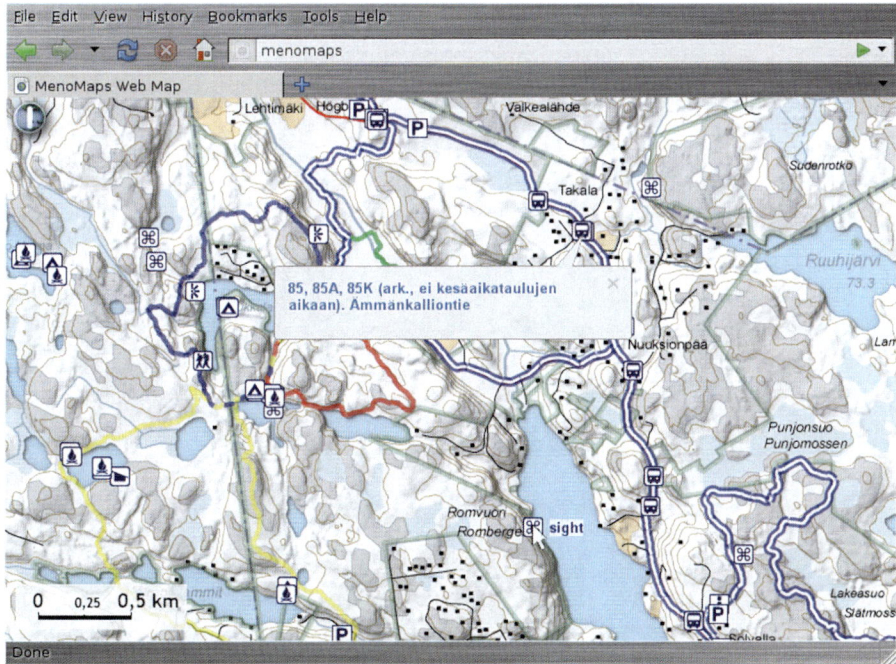

Fig. 12.7 The user can minimize the tool menu. Symbol sizes adapt to the scale level (Selected layers)

12.3.2.2 Map View

The map view is the main usage environment of the MenoMaps web map. Map viewing and map interaction take place within this environment. The map view is based upon the principle of minimalist design. It focuses on the map itself and minimizes the role of menus. The map area is maximized and primary functionalities are embedded within the map following the direct manipulation design paradigm: the user can zoom in and out with the mouse wheel, pan by dragging, get feature tooltips using the mouse-over and open feature pop-ups by clicking on the mouse. The tooltips offer short descriptions of the features and the pop-ups display feature-specific information that can contain text, images and videos (Figs. 12.5, 12.6, 12.7). The map is designed so that it can be used with the mouse, but keyboard controls are also provided for zooming and panning purposes.

The tool menu lies in the upper left corner of the map view. The menu can be minimized to the size of a small icon so that as large a map as possible can be displayed (Fig. 12.7). A transparent scale bar is located in the lower left corner of the map view and it changes dynamically in accordance with the zoom level.

12.3.2.3 Tool Menu

The abstract or complex functionalities of the web map are embedded in an icon-based tool menu in the upper left corner of the map view. The icon menu is built in such a way that it only shows a minimal amount of buttons at the beginning and, as the menu expands, more sophisticated tools can be opened when needed.

The tool menu is made up of three modules that can be opened and closed independently (Fig. 12.8). Initially, the main menu and the overview map are visible and the annotation menu is closed. The usage of the menus is guided by tooltips that come up when the cursor stops for a moment on the button (Fig. 12.8a).

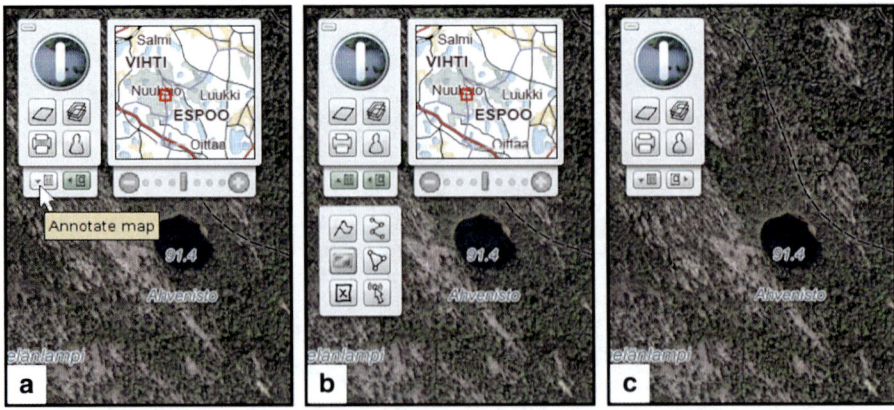

Fig. 12.8 Modularity of the tool menu: (**a**) initial state; (**b**) all menus opened; (**c**) main menu only

The clickable buttons are brightened on mouse-over so that the user can recognize their interactivity.

The overview menu consists of an overview map and zoom buttons. The overview map follows the position of the large map but can also be used for setting the position by clicking on the desired position or dragging the small pointer on the overview map. The zoom buttons are available as an additional means for changing the map scale. The zoom bar denotes the present zoom level among the available levels and can be clicked upon to set a desired view.

The annotation menu gives the user tools for making her/his own markings on the map (Fig. 12.8b). Buttons are available for adding a point of interest or an image, for drawing lines or areas and for editing and removing annotations. The buttons set the annotation mode on "click" so that the annotation can be directly made on the map. For example, the user may denote an interesting area by first clicking the "draw area" button in the right and center of the annotation menu. The map then switches to "draw area" mode and the user is able to draw a polygon on the map. The drawing is finished by double-clicking the location of the last node. The application adds the polygon to the "My content" layer.

The main menu consists of submenus for setting map layers and for communicating with the user database as well as for printing the map view (Fig. 12.9). The background layer is set in the "choose map" submenu and visible overlaid layers are chosen in the "show on map" submenu. The background map selection adapts the related overlaid layers accordingly; for example, ski tracks and winter-only bus stops are hidden when selecting the summer map. The "share" submenu provides a simplistic tool for storing and sharing self-made annotations using the server-side user database. An identifier is entered that is used when saving or loading vector features. The identifier can be used for sharing modifiable annotations with others. The printer button launches the printing functionality of the web browser and applies a printing layout for the present map view. The printing layout is fitted onto the paper, a general title is set on top of the page and the menus are excluded.

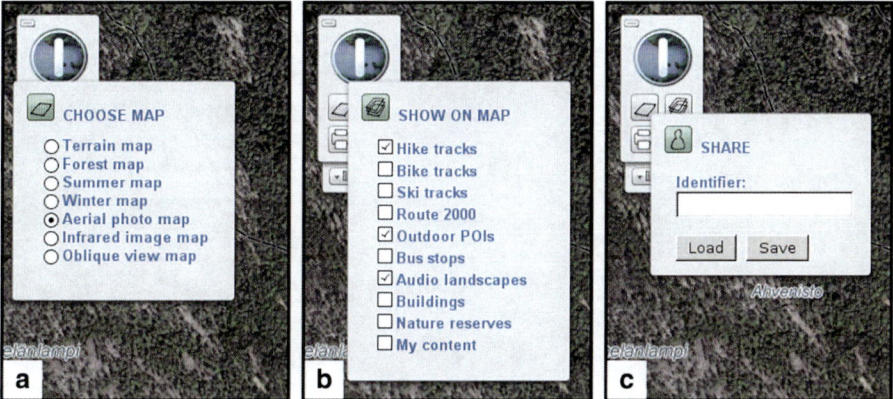

Fig. 12.9 Submenus of the main menu for choosing the visible layers and communicating with the database. The printing button launches the printing function of the browser

12.3.3 Web Map Usability Evaluation

Flink et al., (2011) studied the ease-of-use and usefulness of the MenoMaps multipublishing service by using the methods of thinking aloud and questionnaires. The testing situations were videotaped, and the sessions took about 1–1.5 h each. Six potential users recruited from the Finnish hiking association, Tunturilatu, took part in the usability evaluations. The tasks required the users to use various functions of the web map, such as finding additional information on the trails or drawing their own routes. After the thinking aloud tasks, the participants answered to a System Usability Scale (SUS) questionnaire concerning the tested maps. The SUS is a 0–100 scored Likert measure of the usability of a system as evaluated by users through ten subjective usability questions (Brooke 1996). In the present study, we made use of the preliminary results of the usability evaluations. The evaluations and results are more thoroughly discussed in (Flink et al., 2011).

The usability study guided us to instantly make two simple modifications on the web map functionalities that the participants reported to be problematic. For example, we set the overview map initially visible when entering into the map view the first time. We also added a tooltip to instruct that the drawing should be completed with a double-click, an operation that the participants found difficult to perform without guidance (Fig. 12.10).

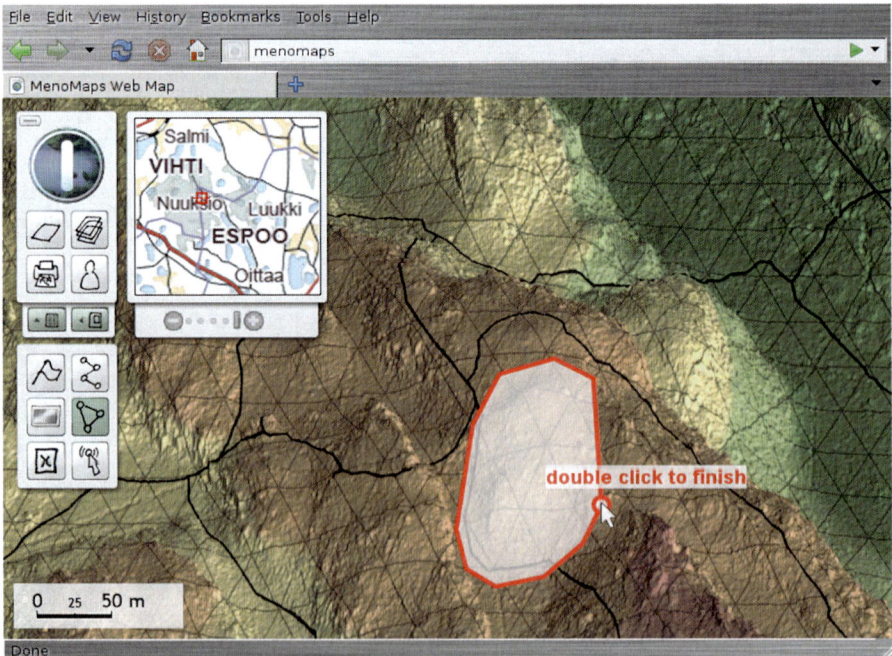

Fig. 12.10 As a result from the usability study, we added guidance so that users know how to complete the drawing action. The large menus and the font size are targeted at elderly users

12.4 Summary and Conclusions

The paper presented on-going research on designing and implementing a web map as one of the channels in a map multipublishing environment to be used for outdoor leisure activities in a national park. The web map client is an end-user application for exploratory cartographic visualization in order to enable creative planning of outdoor activities and to explore detailed information related to the national park.

The design of the web map UI was based upon "minimalist" and "direct manipulation" approaches, aiming at simple usage but versatile functionality. The UI consists of a configuring palette to select display parameters at the home page, a map view for displaying and exploring the map and an expanding tool menu for rich functionalities. The map view is the main usage environment of the web map with a maximized map area that incorporates the basic map functionalities such as panning and zooming. We considered elderly people as a special user group for whom large-size UI components were composed in the design.

The UI design was evaluated by another research group through a usability study. The participants considered the UI usable according to a SUS but encountered difficulties in finding some functionalities and in completing some tasks. Based on the results from the usability study, we made improvements to the UI design.

The web map architecture was fully implemented using FOSS GIS software, which we found to be a functional and flexible solution for creating an effective and robust web map. The FOSS GIS applied thorough implementations of OGC Open Web Services (OWS) for data access in the architecture and offered wide interoperability in terms of data formats and coordinate systems. It was possible to make modifications to the software on both the server and client sides, which was necessary for achieving the design goals of the web map. For example, we replaced the menus and graphical components of the OpenLayers web map client with self-made ones in order to efficiently apply minimalist design.

The MenoMaps web map will be further improved with regards to usability, cartographic design and architecture. The MenoMaps multipublishing service and its channels, including the web map, will be exhibited for the public in the new Nuuksio Nature Centre in southern Finland in the beginning of 2013. This will spur us to conduct additional usability evaluations and, consequently, introduce new UI developments for the web map. In particular, functionalities for sharing content in communities will be advanced.

Acknowledgements This survey is part of research for the MenoMapsI-II projects. These projects, funded by Tekes (the Finnish Funding Agency for Technology and Innovation), are joint ventures by the Finnish Geodetic Institute (FGI), Department of Geoinformatics and Cartography and Aalto University of Helsinki, School of Art and Design.

References

Andrienko GL, Andrienko NV (1999) Interactive maps for visual data exploration. Int J Geogr Inf Sci 13(4):355–374

Brooke J (1996) SUS – a quick and dirty usability scale. In: Jordan PW, Thomas B, Weerdmeester BA, McClelland IL (eds) Usability evaluation in industry. Taylor & Francis, London

Carroll JM (1985) Minimalist design for active users. In: Shackel B (ed) Human computer interaction – INTERACT'84. Elsevier Science, Amsterdam

Cartwright W (1997) New media and their application to the production of map products. Comput Geosci 23(4):447–456

Coltekin A, Heil B, Garlandini S, Fabrikant SI (2009) Evaluating the effectiveness of interactive map interface designs: a case study integrating usability metrics with eye-movement analysis. Cartogr Geogr Inf Sci 36(1):5–17

Dunfey RI, Gittings BM, Batcheller JK (2006) Towards an open architecture for vector GIS. Comput Geosci 32(10):1720–1732

Flink H-M (2009) User centred approach in the concept development of a map-based multi-publishing service. Master's thesis, University of Art and Design Helsinki

Flink H-M, Oksanen J, Pyysalo U, Rönneberg M, Sarjakoski LT (2011) Usability evaluation of a map-based multi-publishing service. In: Ruas A (ed) Advances in cartography and GIScience, vol 1, Lecture notes in geoinformation and cartography. Springer, Berlin, pp 239–257. doi:10.1007/978-3-642-19143-5_14

Gibin M, Singleton A, Milton R, Mateos P, Longley P (2008) An exploratory cartographic visualisation of London through the Google Maps API. Applied Spatial Analysis and Policy 1(2):85–97

González A, Velasco A, Ruiz C, Rubio JM, González J, García Á, Más S (2009) CartoCiudad: national database of all Spanish municipalities available freely including www.cartociudad.es and exploitable through its OGC services. New functionalities implemented. In: Proceedings of the 24th international cartographic conference. The World's Geo-Spatial Solutions, Santiago, 15–21 Nov 2009

Haklay MM, Zafiri A (2008) Usability engineering for GIS: learning from a screenshot. Cartogr J 45(2):87–97

Henrie D (2009) Ordnance survey historic town plans of Scotland (1847–1895): geo-referencing and web delivery with ArcIMS and OpenLayers. e-Perimetron 4(1):73–85

Hutchins EL, Hollan JD, Norman DA (1985) Direct manipulation interfaces. Hum-Comput Interact 1(4):311

Kovanen J, Sarjakoski T, Sarjakoski LT (2009) Studying iPhone as a media for context-aware map-based mobile services. In: Proceedings of the 6th international symposium on LBS & telecartography, CGS, University of Nottingham, Nottingham, 2–4 Sep 2009

Lehto L, Kähkönen J, Sarjakoski T (2001) Multi-purpose publishing of geodata in the web. In: Proceedings of the 4th AGILE conference on geographic information science, Brno, 19–21 April 2001. pp 209–214

MacEachren AM (1994) Visualization in modern cartography: setting the agenda. In: MacEachren AM, Taylor DRF (eds) Visualization in modern cartography. Pergamon/Elsevier Science, Oxford/New York, pp 1–12

MacEachren AM, Kraak M-J (1997) Exploratory cartographic visualization: advancing the agenda. Comput Geosci 23(4):335–343

Moellering H (1984) Real maps, virtual maps and interactive cartography. In: Gaile GL, Willmott CJ (eds) Spatial statistics and models. D. Reidel Publishing Company, Dordrecht, pp 109–132

Nivala A-M (2007) Usability perspectives for the design of interactive maps. Doctoral dissertation, Helsinki University of Technology

Nivala A-M, Brewster S, Sarjakoski T (2008) Usability evaluation of web mapping sites. Cartogr J 45(2):129–138

Oksanen J, Schwarzbach F, Sarjakoski LT, Sarjakoski T (2011) Map design for a multi-publishing framework – case menomaps in Nuuksio National Park. Cartogr J 48(2):116–123

Pulsifer PL, Hayes A, Fiset J-P, Taylor DRF et al., (2008) An open source development framework in support of cartographic integration. In: Peterson MP (ed) International perspectives on maps and the internet. Springer, New York, pp 165–185

Roberts JC (2005) Exploratory visualization with multiple linked views. In: Dykes J, MacEachren AM, Kraak M-J (eds) Exploring geovisualization. Elsevier, Oxford, pp 159–180

Roberts JC (2008) Coordinated multiple views for exploratory geovisualization. In: Dodge M, McDerby M, Turner M (eds) Geographic visualization: concepts, tools and applications. Wiley, Chichester, pp 27–48

Roth RE, Harrower M (2008) Addressing map interface usability: learning from the lakeshore nature preserve interactive map. Cartogr Perspect (Spring) 2008(60):46–66

Sarjakoski LT (2007) Conceptual models of generalisation and multiple representation. In: Mackaness WA, Ruas A, Sarjakoski LT (eds) Generalisation of geographic information: cartographic modelling and applications, series of international cartographic association. Elsevier, Burlington, pp 11–36

Sarjakoski LT, Sarjakoski T, Koskinen I, Ylirisku S (2009) The role of augmented elements to support aesthetic and entertaining aspects of interactive maps on the web and mobile phones. In: Cartwright W, Gartner G, Lehn A (eds) Art and cartography. Springer, Berlin, pp 109–124

Sarjakoski T, Kovanen J, Rönneberg M, Kähkönen J, Sarjakoski LT (2010) Data matrix technology for linking mobile maps in a web-based multi-channel service. In: Proceedings of ubiquitous positioning, indoor navigation and location based service 2010, UPINLBS 2010, Kirkkonummi. IEEE, Seattle, 14–15 Oct 2010. ISBN 978-1-4244-7878-1, CD-ROM

Slocum TA, Blok C, Jiang B, Koussoulakou A, Montello DR, Fuhrmann S, Hedley NR (2001) Cognitive and usability issues in geovisualization. Cartogr Geogr Inf Sci 28(1):61–75

Webster C (1988) Disaggregated GIS architecture lessons from recent developments in multi-site database management systems. Int J Geogr Inf Sci 2(1):67–79

Zipf A (2005) Using Styled Layer Descriptor (SLD) for the dynamic generation of user- and context-adaptive mobile maps – a technical framework. In: Li K-J, Vangenot C (eds) Web and wireless geographical information systems, vol 3833, Lecture notes in computer science. Springer, Berlin, pp 183–193

Chapter 13
User Scalable Graduated Circles with Google Maps

Douglas Paziak

Abstract The circle has long been a useful symbol for displaying quantitative data. Determining the proper size for the circles has remained a problem. If the circles are too small, differences in size are not readily apparent and a pattern is not evident. If the circles are too big, the overlap between the symbols will be excessive or too much of the map will be covered. A method is proposed here that allows the user to interactively adjust both the size and the opacity of the circle shading. The online version of the map, with all relevant code, is available at: http://maps.unomaha.edu/GoogleMapGallery/GradCircles/circles.htm

13.1 Introduction

A major advantage of the graduated circle is that the size can be made in direct proportion to the value that it represents. Multiple methods have been proposed to compute this proportion. One such method is the square root method that takes the square root of the data value and multiplies it by a constant to arrive at a circle radius, from which the circle area can be determined.

Research into how circle area is perceived has shown that the human brain tends to underestimate the area of a circle. In a graduated circle map, this will lead to a misinterpretation of the presented data by the map user. In order to compensate for this size underestimation, another method for calculating the size of graduated circle symbols has been devised, called the psychological scaling method. The psychological scaling method is similar to the square root method but instead using the square root, or ½ power, this method uses the 0.5716 power, as originally devised by J.J. Flannery. This factor has been examined through the years, and although it has found to vary based on the individual and testing method (Chang

---

D. Paziak (✉)
Private Cartographic Contractor, 7528 Pinkney Street, Omaha, NE 68134, USA
e-mail: dpaziak@hotmail.com

1977), circles produced using a non-proportional adjustment that increases the size of the larger symbols are perceived to be generally closer to their correct size than those produced via the standard square root method (Flannery 1971).

The major problem in the use of the graduated circle on maps is determining the proper size for the smallest symbol, as this controls symbol coverage in the map. The use of circles that are too large will lead to an excessive symbol overlap. Circles that are too small will be indistinguishable from each other and will not convey a spatial pattern. Larger circles will also obscure smaller symbols, especially if they are filled with a shading, and the addition of a shading for circles seems to enhance the recognition of a spatial pattern. The problem that is examined here is how to scale proportional circles and change their shading in an interactive mapping environment.

13.2 A Google Maps Implementation

One method to display graduated circles on a map is to use the Google Map service. These maps are displayed using the Mercator projection over a number of scales from small to large. The number of available scales currently is 19 but is updated as higher resolution images and maps become available. For faster map display, these maps are distributed in the form of static raster tiles.

Google has provided a way to customize Google Maps via the use of its API, or application programming interface. The API allows users to access objects created by Google to display fully-functional Google Maps (Google 2010). There is a wide variety of uses for this service, but one of the major applications is the creation of thematic maps in the form of "mashups".

Any interactive mapping environment allows the user greater control and freedom over how a map is presented; as a result, the map can quickly become unusable. For instance, a graduated circle at one map scale may take up the entire screen on a larger scale map, and may not even be visible on a smaller scale map. Changing the map scale is easy to accomplish with online mapping systems. What is viewed as a freedom for the user is actually a map limitation. Various considerations have been approached to combat this loss of map fidelity, such as not allowing the user to change the map scale. However, this reduces the functionality of a Google Map and users do not seem to tolerate any limitations to the expected functionality.

One approach to this circle size dilemma is to allow the user to interactively adjust the size of circles. The psychological scaling factor will still govern the proportion of the circles but the overall circle sizes can be smaller or larger based on user input. Another feature is to make it also possible to adjust the opacity of the circles. To make these two features as easy to use as possible, the user should be able to adjust both at the same time, by dragging a marker on the edge of each circle (see Fig. 13.1).

Fig. 13.1 Mechanism for adjusting circle size and opacity. The circle size is adjusted by moving the triangle on the side of the circle in a horizontal direction. Moving the triangle in a vertical direction changes the opacity of the circle

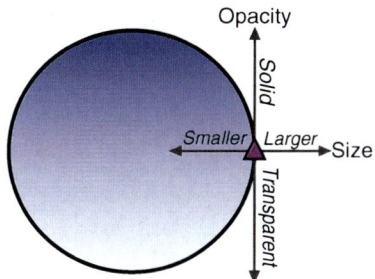

13.3 Implementation

Displaying a graduated circle map using Google Maps is achieved through a number of steps. First, the webpage is set up to use the Google Maps API. A map can then be created with graduated circles based on a given set of data. This involves calculating the radius of each circle that will be displayed on the map.

Point symbols in a Google Map include a functionality called "listening" that can be utilized to help rescale the circles. A small point symbol is placed on the circle that is normally hidden. Each of these point symbols "listen" for a mouse-click action at which point they can be moved and the circle re-scaling sequence starts. This feature is used to allow the user to adjust the size and opacity of the circles, as well as to display a box with information about each circle (see Fig. 13.2). In the example described here, a map is displayed showing graduated circles for the 90 largest metropolitan areas in the United States.

13.3.1 Setting Up the Web Page

In order to set up the webpage properly to use the Google Maps API, access is required to Google's library of code and functions. This is done by obtaining a key from Google, which is available free with certain limitations, such as the restriction that all websites must be freely accessible to the public. Once this key is inserted, the map has access to JavaScript functions and objects to display a Google Map. In the latest iteration of the Google Maps API, a key is no longer required.

13.3.2 Displaying the Base Map

When the web page that is using the map is loaded, a function is called to display the map. A map is created that is centered on a predefined point, at a predefined scale, and using a predefined map type. In addition, some controls are added to give the user the ability to manipulate the map, such zooming in and out, panning the map, among other functions.

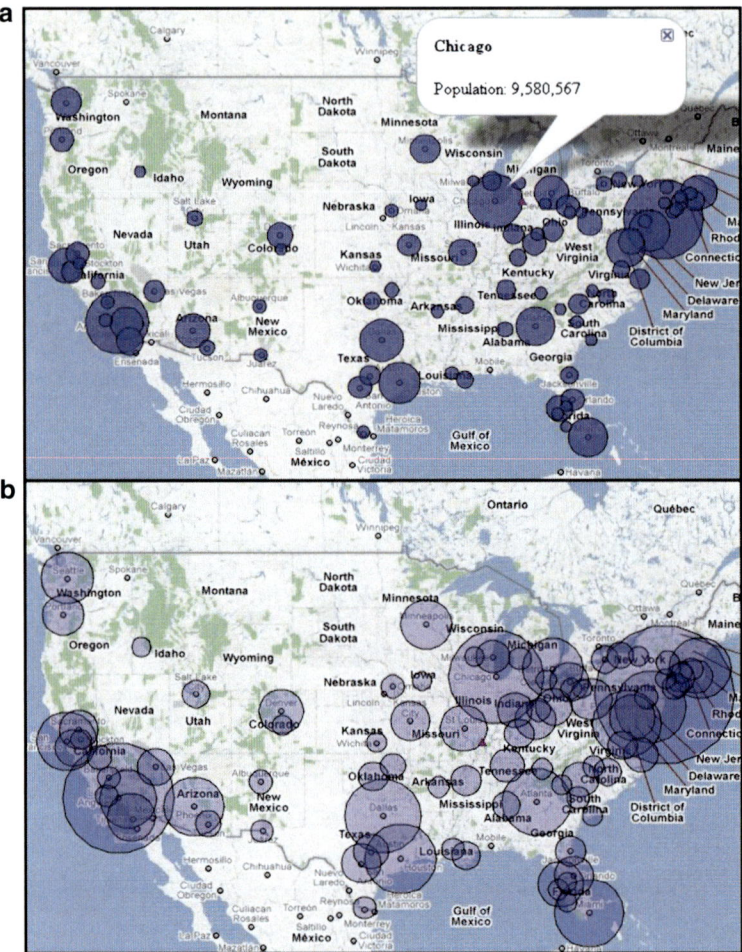

Fig. 13.2 Two examples of alternative circle sizes and opacities: (**a**) shows the original circle sizes with an "info" bubble for Chicago; (**b**) depicts the circles after they have been adjusted in both size and opacity. The small circles in the center of many of the circles are the city markers on Google's base map, not additional circle symbols

Once the map is displayed, a function is called to begin creating the thematic map overlay. This function sets a few variables, such as the initial circle opacity. A new function is then immediately called that begins the retrieval of data. The reason these variables are initialized before this new function is called has to do with the difference between redrawing the circles and resetting the circles.

All of the circles are redrawn every time the user selects a new circle size. In doing so, the user changes the values of the variables that were defined above, so it makes no sense to re-initialize these variables. However, if the user wants to reset the circles to their original sizes, based on the map scale, these variables need to be re-initialized before the circles are redrawn.

13.3.3 Retrieving the Data

Once the map is displayed, the initial variables for the circle symbols are defined, and the data to be displayed is retrieved. These data can be located in pre-populated arrays within the code but it is more functional to obtain the data from an external source, such as an XML file. An XML file contains data structured in such a way that it is easy to parse that data that is required.

The data are parsed using a Google Map function called GXmlHttp. This function opens a connection to the XML data file that allows the data within the file to be freely accessed. The data is then placed within a series of arrays. In this example, the XML file consists of the city name, population, latitude, and longitude.

During the retrieval of data, the latitude and longitude are combined into a Google Maps specific point data type, GLatLng. Also, during this retrieval, the minimum population is calculated, allowing each circle to be correctly proportioned with respect to the size of the smallest symbol. The largest circle is not used as the base size because not all circles may be visible on the map if they are scaled according to its size, as some may be too small.

13.3.4 Calculating the Circle Sizes

The most important part of a graduated circle map is circle size. Each circle must be calculated based on a certain size, usually the smallest circle to be displayed. The basic calculation takes the population of each city and compares it to the population of the smallest city, using a loop. This ratio, which will always be greater than or equal to one, defines the circle radius. In order to adjust the size of the circle, this ratio is multiplied by a minimum circle radius, defined here as five pixels. Thus, the smallest circle will have a radius of five pixels and the largest circle will have some other radius, based on the ratio of its population to the smallest population.

Because of the circle redraw feature of this example, the area of each circle is multiplied by a conversion factor, which is used to resize each circle in proportion to amount by which the user adjusted one circle. For example, if the user doubled the circle size area, this conversion factor would be two. This conversion factor is initially one, and by resetting this variable to one, the symbols will display at their original sizes. Using this conversion does not compromise the use of the psychological scaling factor because all circles are resized by the same proportion after they have been scaled using the psychological scaling factor.

The next step of the loop is to calculate the location of the marker point that will be used by the map user as a handle to adjust both the size and opacity of the circle symbols. The opacity of the circle is defined by the angle of the marker point in relation to a horizontal line through the center of each circle, with 90° being solid (or an opacity of 1) and $-90°$ being transparent (an opacity of 0). Initially, this angle is zero, but it is changed every time the symbol opacity is adjusted.

The location of the marker point for each circle that is used for re-scaling is calculated by combining the radius of the circle with the angle that defines the opacity. This calculation will display a marker point on the edge of the circle symbol and, as a result of the distortion of circle by the Mercator projection, the location of the marker should be calculated in pixel coordinates and then converted to geographical coordinates. This is done by invoking the fromLatLngToPixel function and fromPixelToLatLng function, when necessary. The location of the marker point is calculated using sine and cosine trigonometric functions.

13.3.5 Displaying the Circle Overlays

Once all of the circle radii and marker points are calculated, the circles need to be displayed, along with the marker points. Circles are drawn as a many-sided polygon. The number of sides of the polygon defines how realistic the circle looks; a greater number of sides will result in a smoother looking circle, a lesser number of sides will produce a rougher circle. A circle with fewer sides will be drawn faster than a circle with many sides. In this example, the number of sides of the circle is defined by the circleQuality variable. This variable sets the angular distance, in degrees, between each point on the circle edge.

To draw the circles, the center point is converted to pixel coordinates, and the sine and cosine trigonometric functions produce the edge point. However, to calculate all of the points on the circle, a loop is run through all of the angles of the circle, using the circleQuality variable as the increment. So, using the circleQuality variable of 20 will cause the loop to run 18 times, or 360 divided by 20. The first time through the loop, $0°$ is used to calculate the circle edge point; the second time, $20°$ is used; the third time, $40°$ is used, and so on.

These points are placed into an array that contains all of the points of the circle. Following this, a polygon is defined using the points array, as well as some of the circle constants, such a color and line thickness, and the opacity variable. As each circle symbol is calculated, it is added to the map. At this time, a listener is added to each circle to display an info box when the circle is clicked. Also, a listener is added to check to see if the mouse enters or leaves a circle. This listener displays the handle marker if the mouse enters the circle, and hides the marker once the mouse leaves. The marker is used to rescale the circle.

13.3.6 Adding Functionality

Adding the info box, mouse in, and mouse out listeners to the circle polygon adds some basic functionality to the map, but to allow the user to resize the circles takes a few more steps. These steps are implemented when the individual circle edge points are being defined. For each circle, every time the loop calculates the first point of the circle, the handle marker point is added to the map. This marker point is displayed by a custom icon using Google's GIcon function. This draggable marker is added to the map but immediately hidden so it is not intrusive to the appearance of the map; it will be displayed later using the mouse listener for the actual circle polygons.

To allow the map to change when the marker is moved, listeners must be added to this marker. The first listener results in a function that runs while the marker is being moved. This function draws a temporary, constantly changing circle that defines what the size and opacity of the circles will be, based on the current position of the marker. This is basically the same function that draws the actual circle symbols; however, instead of using a pre-defined radius, the radius is calculated based on the current position of the marker.

To calculate the radius of the temporary circle, the marker point and center point are first converted to pixel coordinates. The horizontal and vertical components of the radius are defined as the difference between the horizontal and vertical portions of the center and marker point coordinates. The radius is calculated by the Pythagorean Theorem, using the horizontal and vertical components as the legs of the triangle and the radius as the hypotenuse. Then, the area of the temporary circle is calculated and compared to the original area of the circle. This comparison defines the conversion factor that is used to initially calculate the circle radius. The temporary circles are drawn using the same method that produced the actual circle symbols.

The last part of this listener calculates the opacity based upon the angle of the marker point to the center point of the circle. First, the vertical distance between the center point and the marker point is calculated; if the marker point is above the center point, this distance will be positive, if below, this distance will be negative. The ratio of this distance to the radius of the temporary circle is used to determine the opacity. This ratio is always between −1 and 1, and by adding 1 and dividing by 2, a value for the opacity is achieved, which must be between 0 and 1.

The angle of the marker point is then calculated using the inverse tangent trigonometric function. This angle is important to ensure the marker point for each circle is placed in the same relative location. However, this function only calculates the angle between the marker point and the center point vertically; separate lines of code need to be run to calculate if the marker should be placed on the left or right side of the circle. The opacity, conversion factor, and all of the marker point angle variables are global variables, meaning they will still hold their value once the CircleFill function is ended.

The last listener of this marker calls the function that calculates the circle radii and then draws the circles when the marker is dropped. Because the global variable for the conversion factor and opacity have been changed, the circles will all be redrawn based on the prototype of the temporary circle.

13.3.7 Circle Size Reset

The final part of the process is to put a button within the webpage that allows the user to reset the circle sizes and opacity to their original values. Once this button is clicked, all of the variables are reset to their original values and the circle symbols are redrawn. The minimum circle size and the proportion of each circle size to one another will still be the same as the original; however, because the map scale may have changed, the circles may appear to cover less area on the map. If a user zooms in on the map, the circles may fill up the entire screen (see Fig. 13.3). This feature gives the user the instant ability to reset the circle in such a way that no circle will

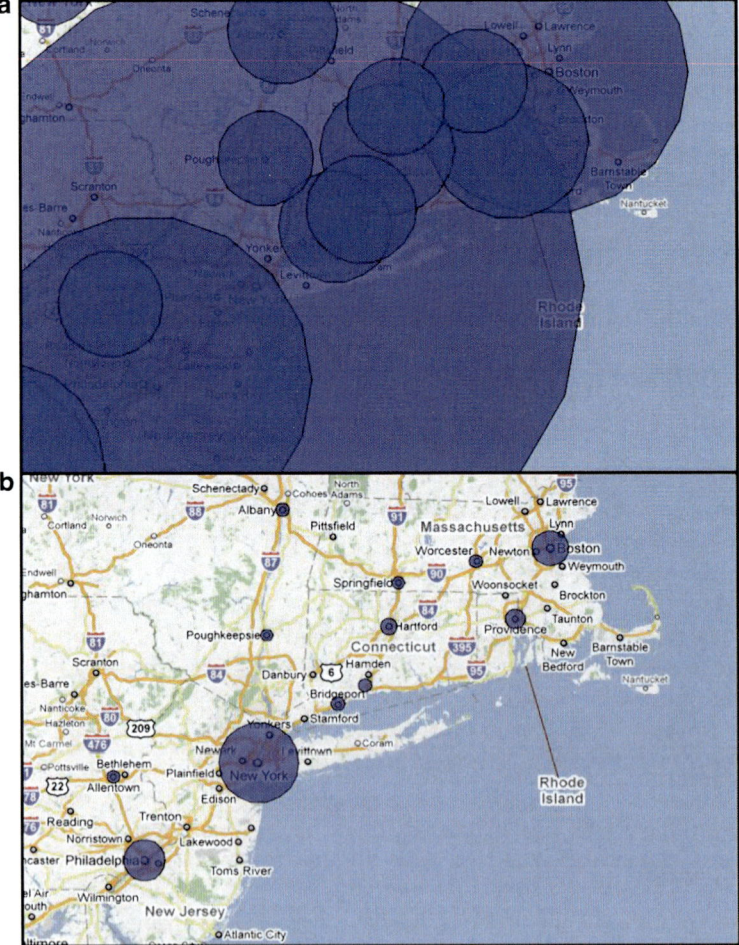

Fig. 13.3 Circle Size Reset: (**a**) shows the result of selecting a larger map scale; (**b**) depicts the circles after the reset button has been selected. Vertices are clearly visible in the larger circles in (**a**)

cover the whole map, because the circle sizes are based on the visible map area via the pixel coordinates that change every time the zoom level changes, instead of static geographical coordinates.

13.4 Conclusion

Graduated circles are an excellent way to display quantitative data at points, and the Google Map API is a powerful tool for producing maps through the Internet. The method discussed here combines these two to give the user unprecedented control over the display of graduated circle symbols, control that makes a map that is more useful and relevant to the user, which is the ultimate goal of cartography.

References

Chang KT (1977) Visual estimation of graduated circles. Can Cartogr 14(2):130–138
Flannery JJ (1971) The relative effectiveness of some common graduated point symbols in the presentation of quantitative data. Cartogr Int J Geogr Inf Geovis 8(2):96–109
Google. Google maps API. http://code.google.com/apis/maps/documentation/reference.html. 19 Apr 2010

Chapter 14
Webservices for Animated Mapping: The TimeMapper Prototype

Barend Köbben, Timothée Becker, and Connie Blok

Abstract Within a larger aim of improving automated vector animated mapping, the main objective of this research was to look into the possibility of combining two technologies: distributed webservices and animated, interactive vector maps. TimeMapper was developed as a prototype for an OGC-compliant Web Map Service implementation that serializes spatio–temporal data from a database backend as Scalable Vector Graphics. The SVG is used in a web browser to show animated maps with a built-in advanced user-interface. This interface allows the user to interact with both the spatial and the temporal dimensions of the data. The potential and limitations of the TimeMapper prototype were explored using Antarctic iceberg movement data. The prototype can be explored on the TimeMapper website (http://geoserver.itc.nl/timemapper/).

14.1 Introduction

The motivation for building the TimeMapper prototype was the improvement of automated animated vector mapping. We are interested in animation, because interactive animated mapping has been pointed out, by Andrienko et al., (2003) among others, as the only technique to be generically applicable to visually analyze the dynamic nature of real world phenomena.

We want to facilitate the production of animated maps from spatio–temporal data to a format suitable for Internet dissemination, *automatically* and *directly*. To achieve that, we looked specifically into the possibilities of the loose coupling of distributed webservices with animated, interactive vector maps.

B. Köbben (✉)
Faculty of Geo-Information Science and Earth Observation, ITC – University of Twente, Enschede, The Netherlands
e-mail: kobben@itc.nl

14.1.1 Animated Mapping in a Webservices Environment

The principle of disseminating maps in a webservices environment is depicted in Fig. 14.1. This general set-up is being used in many of today's webmapping efforts, with considerable variation in the choice of technology for the mapping service (depicted as A) and the subsequent map formats (B). This choice to a large extent defines the possibilities of the system as a whole to achieve what we called previously the *automatic* and *direct* production of animated maps.

We consider "automatic" to mean that the animated maps will be generated from the spatio-temporal data by the system "working by itself with little or no direct human control" (which is how "automatic" is defined in Fowler et al., (1976)). It is important to note that this automation will not include the cartographic decisions as to what type of map to use for different data-types and data-instances. As in most current systems, this level of automation in cartography will not be achieved. The link between data- and visualisation-type has to be made by a human (the cartographer in Fig. 14.1), setting up the appropriate configuration parameters beforehand.

By "direct" we mean that the maps are generated on-the-fly from the data, without conversion or pre-processing needed for the purpose of visualisation only. This is important because we want our system to fit in an interoperable Spatial Data Infrastructure, where it would be an SDI node, as defined in de By et al., (2009): "a single system node in an SDI network, which archetypically has data, catalog or portrayal services." In such a context, our system would be just another service, that should be able to consume data from any other SDI node, and whenever that data changes, the maps generated from that data would change. To achieve this, the visualization functionality should be loosely coupled to the other parts of the system. The Open Geospatial Consortium's (OGC) Web Map Service (WMS) and related specifications (see OGC 2002b, 2006) are especially useful for this. These webservices are designed to take their input from a variety of distributed sources and generate output meant for Internet dissemination.

Current WMS implementations mostly generate so-called "picture" output, such as the raster formats PNG, GIF and JPEG. These picture formats are fine for

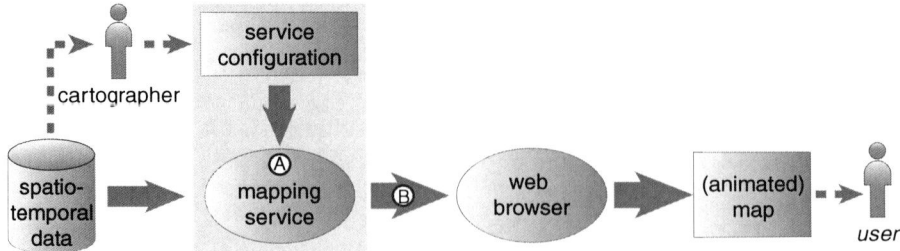

Fig. 14.1 General principle of dissemination of maps in a webservices environment. Note the important influence, discussed in Sect. 14.1.1, of the choice of technology and the subsequent formats in *A* and *B*. In this research *A* is the TimeMapper WMS prototype, and *B* is Scalable Vector Graphics XML, including SMIL animation and interactivity

delivering static map images over the web. For data that has a temporal component, a series of such static images, packed into some sort of animation format (e.g., an animated GIF or a QuickTime movie) can be generated; this is supported by several existing implementations of WMS. These raster animations, comparable to movies, can be useful for certain types of temporal datasets, especially satellite image series.

However, this type of animation lacks the possibility to express many of the aspects of temporal changes that are connected to specific *objects*, not fields. In geographic theories on change, different types of change are usually recognized. Based on the work of Blok (2005), Andrienko and Andrienko (2006) describe three categories of change: existential change (e.g., appearance/disappearance), change of spatial properties (e.g., location, shape) and change in attribute (e.g., ownership change). These changes occur over time, to selected geographic objects, that are in many cases modeled in the data as polygons, lines and points.

To be able to express these changes on the object level, we are interested in *vector* animations. The WMS specification also allows for "graphic element" output, i.e., using vector formats with added possibilities, such as scripting and animation. The added benefit of this is that such formats allow the inclusion of attributes within the objects, providing the client-side mapping application access to data attributes without the need for a round-trip to the server.

At the time of our research, two graphic element formats were widely used to produce vector animations for the Internet: Scalable Vector Graphics (SVG) and Adobe Flash. SVG is our format of choice: firstly because it is an XML-based Open Standard (W3C 2010), whereas Flash is a binary format and a proprietary technology. Secondly, because we have good experiences with a previously built SVG-based WMS for static maps. And thirdly, because the SVG specification allows the use of the Synchronized Multimedia Integration Language (SMIL), a declarative XML-based language well-suited for building data-driven animations (W3C 2008).

14.1.2 Objective and Set-Up

The **objective** of our research was therefore *to design an animated mapping system which can generate SVG SMIL animations, automatically and directly, from spatio–temporal data.* The scope of vector animated mapping is broad and we therefore limited our initial goal by developing a prototype system for one specific type of real-world phenomenon only: the movement of objects in geographic space.

To achieve our objective, we extended our existing SVG-based Web Mapping System for static maps with the animated mapping functionality. We briefly describe this existing system, called RIMapperWMS, as well as other related work, in Sect. 14.2. We then explain the extensions necessary to create animated maps and their set-up and introduce our prototype, called TimeMapper, in more detail in Sect. 14.3. We evaluate the prototype in Sect. 14.4 and, based on that, draw some conclusions and give an outlook on further work in Sect. 14.5.

14.2 RIMapperWMS and Other Related Work

Our starting point for the TimeMapper prototype was RIMapperWMS, which in turn was a further development of RIMapper. This was a non-standardised webservice, that started out as the visualisation part of an urban risk management system, hence the name, an acronym for Risk Indicator Mapper. RIMapperWMS, described in more detail in Köbben (2007), was an effort to make the software compliant with OGC's Web Map Service (WMS) specification. Like most of the software we use, it fits within the so-called SDI^{light} approach, that we have been using for several years in our teaching and research at ITC. It has been discussed in detail elsewhere, most recently in Köbben et al., (2010). In short, it is a down-to-earth approach towards Spatial Data Infrastructures, using Open Standards whenever available, and open source solutions where possible. This approach provides researchers, students and partners with a platform for relatively simple, low-cost, yet powerful ways of sharing data amongst various stakeholders.

The main components of the platform are firstly a spatial database back-end (we most often use the object-relational DBMS PostgreSQL and its spatial extension PostGIS). Secondly, there are middleware web applications that interface with the database back-end and with each other, and fulfill tasks such as delivering maps, data and processing services. We use existing open source applications (mostly MapServer and GeoServer), but also develop our own components. The RIMapperWMS and TimeMapper discussed here are examples of those. And, finally, there are open source desktop clients (such as QuantumGIS and ILWIS), as well as browser-based clients (such as OpenLayers and the SVG GUI part of TimeMapper), that enable access to the maps and data.

There are several established WMS implementations that serve SVG. However, based on our own experiences and the capabilities as described on the OGC implementation webpage (OGC 2002a), they treat the map output as just another graphics-only format. So, just like the raster format output in PNG and JPEG and such, the maps are basically pictures only, with no interactivity or "intelligence". To have the map behave like a *mapping application*, it needs to be wrapped in a Graphical User Interface (GUI) that takes care of zooming, panning and other interactions. There are many such GUI's available, as dedicated server-side applications, as part of general GIS clients (such as QuantumGIS), or as client-side API's for browser environments (such as the popular OpenLayers Javascript API). But, because SVG is more than a graphics format only, it also offers interactivity through built-in scripting and full access to its XML Document Object Model. It is therefore possible to deliver straight from the WMS an *SVG application*, that includes its own GUI. This is the strategy we use in RIMapperWMS, and its output can therefore be used without a dedicated mapping client, any SVG-enabled client is sufficient. At present, this includes most major web browsers: Firefox, Opera, and WebKit based browsers (Chrome, Safari, etc.). Until recently one needed to use a plugin for Internet Explorer but the latest IE 9 release now supports SVG.

Since introduced in 2007 (Köbben 2007), the RIMapperWMS software has been further developed to version 1.0, and is now a fully functional and compatible version 1.1.1 Basic Web Map Service. The GUI has been improved by including a layer switcher and an info tool, and is now draggable. A transcoder has been added for the ability to output PNG and JPEG, for increased interoperability and cascading WMS ability. RIMapperWMS is published under an open source license and the components, their source, as well as some documentation can be found on the RIMapperWMS website (ITC 2010a).

Other researchers have also looked into the added possibilities of SVG in a WMS environment. In the work presented by Ostländer et al., (2005), the SVG is only used to build an interactive mapping client, the maps are traditional WMS raster output wrapped in the SVG client-side. Behr et al., (2006) present a WMS with dedicated SVG-based clients (both for browsers and for mobile clients), but these client GUIs are not included in the SVG output of the WMS. The system they describe actually stores its data in SVG format. This does diminish the possibility for interoperability and loose-coupling of the WMS in a SDI environment as described in Sect. 14.1.1. Their work has been further developed into the SUAS Mapserver (see EasyWMS 2011), a project with a much broader focus than WMS only, but which does not include animated mapping.

Professor Yan Li and a group of researchers at the Spatial Information Research Center of the South China Normal University are working on a complete GIS solution based fully on XML–GML–SVG (Li et al., 2009). It is a wide-ranging project involving much more than just map services, but like Behr's prototype, it is not intended to function loosely-coupled in an SDI environment and does not include animated mapping.

Brinkhoff (2007) describes an interesting use of SVG in combination with GML and SLD for what he calls "WMS mashup services". Brinkhoffs' architecture is actually focussed on a *tight* coupling to allow users to add their own geospatial objects to WMS *images*. Although he mentions the use of SMIL for image maps, it is not used for animating the WMS layers.

We therefore feel that the prototype described in the next section is the first effort to build an OGC-compliant Web Map Service implementation that serializes spatio–temporal data from a database backend as Scalable Vector Graphics with SMIL animation.

14.3 Animated SVG Maps from a WMS: The TimeMapper Prototype

The TimeMapper prototype was designed to add animated SVG and associated GUI extensions to the existing RIMapperWMS architecture. To focus on the animated movement of objects in geographic space, we chose the visualisation of the dynamics underlying the calving and spread of icebergs. The data used is the U.S. National Ice Center (NIC) Antarctic icebergs dataset.

14.3.1 Test Case Data: Antarctic Icebergs

This data consists of movement data, gathered since 1978, of 301 icebergs (with currently 15,737 records) that can be downloaded from the NIC website (National Ice Center 2010). Besides the iceberg position in latitude and longitude, and the time of recording, the dataset contains basic descriptive information (name, size, etc.). The temporal resolution is quite irregular, ranging from 1 day between consecutive positions, to up to 50 days. The iceberg dataset has been used by different researchers with various objectives. Many have described the importance of visualising the dynamics of movement as well as calvings or splits, appearances and disappearances, of the icebergs. For example, Schmittner et al., (2002) used these as climate change indicators, and Benn et al., (2007) looked at the relationship with external factors, e.g., winds and ocean currents. To be useful for visualising the properties and relations of the data, the TimeMapper prototype would need to show animated point symbols (depicting icebergs), together with relevant topographic data, to provide an appropriate spatial reference, as well as other relevant data, such as ocean currents.

14.3.2 Extending RIMapperWMS

To extend the RIMapperWMS system to achieve the mapping of temporal data as SMIL SVG animations, we addressed the following points:

- Firstly, we designed SVG SMIL animation types we could use to represent the real-world dynamics;
- Secondly, we needed to develop an algorithm for converting the temporal component of the data, compliant with relevant standards, to the temporal designations in the SMIL format within SVG;
- Thirdly, we determined, based on animated mapping literature and experiments, which user-interface elements and which interactivity we would offer to the user, and how those could be implemented in an SVG environment;
- Finally, an appropriate architecture for the overall system was designed.

14.3.2.1 Designing the SVG SMIL Animations

In our present test case, we focus on *point* objects, i.e., the icebergs, and their changes in *position* over time. To show the movement of the objects, two main types of animations can be used, *discrete* animations and *interpolated* animations. Such animations are easy to declare using SVG SMIL. The following code snippet shows an example of how SMIL declaration works, in this case for an animation for the step-wise movement of an iceberg symbol:

```
<circle id="IB_A22B" cx="0" cy="600" r="25" >
  <animate id="XanimIB_A22B_0" attributeName="cx"
     repeatCount="none" fill="freeze" begin="0s"
     from="0" to="200" dur="2s" calcmode="discrete" />
  <animate id="YanimIB_A22B_0" attributeName="cy"
     repeatCount="none" fill="freeze"
     from="600" to="550" dur="2s" calcmode="discrete" />
</circle>
```

SMIL animations are defined by including them as children of SVG elements, in this case of the circle element depicting iceberg IB_A22B. The first animate element will lead to movement along the x-axis, the second along the y-axis. The begin attribute tells the animations when to start, and the dur attribute sets its duration and therefore its apparent speed. The calcMode attribute is set to "discrete", which means we will see a stepwise animation. In brief, the effect of these two animations is that the object is going to jump from 0 to 200 on the x-axis and at the same time from 600 to 550 on the y-axis after a count of 2 s. For a linearly interpolated animation, the calcMode attribute would have to be changed to "linear", and then the movement would be gradual over the course of the 2 s.

In our moving-object prototype, the interpolated animation type is offered to the users for them to be able to appreciate the movement of the objects taken individually or in small groups. In other words, we are interested in the dynamics of the *trajectories*. To help visualise this, we added animated tracks. SVG SMIL doesn't have a standard way of animating movement with line segments. However, members of the SVG community have developed a workaround by animating the stroke-dasharray attribute of lines, resulting in lines that are "growing" over time. The result can be seen in Fig. 14.2 as the thin lines trailing the circles depicting the icebergs. As can be seen in the code example above, the SMIL animations use a timing mechanism based on display time, that is defined in seconds. Therefore the real-world time in the spatio–temporal data has to be converted to appropriate values.

14.3.2.2 Converting the Temporal Dimension of the Data

For loose coupling of the WMS with data from arbitrary SDI nodes, we have to assume the temporal component of the data is made available in a standardised format. We therefore followed the relevant OGC specifications, which in turn are based on the ISO 8601 standard. This defines the encoding of time stamps and time periods in single strings, following the format ccyy-mm-ddThh:mm:ss.sssZ: Up to 14 digits specify century, year, month, day, hour, minute, seconds, and optionally a decimal point followed by zero or more digits for fractional seconds, with non-numeric characters to separate each piece. All times are expressed in Coordinated Universal Time (UTC) as indicated by the suffix Z (for "zulu"). Thus the timestamp 2009-10-04T14:50:58Z indicates October 4, 2009 at 2:50 pm and 58 s. In the

Fig. 14.2 Screenshot of the prototype of interactive animation of 19 icebergs in the Weddel Sea (Antarctica) from January 1, 1999 to December 31, 2004. An interactive version can be loaded from the TimeMapper website (http://geoserver.itc.nl/timemapper/)

previous section, we showed how display time is used in SMIL animation. To go from real-world time in the data to display time in the animations, the first step is to convert the ISO 8601 time stamps to a single unit time format, in our case, the number of seconds since an arbitrary fixed starting point. Next comes the scaling of time, by multiplying all time-stamps by a ratio that we call the *temporal scale*. Just as the traditional cartographic spatial scale is the ratio between map distances and real-world distances, the temporal scale in animated cartography is the ratio between display-time and real-world time (Harrower and Fabrikant 2008). Most often, as in our prototype, time will be shrunk for the viewer to visualize in a short time the events which happened during a comparatively long period.

14.3.2.3 Developing the Animated Mapping User Interface

Most theorists agree that the users of animated maps should be helped in their understanding of the information by two elements: temporal legends and ways to interact with the temporal dimension of the maps. Kraak et al., (1997), among others, state the importance of temporal legends and recognise three main types of temporal legends: digital clocks, time bars and cyclic temporal legends. All three of these legend types were developed for our prototype.

Various interactive functionalities were developed for the graphic user interface (GUI): to enable the user to choose between stepwise and interpolated animations, to add or remove interpolated tracks, and to choose which cyclic legend to view (if any). To control the temporal dimension, both a time scaling mechanism and time control tools were devised. The mechanism behind these interactive tools is a series of Javascript functions embedded in the SVG, changing the attributes of the various elements that make up the GUI and the map.

Various elements of the GUI are presented in Fig. 14.2: the *Big Hand* cyclic temporal legend has been chosen, and it is visible as a transparent circle over the whole map image with a red bar that travels the full 360° in 1 year of real-world time. By selecting *Motion Dynamics*, the user has chosen interpolated animations. The *View Tracks* option has resulted in the moving point symbols depicting the icebergs (the white circles) having their tracks traced with narrow lines. The *Speed control* on the lower-right side lets the user adjust the temporal scale, and the *time control bar* on the lower-left is for stopping, starting and brushing time.

There is limited space here to go into details on the choices for, and set-up of the various GUI elements. GUI design was based on a literature review and experiments that were part of an ITC MSc research and are described in more detail in Becker (2009).

14.3.2.4 System Architecture

To bind all elements together we used an overall system architecture that is basically an extension of RIMapperWMS (see Sect. 14.2). We therefore provide only a general overview here, as depicted in Fig. 14.3.

The core of the application tier is a set of Java Servlets that can be deployed in any J2EE compatible servlet container. The servlets do recurring tasks like extracting OGC features and attribute data from the database, translating these into fragments of SVG, SMIL and JavaScript, collecting and structuring these fragments into valid output and delivering this output to the clients. The web interfaces of the servlets comply to OGC's Web Map Service 1.1.1 standard (OGC 2002b): Firstly, *GetCapabilities*, returning an XML description of the WMS information content and acceptable request parameters and secondly, *GetMap*, returning the map itself.

The WMS is configured using a database back-end. In our prototype, this database is a set of PostGIS tables, and it can also include the spatial and attribute data to be mapped, stored according to the OGC "Simple Features for SQL" specifications (OGC 1999). Other spatio–temporal data can be retrieved from external SDI-nodes, but this has not been implemented in the current prototype.

Based on a WMS GetMap request, the servlets generate an SVG file that defines the setup of the map and its GUI: the necessary scripts for the basic WMS GUI as well as the animation GUI are loaded, and the map layers are defined as a set of nested SVG elements.

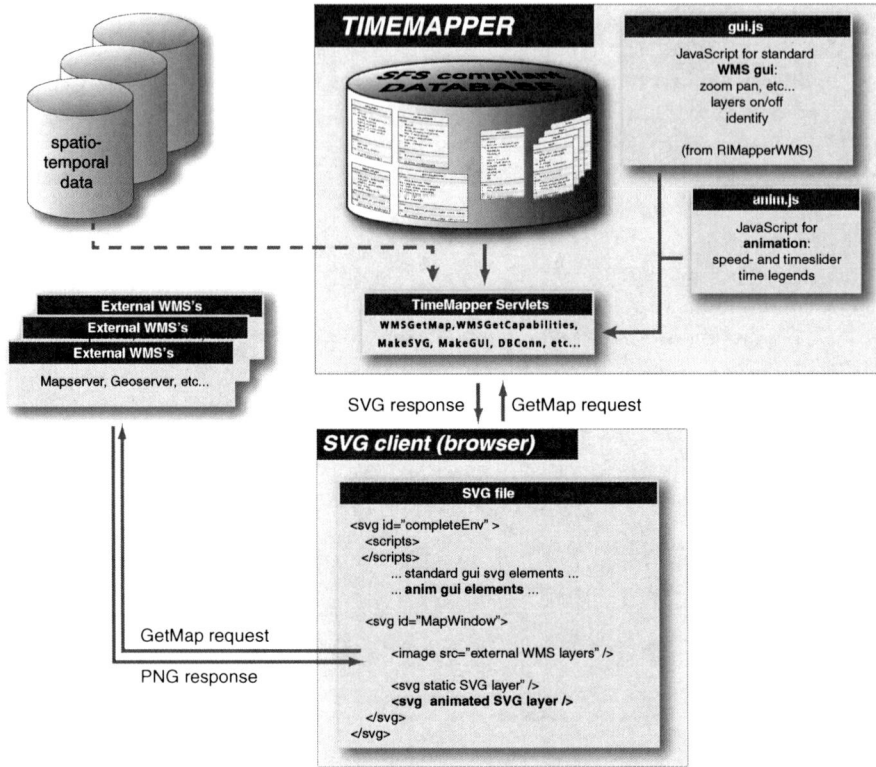

Fig. 14.3 TimeMapper system architecture overview. Note that the *dashed* link with external spatio–temporal data is not used in the current prototype

Note that the content of the map layers is not limited to the animated SVG generated by TimeMapper. External WMS layers can also be included, for instance, in the example in Fig. 14.2, the land and ice shelf are raster WMS layers from the Antarctic Cryosphere Access Portal OGC services (http://nsidc.org/agdc/acap). The GetMap requests for these external WMS's are encoded in the src attribute of an SVG image element, and subsequently retrieved asynchronously by the browser. And, of course, additional non-animated SVG layers can be retrieved from RIMapperWMS services.

14.4 Evaluation of the Prototype

With our TimeMapper prototype, a series of three test-cases was generated, all depicting the same 5 year period (1999–2003), with an increasing number of icebergs displayed. The *small* test case shows three icebergs, with a total of 218 separate data entries, or occurrences. The *middle* test case has 19 icebergs, i.e., all the icebergs present in the Weddell Sea region during the 5 years, accounting for 811 occurrences. The *big* test-case shows 99 icebergs, including all icebergs in the whole Antarctic Ocean during the 5 years, totalling 4,825 occurrences.

Testing was done using the Opera web browser, as the only web browser supporting the full SVG SMIL specification at the time of testing. Others browsers are nearing full compatibility at the time of writing, Mozilla Firefox and browsers based on the Webkit engine (e.g., Safari and Chrome) support almost all of TimeMapper's functionality. Microsoft recently announced that it will implement SVG natively starting with Internet Explorer 9, although it currently looks like SMIL animation will not be supported in the first version.

While testing the prototype with the small test case, we found all the features working as planned. The speed setting system (setting the time scale) as well as the time control slider work particularly well, and are very responsive. The user can interact with the temporal settings of the animation with hardly any delay, creating a true time-brushing effect.

However, with larger test-cases, there are important responsiveness limitations, that mostly affect the time scaling mechanism and the time control slider. Changing the speed of the animations with the middle test case takes a few seconds, after which the animations themselves work fine. But, although interacting with the time slider while the animation runs still works, it is too slow to constitute a usable exploratory tool. Finally, the big test case takes a long time (over 60 s) to load. Setting the speed of the animations and interacting with the time slider hardly works at all, although once it is finally running, the animation is still reasonably smooth.

In recent meetings at the SVG Open developers conference (http://svgopen.org), the performance issues were discussed with SVG implementors. The lack of performance in interactivity seems to be mainly due to the fact that the browser needs to manipulate large amounts of objects in the DOM tree that represents the SVG in memory. DOM manipulation is a recognized performance problem in current browser implementations. With our present animation set-up, each segment travelled by an object leads to two animation objects (one x and one y) per segment travelled. Testing is planned using an alternative coding method for the animation, such as keyTimes and KeyValues. With these SMIL methods, the whole trajectory (i.e., the sum of the segments) travelled by an object would lead to only two animation objects and we feel this may improve the responsiveness of the animations to the time slider and the speed-setting mechanism. With this change the amount of DOM manipulations necessary would be reduced, resulting in an expected performance gain. There is always going to be a limit to performance set by the total amount of objects to be animated. However, this limit is expected to be relatively high, as we observed that the large test-case of 4,825 occurrences did run fairly smoothly once the interactive setting of the parameters (and thus the DOM manipulation) was done.

14.5 Conclusion

We can conclude that combining distributed webservices and animated, interactive vector maps in a service-oriented, distributed architecture is possible with the use of existing interface, service and data standards with smaller files. We have built an OGC-compliant Web Map Service implementation that serializes spatio–temporal data from a database backend as animated Scalable Vector Graphics. The SVG is used in a web browser to show animated maps with a built-in advanced user-interface. This interface allows the user to interact with both the spatial and the temporal dimensions of the data.

A use case for the visualization of the movements of Antarctic icebergs was developed and our test-cases show that the system works as planned. There were clear performance problems in the interactivity, which makes this particular implementation less useful for exploratory purposes, where time-brushing and similar interactive features are needed. But, for less demanding tasks, where the speed of the animation can be fixed and brushing is not needed, the system could be very useful. In addition, we have identified the cause of the interactivity performance limitations, and feel we have pointers to solving them in the longer run. These results make us think that the TimeMapper system could be the starting point for a complete animated mapping system within an SDI context.

Our objective to design an animated mapping system which can generate SVG SMIL animations, automatically and directly, from spatio-temporal data, has thus been reached. Although our prototype only uses one specific data set, the employment of OGC standards and services in the system does make us confident that any other SDI-type of data could be used directly. It also works automatically, albeit in the more narrow sense explained in Sect. 14.1.1. The broader sense of fully automatic mapping from data, with cartographic design decisions included, would make a system "...simulate human action..." (from the more extensive definition in Simpson and Weiner (1989)). This still remains as an interesting research challenge that we hope to work on in the future.

References

Andrienko N, Andrienko G (2006) Exploratory analysis of spatial and temporal data: a systematic approach. Springer, Berlin

Andrienko N, Andrienko G, Gatalsky P (2003) Exploratory spatio–temporal visualization: an analytical review. J Vis Lang Comput 14:503–541

Becker T (2009) Visualizing time series data using web map service time dimension and SVG interactive animation. MSc thesis, ITC, Enschede

Behr F-J, Li H, Cheng H, Weldeslasie F, Chen Z (2006) PHPMyWMS – an open source based, SVG oriented framework for extended web map services. In: Wu H, Zhu Q (eds) Geospatial information technology – proceedings of SPIE, vol 6421. SPIE, Bellingham

Benn DI, Warren CR, Mottram RH (2007) Calving processes and the dynamics of calving glaciers. Earth Sci Rev 82(3–4):143–179

Blok CA (2005) Dynamic visualization variables in animation to support monitoring of spatial phenomena, vol 328, Netherlands Geographical Studies. KNAG/Universiteit Utrecht/ITC, Utrecht/Enschede

Brinkhoff T (2007) Increasing the fitness of OGC-Compliant Web Map Services for the Web 2.0. In: Fabrikant S, Wachowicz M (eds) The European information society, Lecture notes in geoinformation and cartography. Springer, Berlin, pp 247–264

de By RA, Lemmens R, Morales J (2009) A skeleton design theory for spatial data infrastructure. Earth Sci Inf 2(4):299–313

EasyWMS (2011) Easywms open source webgis projects. URL: http://www.easywms.com/

Fowler H, Fowler F, Sykes J (eds) (1976) Concise Oxford dictionary of current English, 6th edn. Clarendon, Oxford

Harrower M, Fabrikant S (2008) The role of map animation for geographic visualization. In: Dodge M (ed) Geographic visualization: concepts tools and applications. Wiley, Chichester, pp 44–66

ITC (2010a) RIMapper project pages. URL: http://kartoweb.itc.nl/rimapper/

ITC (2010b) TimeMapper project pages. URL: http://geoserver.itc.nl/timemapper/

Köbben B (2007) RIMapperWMS: a Web Map Service providing SVG maps with a built-in client. In: Fabrikant S, Wachowicz M (eds) The European information society – leading the way with geo-information, Lecture notes in geoinformation and cartography. Springer, Berlin, pp 217–230

Köbben B, de By R, Foerster T, Huisman O, Lemmens R, Morales J (2010) Using the SDIlight approach in teaching a geoinformatics master. Trans GIS 14(s1):25–37

Kraak MJ, Edsall R, MacEachren AM (1997) Cartographic animation and legends for temporal maps: exploration and/or interaction. In: Ottoson L (ed) Proceedings of the 18th ICA international cartographic conference, vol 1. International Cartographic Association, Stockholm, pp 253–260

Li Y, Dong X, Chi G (2009) Towards spatial information service with applied extension to SVG. In: Proceedings of new trends in information and service science, 2009 – NISS 09, Beijing, pp 474–480

National Ice Center (2010) Antarctic icebergs. URL: http://www.natice.noaa.gov/products/iceberg/

OGC (1999) OpenGIS simple features specification for SQL revision 1.1. Technical Report OGC 99-049, Open Geospatial Consortium

OGC (2002a) Implementations by specification. URL: http://opengeospatial.org/resource/products/byspec/

OGC (2002b) Web map service implementation specification (1.1.1). Technical Report OGC 01-068r3, Open Geospatial Consortium

OGC (2006) Web map server implementation specification 1.3.0. Technical Report 06-042, Open Geospatial Consortium

Ostländer N, Tegtmeyer S, Foerster T (2005) Developing an SDI for time-variant and multilingual information dissemination and data distribution. In: Proceedings of 11th EC-GI & GIS Workshop ESDI: setting the framework, Alghero, Sardinia, European Commission

Schmittner A, Yoshimori M, Weaver AJ (2002) Instability of glacial climate in a model of the ocean–atmosphere-cryosphere system. Sci Exp 295(5559):1489–1493

Simpson J, Weiner E (eds) (1989) Oxford English dictionary, 2nd edn. Clarendon, Oxford

W3C (2008) Synchronized multimedia integration language (SMIL 3.0). URL: http://www.w3.org/TR/SMIL/

W3C (2010) Scalable vector graphics (SVG) 1.1, 2nd edn. URL: http://www.w3.org/TR/SVG11/

Chapter 15
The Possibilities of Globe Publishing on the Web

Mátyás Gede

Abstract This paper discusses the different techniques one can use to publish digital models of real globes on the web.

The described solutions form two groups. The first one includes various KML/KMZ models viewable in geo-browsers or using the Google Earth plug-in, while the second one consists of VRML/X3D models. The main advantages and disadvantages of each solution are also discussed.

Although the main purpose of these techniques is to publish digital models of old globes, any other kind of global raster data can be visualized in the same way.

15.1 Introduction

Old globes are very peculiar things. These objects not only show the geographic knowledge of their time but they are artistic works. Unfortunately, globes are rather vulnerable objects; therefore, most of the older ones are kept in museums, where visitors can observe them only from a secure distance. But what is the use of a globe if one cannot spin it around or have a closer look to examine its details?

The Virtual Globes Museum project (Márton 2007) was launched to solve this problem. On its web page, steerable-zoomable three-dimensional models of near 100 globes are displayed forming a "virtual exhibition".

The following pages describe the techniques the author used to bring these models to the computer screen.

To realize these solutions, a digitised globe map in Plate Carrée is required. The methods of producing such a map is not described here; for detailed information on the topic see the references (Adami and Guerra 2008; Gede 2009, 2010; Hruby et al., 2006).

M. Gede (✉)
Department of Cartography and Geoinformatics, Eötvös Loránd University, Budapest, Hungary
e-mail: saman@map.elte.hu

15.2 Required Materials, Software

The most important material needed is a globe map in Plate Carrée (Fig. 15.1) or another global raster, e.g., a relief map (Fig. 15.2).

Sample globe maps and most of the examples can be downloaded from the following web site: http://mercator.elte.hu/~saman/webglobes/.

A text editor is essential if one tries the following examples. The author prefers *Notepad++*, which is free software. This program can be extended with additional language files to support syntax highlight for several file formats, including VRML and X3D as well.

There will be need of gridding large images to smaller parts, and eventually to perform a projection transformation. A very useful tool for this and several other geographical data-transforming tasks is *Global Mapper* (Global Mapper 2011), which is unfortunately commercial software. Naturally, it is possible to find free tools also on the web such as utilities of the *GDAL* library (GDAL 2011).

To view VRML/X3D models an appropriate player or plug-in is needed. See the references for the details.

15.3 KML/KMZ "Globe Layers"

If one hears the "globe" and "internet" phrases together, the first thought will be one of the well-known geo-browsers like Google Earth. An easy but impressive solution of globe publishing is to use one of these programs for visualizing. The big advantage of this solution is that the content of the digitised globe can be examined together with other geodata by using the program's predefined layers or loading other data files.

Although one can usually load a globe map image into these programs using menu commands (e.g., "Add Image Overlay" command in Google Earth, see Fig. 15.3), it is advisable to create a KML/KMZ file to define the bounding rectangle of the image to help those who cannot or do not want to do this task.

15.3.1 About the KML/KMZ Format

The KML (Keyhole Markup Language) is an XML-based file format to describe different types of geographic data. It is recognized by all geo-browsers and many GIS software. Only a small subset of its features is discussed here, but the full documentation of the format can be found on the web (Google 2011a). KMZ is the compressed version of KML; it can be created by simply zipping the KML file and all the referenced files together, and change its extension to ".kmz".

15 The Possibilities of Globe Publishing on the Web 221

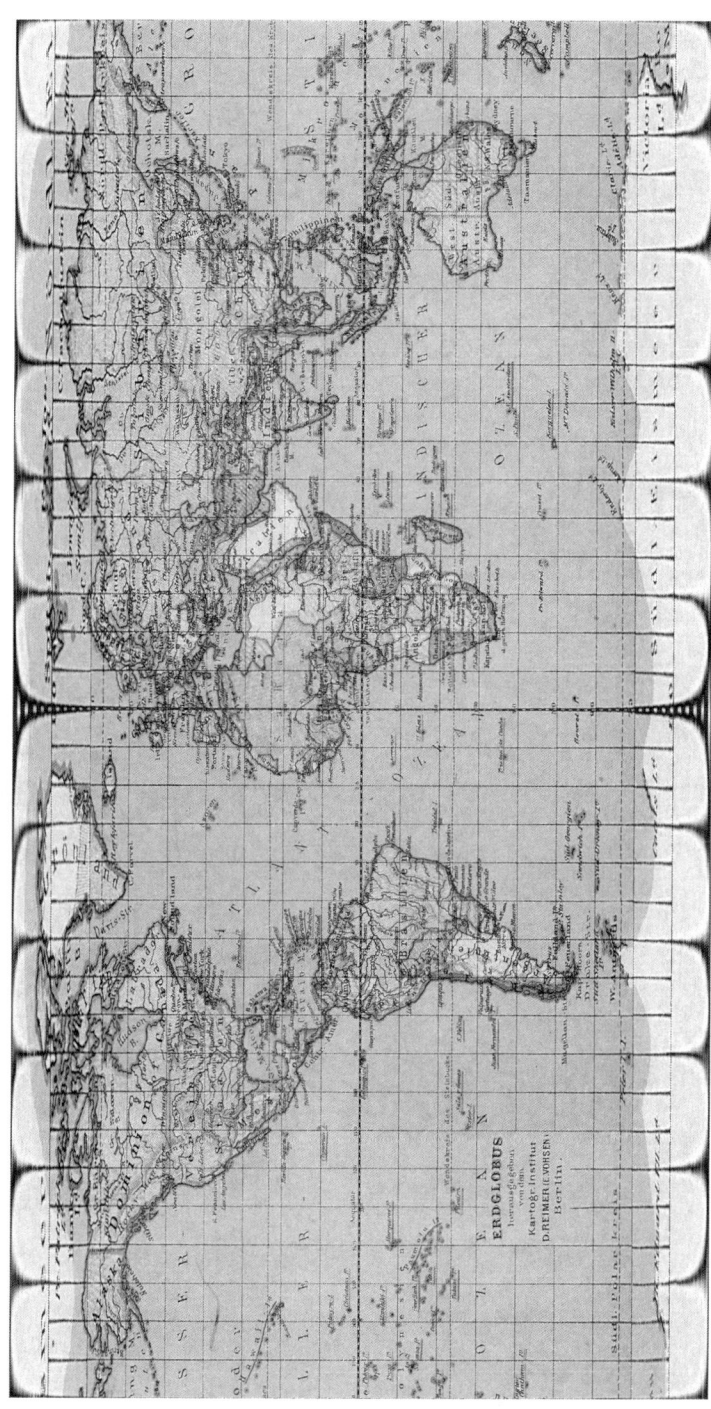

Fig. 15.1 Plate Carrée map of an old globe

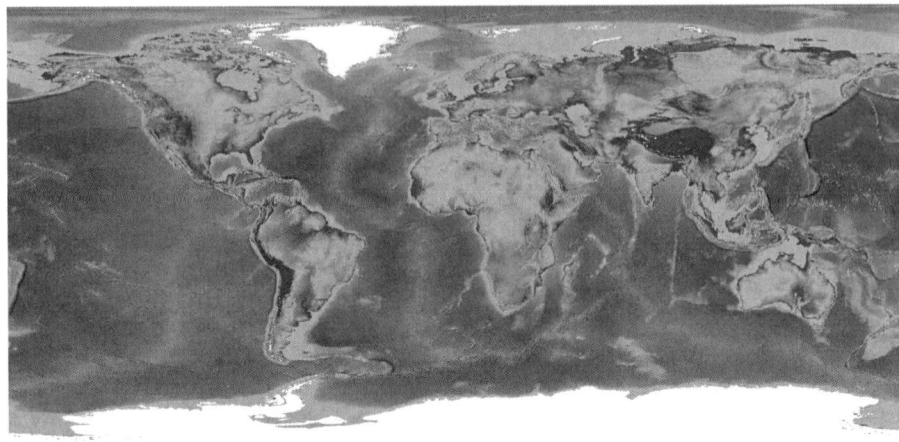

Fig. 15.2 Plate Carrée relief map

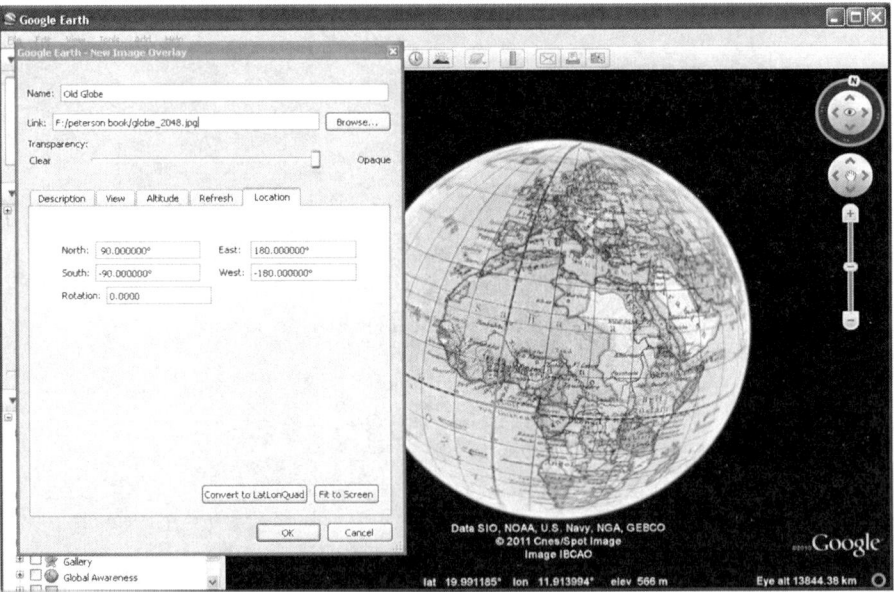

Fig. 15.3 Using the "Add image overlay" dialog of Google Earth to add a globe map

15.3.2 The Simple Solution

If the map image is not too large, we can simply stretch it to the globe using the following KML code:

15 The Possibilities of Globe Publishing on the Web

```xml
<?xml version="1.0" encoding="UTF-8"?>
<kml xmlns="http://earth.google.com/kml/2.2">
 <Document>
  <name>Old globe</name>
  <GroundOverlay>
   <Icon><href>globe_map.jpg</href></Icon>
   <LatLonBox>
    <north>90.00</north>
    <south>-90.00</south>
    <west>-180</west>
    <east>180</east>
   </LatLonBox>
  </GroundOverlay>
 </Document>
</kml>
```

Let's save this code with ".kml" extension and place it into the same folder with the referenced image file (here *globe_map.jpg*). Finally, let's create a KMZ file by zipping the KML and the image file together into an archive named e.g., "globe.kmz". Now it can be loaded with a geo-browser to see the globe (Fig. 15.4). It is advisable to turn off atmospheric effects to clearly see the globe.

If our globe map was made using other prime meridian than Greenwich (European globes older than 1884 often use the Ferro prime meridian), we should slightly modify this code, using the appropriate values for the *west* and *east* fields. These values can be less than $-180°$ or more than $180°$ if necessary. (This is the only solution if a bounding box stretches over the $\pm 180°$ meridian.)

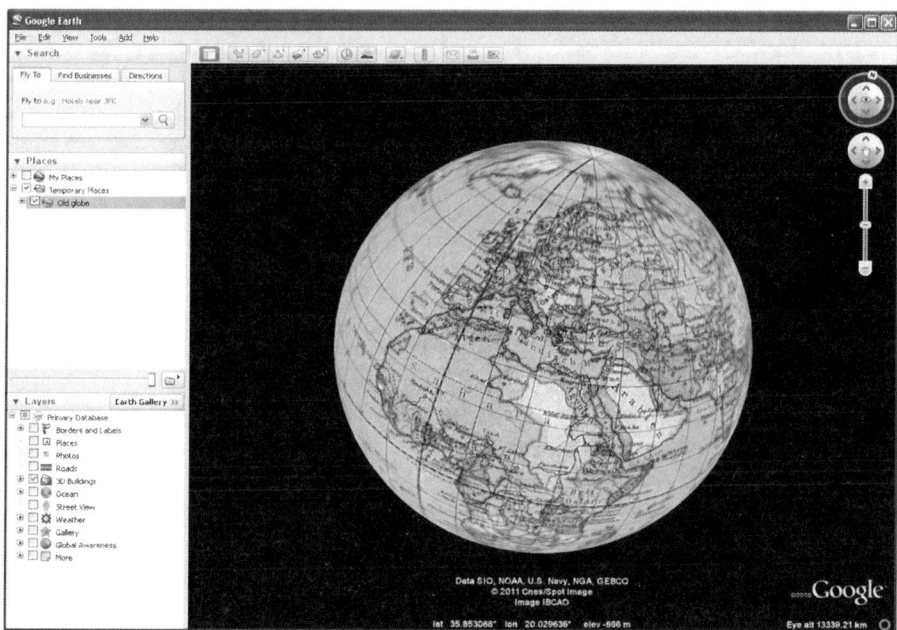

Fig. 15.4 KMZ "Globe layer" in Google Earth

15.3.3 Using Gridded Files

If the globe map size is larger than 2,048 × 2,048 pixels, the image is needed to be gridded, or the browser may reduce its resolution. The KML file for a gridded image file set looks similar to the previous one, but there are multiple *GroundOverlay* nodes (one for each grid cell), each one with an appropriate bounding *LatLonBox* field.

If there are more than two rows in the grid, the total file size can be reduced by taking advantage of the fact that the longitude lines are shorter at higher altitudes. Therefore, the images may have lower horizontal resolution. For example, we can reduce the horizontal resolution of a map part depicting an area between the pole and the 60th latitude to the half of its vertical resolution. The reduction however, should be used cautiously as resampling the images may cause undesirable interference (Moire effect).

15.3.4 Using Multiple Levels of Detail

The KML format allows the use of multiple levels of detail (LOD). This phrase means that different resolution images are used when an area is viewed from a longer or a shorter distance. The advantage of this solution is that there is no need to load a high-resolution image for the whole globe, and rendering can be faster.

To define LODs in KML, a hierarchical file structure is needed, as LODs can be defined to *Region* nodes, and have effect on the sibling geometry of those *Region*s (i.e., nodes that are in the same grouping node). The following code instructs the viewer to show the image only if it occupies an area wider than 400 pixels on the screen:

```
<Folder>
 <Region>
  <LatLonAltBox>
   <north>90</north>
   <south>78</south>
   <west>-137.66</west>
   <east>-77.66</east>
  </LatLonAltBox>
  <Lod>
   <minLodPixels>400</minLodPixels>
  </Lod>
 </Region>
 <GroundOverlay>
  <Icon><href>g_1_2.jpg</href></Icon>
  <LatLonBox>
   <north>90</north>
   <south>78</south>
   <west>-137.66</west>
   <east>-77.66</east>
  </LatLonBox>
 </GroundOverlay>
</Folder>
```

Global Mapper provides an easy way of creating KML globe overlays with multiple levels of detail: the gridded image set and the describing KML file can be created with one simple menu command.

Another utility for this task is the open source *gdal2tiles*, part of *GDAL*.

15.3.5 Adding Viewpoints

We can add predefined viewpoints to the KML globe model. This is useful if we want to draw attention to details that are not conspicuous e.g., the colophon (title field) of the globe. The viewpoints are realized as *Placemark* nodes without any geometry but having a *LookAt* node:

```
<Placemark>
 <name>Europe</name>
 <LookAt>
  <longitude>20</longitude>
  <latitude>50</latitude>
  <range>4000000</range>
  <heading>0</heading>
 </LookAt>
</Placemark>
```

15.3.6 Network Links

Network links provide the possibility of auto-updating the published geographic information. The idea is simple: the KML document contains references to on-line geodata stored in other KML or KMZ files. Even if the KML file is downloaded and stored locally (or e.g., added to *My Places* in Google Earth), the referenced information is refreshed at every load (or periodically, if defined so).

A simple example to illustrate the usage of network links:

```
<?xml version="1.0" encoding="UTF-8"?>
<kml xmlns="http://earth.google.com/kml/2.0">
 <NetworkLink>
  <name>Old Globe</name>
  <Link>
   <href>http://mercator.elte.hu/~saman/webglobes/globe_grid.kmz</href>
  </Link>
  <visibility>1</visibility>
 </NetworkLink>
</kml>
```

Network links can be refreshed periodically (e.g., once a day or in every 5 s), which can be useful for quickly changing data. Detailed information can be found in the online KML reference (Google 2011a).

15.3.7 Taking Advantages of Google Earth Plug-In

A very convenient way of visualizing a KML globe layer is to embed a Google Earth globe into a HTML page using the software's plug-in version (Google 2011b). The JavaScript API of this plug-in allows developers to add rich functionality to the globe. It is possible, for example, to add a searchable place name index where clicking on a name turns and zooms the globe to the specified feature.

The following example simply embeds a virtual globe and adds a select control with a few predefined viewpoints (see the result in Fig. 15.5):

```
<!DOCTYPE HTML>
<html>
 <head>
  <title>KML Globe layer visualized on Google Earth Plugin</title>
  <script src="http://www.google.com/jsapi?key=key_comes_here"> </script>
  <script type="text/javascript">
   var ge,vp;
   google.load("earth", "1");

   function init() {
     google.earth.createInstance('map3d', initCB, failureCB);
   }
```

Fig. 15.5 KML globe layer visualized on Google Earth plug-in

15 The Possibilities of Globe Publishing on the Web

```
    function setVP(latLon) {
      vpt=latLon.split(',')
      vp.set(vpt[0]*1, vpt[1]*1, 6371000,
             ge.ALTITUDE_RELATIVE_TO_GROUND, 0, 0, 0);
      ge.getView().setAbstractView(vp);
    }

    function initCB(instance) {
      ge = instance;
      ge.getWindow().setVisibility(true);
      // add a navigation control
      ge.getNavigationControl().setVisibility(ge.VISIBILITY_AUTO);
      // no atmosphere
      ge.getOptions().setAtmosphereVisibility(false);
      // no terrain
      ge.getLayerRoot().enableLayerById(ge.LAYER_TERRAIN, false);
      function finished(object) {
        if (!object) {
          // wrap alerts in API callbacks and event handlers
          // in a setTimeout to prevent deadlock in some browsers
          setTimeout(function() {
            alert('Bad or null KML.');
          }, 0);
          return;
        }
        ge.getFeatures().appendChild(object);
        setVP("50,20"); // set the viewpoint to Europe
      }
      // fetch the KML
      google.earth.fetchKml(ge, 'http://**HERE_COMES_THE_KMZ_URL**', finished);
      vp=ge.createLookAt('');
    }

    function failureCB(errorCode) {
    }

    google.setOnLoadCallback(init);
  </script>
 </head>
 <body>
  <h1>KML Globe layer visualized on Google Earth Plugin</h1>
  <b>Viewpoint:</b>
  <select onchange="setVP(this.value)">
   <option value="50,20">Europe</option>
   <option value="90,0">North Pole</option>
   <option value="0,20">Africa</option>
   <option value="40,-90">North America</option>
   <option value="-30,-40">South America</option>
   <option value="-90,0">South Pole</option>
   <option value="-25,150">Australia</option>
   <option value="45,90">Asia</option>
  </select>
  <hr/>
  <div id="map3d" style="height: 600px; width: 700px;"></div>
 </body>
</html>
```

To use the Google Earth plug-in we have to obtain a key for every domain. It is free of charge and can be done via the website of the plug-in.

An important note: Google Earth API currently does not support local KML files; the URL must be absolute, e.g.: http://some.url/globe.kml.

This solution supports KMZ format also, therefore it is advisable to create KMZ packages instead of a set of separate images with a KML file.

15.4 VRML/X3D Globes

Although the use of KML globe layers on virtual globes is very impressive, this solution has a few disadvantages. The equirectangular maps as textures are usually distorted at polar regions due to the fact that a pole line is mapped to a pole point on the globe (Fig. 15.6). Another problem is that special non-spherical objects (e.g., a detachable Earth-model, Fig. 15.12) cannot be visualized in this way. Furthermore, to visualize the KML globes on a web page, the Google Earth plug-in and its JavaScript API are needed, which requires a rather high network bandwidth to work properly, and cannot operate in off-line mode.

A solution to these problems is to create the globes using a 3D modeling language. The author uses VRML and X3D as these are widely supported formats.

15.4.1 About the Formats

VRML (Virtual Reality Modeling Language) is a standard 3D modeling format. It was designed particularly with the WWW in mind. Although it has been superseded

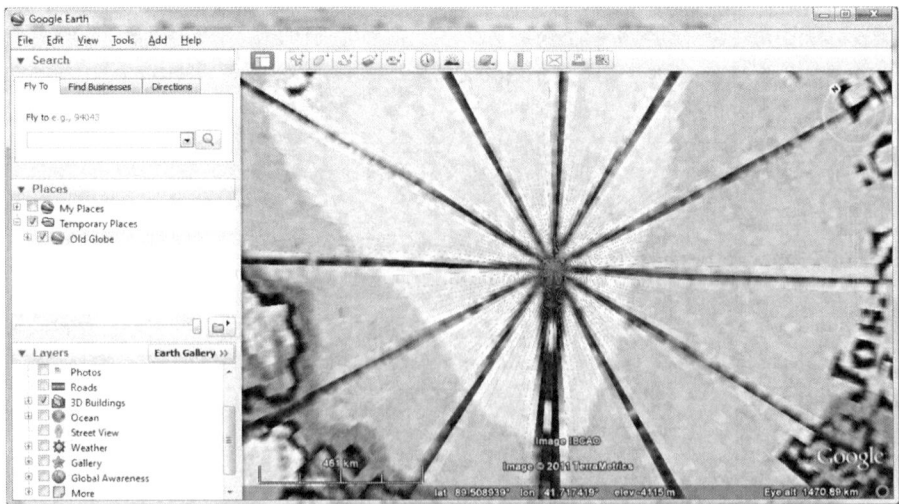

Fig. 15.6 Imperfect texture mapping at the poles in Google Earth

with its XML-based successor, X3D (Web3D 2011), nowadays both formats are still used. As all the features of VRML can be realized in X3D also, only the first code example is shown in both languages. All the latter VRML codes can be easily converted to X3D with an automatic converter tool, e.g., Instant Labs online X3D encoding converter (InstantLabs 2011).

This text does not intend to provide a comprehensive VRML guide. For those who need more background knowledge, there are very good on-line tutorials e.g., (VapourTech 1999).

15.4.2 Using the Built-In Sphere Node

The easiest way to create a "virtual globe" in VRML or in X3D is the use of the built-in Sphere node:

```
#VRML V2.0 utf8

Shape {
  geometry Sphere { }
  appearance    Appearance {
    texture ImageTexture {url "globe.jpg"}
  }
}
```

```
<?xml version="1.0" encoding="UTF-8"?>
<!DOCTYPE X3D PUBLIC "ISO//Web3D//DTD X3D 3.0//EN"
  "http://www.web3d.org/specifications/x3d-3.0.dtd">
<X3D xmlns:xsd='http://www.w3.org/2001/XMLSchema-instance' profile='Full' version='3.0'
     xsd:noNamespaceSchemaLocation='http://www.web3d.org/specifications/x3d-3.0.xsd'>
  <Scene>
    <Shape>
      <Appearance>
        <ImageTexture url="globe.jpg"/>
      </Appearance>
      <Sphere/>
    </Shape>
  </Scene>
</X3D>
```

The globe.jpg file contains the globe map in Plate Carrée. Luckily, the texture mapping rules of the Sphere node fit for this solution, and the image will be mapped correctly (Fig. 15.7).

To embed a VRML or X3D model into a HTML page we must use the *object* tag:

```
<object data="globe.wrl" type="model/vrml" width="600" height="600">
 <param name="src" value="globe.wrl">
</object>
```

The parallel use of the *data* attribute and the *src* parameter is required to work properly in different browsers.

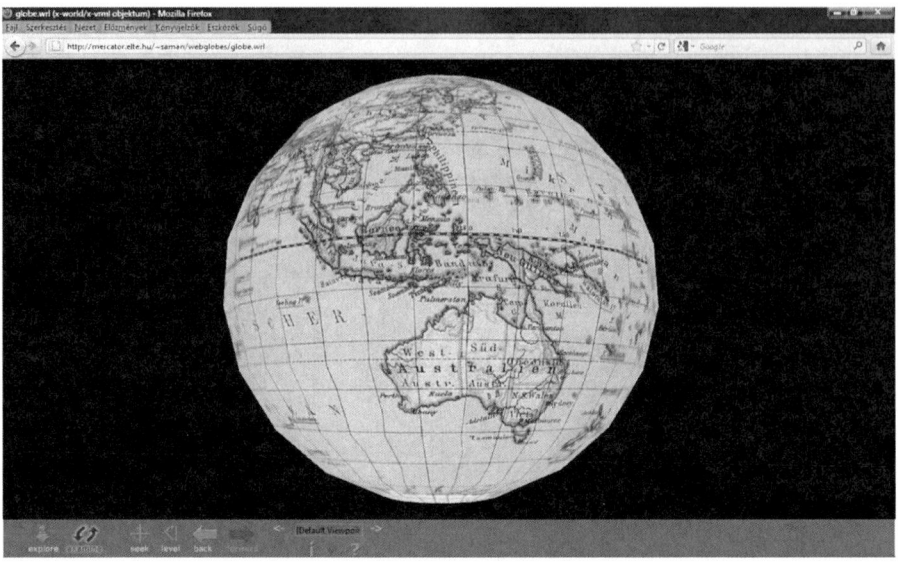

Fig. 15.7 Simple VRML globe using the Sphere node

15.4.2.1 Transformations

A minor problem of this solution is that the axis of the globe is not along the Z axis of the coordinate system but along the Y one. Although it is not very important while the globe stands alone, it is not convenient when additional viewpoints are defined or if the globe is only a part of a bigger model.

The solution is to nest the shape in a Transform node as follows:

```
Transform {
  rotation 1 0 0 1.57
  children [
    Shape {
      geometry Sphere { }
      appearance Appearance {
        texture ImageTexture {url "globe.jpg"}
      }
    }
  ]
}
```

This transformation rotates the sphere around its X axis by 90° (1.57 rad), causing the globe axis coincidence with the global Z axis.

15.4.2.2 Problems with the Sphere Node

There are a few problems with the solution described above. First, the texture image size: computers with lower graphical performance may not support textures larger than 2,048*2,048 pixels, which is insufficient for the detailed presentation of a larger globe. Another problem is that the Sphere node is not a real sphere but a polyhedron. The interference between the face edges of this polyhedron and the grid lines of the globe map results in zig-zagging lines at higher latitudes. Finally, the problem of mapping the pole line to a point cause problems at the poles again (Fig. 15.8).

15.4.3 Building Your Own Globe Using IndexedFaceSet

Due to the problems of the Sphere node it is advisable to define a new spherical shape which fits better to our aims.

15.4.3.1 How to Define a Sphere?

Naturally, the globe is modeled by a polyhedron, but with flat faces matching 10°*10° geographic quadrangles (or triangles at the poles) to prevent the

Fig. 15.8 Texture mapping problems of the Sphere node at higher altitudes and the poles

interference of face edges and grid lines. This solution grants that the grid lines run along the edges of the shape as most globes have a graticule density of 10°.

A shape can be defined using the IndexedFaceSet node. This node requires a list of vertices, described by 3D Cartesian coordinates, a list of faces, and optionally a list of texture point coordinates (2D Cartesian), one texture point for each vertex for adequate texture mapping. The faces are defined by the indexes of vertices forming the face.

For example, the definition of an octahedron is the following (Fig. 15.9 shows the resulting shape):

```
geometry IndexedFaceSet {
  coord Coordinate {
    point [ 0 0 1, 1 0 0, 0 1 0, -1 0 0, 0 -1 0, 0 0 -1 ]
  }
  coordIndex [ 0 1 2 -1, 0 2 3 -1, 0 3 4 -1, 0 4 1 -1,
               5 2 1 -1, 5 3 2 -1, 5 4 3 -1, 5 1 4 -1]
}
```

The globe definition is similar to this, only there are more points and faces. The Cartesian coordinates of a specific point of a sphere described with its latitude/longitude coordinates (φ, λ) are as follows:

$x = R \cos \varphi \cos \lambda$
$y = R \cos \varphi \sin \lambda$
$z = R \sin \varphi$

where R means the radius of the sphere, which is practically equal to 1. It is advisable to write a small script to generate the point coordinates, as it would be a hard work to do it manually.

In order to hide the face edges, it is advisable to set the creaseAngle property of the IndexedFaceSet to an appropriate value (e.g., 1). If the angle of faces is smaller than this value, the renderer smoothes the edge between them (Fig. 15.10).

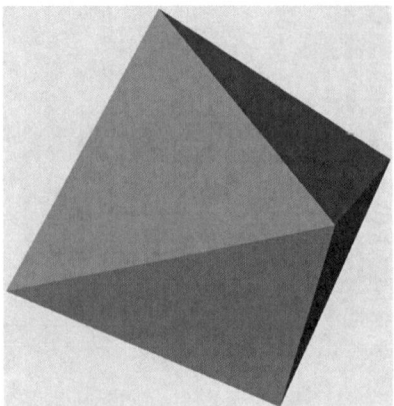

Fig. 15.9 An octahedron defined by IndexedFaceSet

Fig. 15.10 The same half-sphere without (*left*) and with *creaseAngle* value set (*right*)

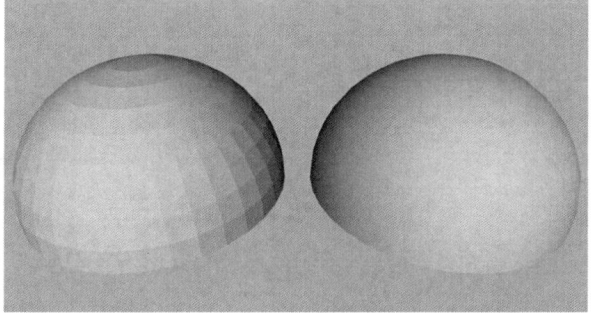

15.4.3.2 Dividing the Globe to Sub-surfaces

Texturing a globe with only one image always results in line-to-point texture-mapping problems i.e., there will be at least one line on the texture image which is mapped into one single point. This is rendered usually imperfectly (Fig. 15.8).

To prevent this problem, the sphere has to consist of two or more parts. This solution has a positive side-effect also: as some of the parts will be identical, differing only in position and texture, the VRML (or X3D) code can be shorter.

Two Hemispheres

The easiest solution is to define two hemispheres and draping them with images in polar azimuthal projection.

Two Polar Caps and Four Equatorial Quadrangles

A more sophisticated solution is to define six parts: two polar caps delimited by specified latitude lines (textured with polar azimuthal projection images), and four identical quadrangles (textured with Plate Carrée maps) together forming the equatorial belt around the Equator (Fig. 15.11).

The benefit of this model is that more texture is used to drape one globe so the size of one texture can be smaller (or the resolution can be higher using the same size). If the delimiter latitudes are set to $\pm 45°$, all the images are rectangular and they depict areas of identical size ($90°*90°$ quadrangles and polar caps with $90°$ diameter), so if there is any software or hardware limit of the texture sizes, this limit means the same resolution of each image.

As the connecting lines of the different globe parts are not always rendered perfectly, it is advisable to set the boundary latitude to match a latitude line of the globe map. Due to this reason, this latitude is usually set to $\pm 50°$.

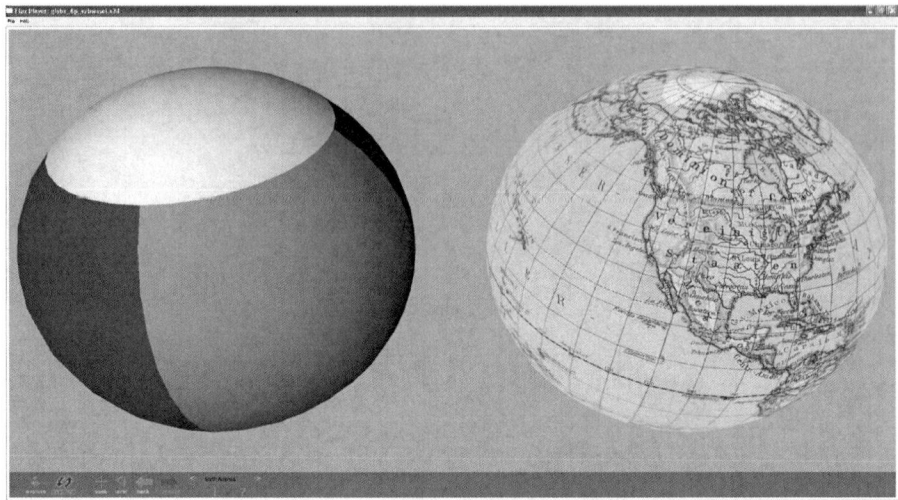

Fig. 15.11 Globe model made up of six parts without (*left*) and with (*right*) textures

15.4.4 Adding Viewpoints

As the navigation with the mouse in the virtual space requires some skill and practice, it is very useful to define some pre-programmed viewpoints to the globe models. This can be done by adding Viewpoint nodes by the following code:

```
transform {
  rotation -x -y -z -π/2 -λ
  children [
    viewpoint {
      position x y z
      orientation -y x 0 π/2-φ
      fieldOfView 0.8
      description "name of viewpoint"
    }
  ]
}
```

where (φ, λ) are the geographical coordinates of the centre of the viewed area (in radians), and

$x = R \cos \varphi \cos \lambda$
$y = R \cos \varphi \sin \lambda$
$z = R \sin \varphi$

R is the distance of the viewpoint from the globe centre.

The 0.8 value of the *fieldOfView* property can be modified, but the author had found this to be the perfect value from aesthetic aspects. The best value of R depends on the area to be visualized. R = 2.5 fits for viewpoints of whole continents. When modeling other planets or the Moon, high R and low *fieldOfView* value pairs will result in views similar to the ones when the planets are observed from the Earth.

As the calculations required by the viewpoint codes have to be done for each predefined viewpoints, it is advisable to create a simple program to generate these codes automatically for any given parameter sets.

15.4.5 Special Objects

The possibilities of VRML and X3D allow us to create special shapes also. A good example is the detachable structural earth model of the Hungarian Cartographia company (Fig. 15.12).

The modeling languages also provide tools to make models interactive. Using these features, it is possible to create such complex models like a combined terrestrial-celestial globe, which can be opened and closed by clicking its appropriate point (Fig. 15.13).

15.4.6 Integrating Our Models to Web Pages Using X3DOM

A very fresh development allows us to integrate our models into web pages without the need of VRML or X3D plug-ins. The X3DOM open source framework of the Fraunhofer Institute (X3DOM 2011) provides a JavaScript library which enables the insertion of X3D code into HTML codes and even makes possible to manipulate X3D nodes as a part of the document object model in browsers supporting WebGL.

Although this tool is very promising, currently there are severe limitations: by this time (March, 2011) only the most decent browsers (like Firefox 4 or Chrome 9) support WebGL natively. The texture images must be $2^n * 2^n$ in size, and there is a limitation in the number of vertices too.

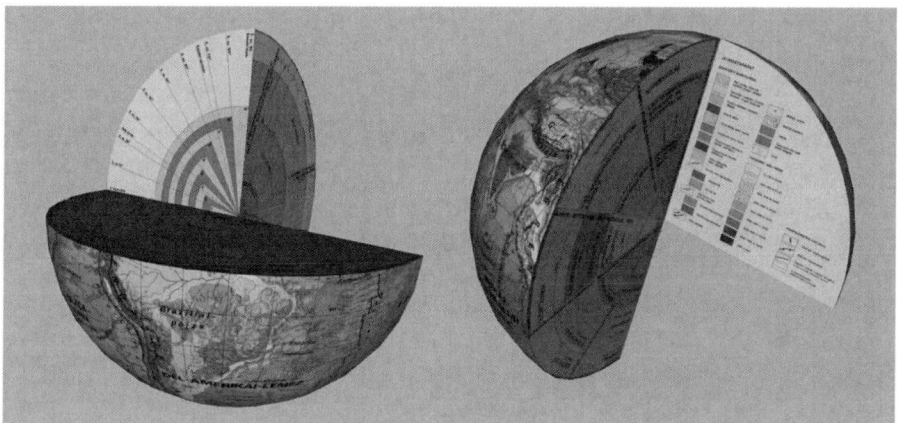

Fig. 15.12 The two parts of the detachable structural earth model

Fig. 15.13 Combined terrestrial-celestial globe (*left*) and its fully functional VRML model (*right*)

15.4.6.1 Defining Globes in X3DOM

The simplest HTML page containing a globe model in X3DOM (similar to the first VRML/X3D example) has the following code (see the resulting web page in Fig. 15.14):

```
<!DOCTYPE HTML>
<html>
 <head>
  <title>X3DOM globe model</title>
  <link rel="stylesheet" type="text/css"
        href="http://www.x3dom.org/x3dom/release/x3dom.css">
  </link>
  <script type="text/javascript"
        src="http://www.x3dom.org/x3dom/release/x3dom.js">
  </script>
 </head>
 <body>
  <h1>X3DOM globe model</h1>
  <x3d width="600px" height="400px">
    <scene>
      <shape>
        <appearance>
          <imageTexture url="globe.jpg"/>
        </appearance>
        <sphere/>
      </shape>
    </scene>
  </x3d>
 </body>
</html>
```

15 The Possibilities of Globe Publishing on the Web

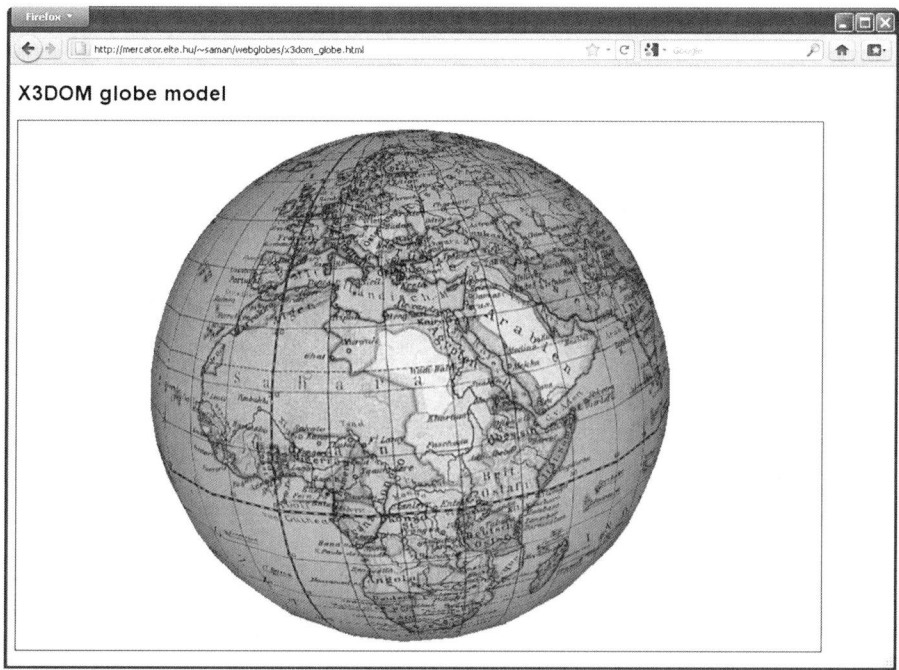

Fig. 15.14 A simple web page using X3DOM

15.4.6.2 Adding Interaction

The X3DOM framework allows developers to handle X3D nodes just like HTML DOM elements. This makes things rather easy as any interactivity can be driven by simple Javascript codes. The following example defines a JavaScript function, which controls the settings of a Transform node specified by its *id* attribute:

```
...
<script>
 function setTransformParameters(tx,ty,tz,rx,ry,rz,ra)
 {
   with(document.getElementById('tr1'))
   {
     setAttribute('translation',[tx,ty,tz]);
     setAttribute('rotation',[rx,ry,rz,ra]);
   }
 }
</script>

...
<transform id="tr1">
...
</transform>
```

As the viewing canvas of X3DOM framework does not support manual viewpoint selection, the handle of viewpoints can be realized in similar ways: changing the attributes of a Viewpoint node by JavaScript code.

Naturally, manipulating existing nodes is not the only possibility. X3D nodes can be created and removed dynamically. These features give the opportunity to create highly interactive models integrated to web pages. Those who are interested in this topic can find more detailed information on the web page of X3DOM (see references).

Acknowledgements The European Union and the European Social Fund have provided financial support to the project under the grant agreement no. *TÁMOP 4.2.1./B-09/1/KMR-2010-0003*.

References

Examples in this text and more resources on the topic including links to VRML/X3D players and plug-ins are available at the following URL: http://mercator.elte.hu/~saman/webglobes/

Globe Digitising

Adami A, Guerra F (2008) Coronelli's virtual globe. E-Perimetron 3(4):243–250, http://www.e-perimetron.org/Vol_3_4/Adami_Guerra.pdf
Gede M (2009) Publishing globes on the internet. Acta Geod Geophys Hung 44(I):141–148
Gede M (2010) The use of the Nelder–Mead method in estimating projection parameters for globe photographs. Acta Geod Geophys Hung 45(I):17–23
Hruby F, Plank I, Riedl A (2006) Cartographic heritage as shared experience in virtual space: a digital representation of the earth globe of Gerard Mercator (1541). E-Perimetron 1(2):88–98, http://www.e-perimetron.org/Vol_1_2/Hruby_etal/Hruby_et_al.pdf
Márton M, (2007) Virtual globes museum. http://vgm.elte.hu

File Formats, Software

GDAL (2011) Geospatial data abstraction library. http://www.gdal.org/
Global Mapper (2011) http://globalmapper.com/
Google (2011a) KML reference. http://code.google.com/intl/hu-HU/apis/kml/documentation/
Google (2011b) Google Earth API. http://code.google.com/intl/hu-HU/apis/earth/
InstantLabs (2011) X3D encoding converter. http://doc.instantreality.org/tools/x3d_encoding_converter/
VapourTech (1999) Floppy's VRML guide. http://rvirtual.free.fr/programmation/VRML/tuto_eng/
Web3D (2011) X3D for developers. http://www.web3d.org/x3d/
X3DOM (2011) X3DOM 1.0 – Instant 3D the HTML way! http://www.x3dom.org/

Part IV
Applications

Chapter 16
Mapping Social-Network Interactions

James O'Brien and Kenneth Field

Abstract Blogs, micro-blogs and online forums are fundamental building blocks of an interconnected world. They provide a mechanism for people to communicate details of their lives and the spatial locations of their activities. Desktop, online and mobile mapping APIs have never been so rich yet this presents challenges to build applications that blend meaningful content with visual appeal.

Here, we begin by examining the series of APIs needed to collect this spatial expression of micro-blogging from the social networking tool Twitter. To create cartographically appropriate and semantically relevant 'twitter maps' we blend functionality and data from APIs by Esri, Google, Twitter and others.

We then demonstrate how to leverage the available APIs to create an interactive application enabling real – time mapping of students undertaking mobile data collection exercises. Two examples are presented: a "race" monitoring application focussing on extracting and mapping temporal variables and a category building, asynchronous collaborative land – use mapping exercise where the semantic content and location of tweets is emphasized.

16.1 Introduction

There have been many recent developments in online mapping environments due to several factors. The main driver has been the rapid increase in the availability of online map and data services (e.g., Googlc Maps, Yahoo Maps, Microsoft Bing and Esri ArcGIS Server) and the ease with which non-experts can produce online spatial content using Web 2.0 tools. A parallel period of growth in social networking such as blogs, micro-blogs (e.g., Twitter posts and Facebook status updates) and online forums has created an increasingly interconnected world; where place is often at the

J. O'Brien (✉)
Risk Frontiers, Macquarie University, NSW 2109, Australia
e-mail: james.obrien@mq.edu.au

forefront of the discussions taking place in these virtual worlds. This increasing recognition of place is highlighted by the implementation of geolocation in Twitter and Facebook Places and events like O'Reilly's Where 2.0.

These social networking location sharing features reflect a growing desire for maps depicting our location, navigating us, providing spatial context for our activities and defining our online presence. Instead of remaining spatially anonymous in online environments, many people choose to reveal their locations (Gibin et al., 2008).

Finally the increasing availability of free data in many countries creates opportunities for mapping previously unavailable data. Coupled with the widespread use of social networks, the variety of internet mapping services and the abundance of programming interfaces for interacting with all three (data, social networks and map services) it is not surprising that "mashups" of spatial locations and aspatial data (such as conversations, images and video) are popular. Georeferencing of photos and video (available from approximately 2005) is not a new concept but the mapping of contextual information derived from blogs and microblogs (such as Twitter) is more recent.

This online spatial presence mirrors the move of GI Systems to the Internet. As Chow (2008) notes the use of GIS has shifted from an isolated, proprietary single user implementation to an interoperable, distributed, open, Internet environment. This increase in visibility and access to GIS has accompanied technological advancements in location aware mobile devices and has raised the awareness of GIS among the general public (Simon et al., 2007). The wide use of spatially aware devices supporting applications such as Twitter, Facebook, SMS, photo and video capture and sharing allows a greater proportion of the population to spatially reference the data they collect (volunteered geographic information).

This chapter builds on these trends of spatially enabled, interacting, data sharing user communities and briefly examines the programming interfaces of each of these silos of data (open resources, social networks and internet map services) in isolation before examining approaches for linking them together. We develop a multi-purpose, extensible Web 2.0 application through iterative task focussed exercises (land use mapping, conference discussions, a return to the land use mapping and a "race" exercise) and these exercises are then used to demonstrate how these applications can be developed to share meaningful (in the context of the exercises) content in a visually appealing manner.

16.2 Background

The creation and use of maps has changed profoundly in recent years. Today, a large proportion of maps are screen maps authored by users leveraging web services as a result of advances in Web 2.0 and online mapping environments (Haklay et al., 2008; Graham 2009). The production of map mashups using spatial and aspatial data from several sources often results in a tendency for maps to become overly

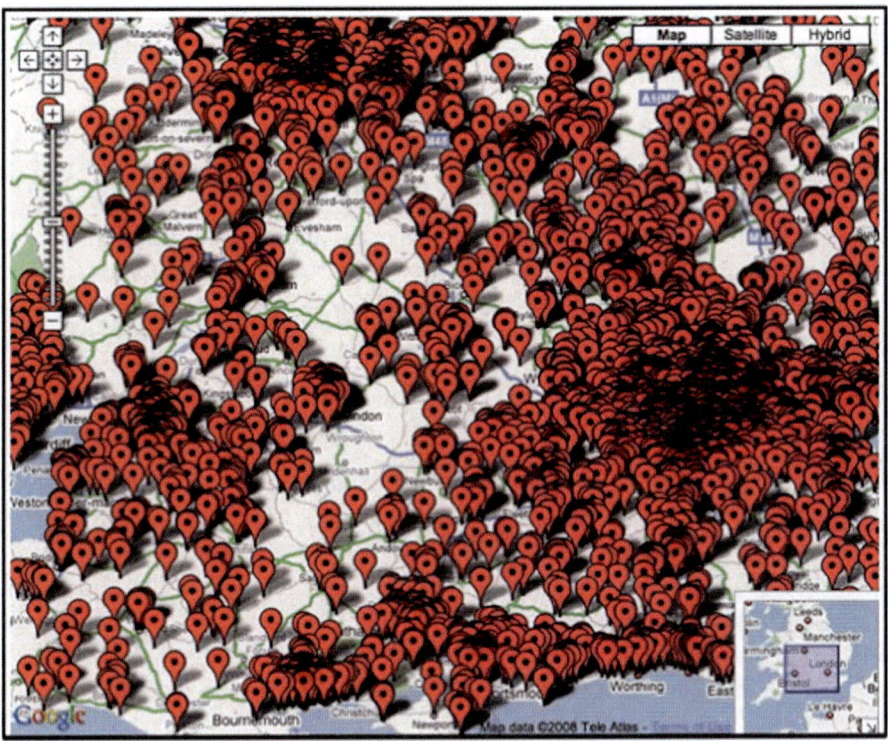

Fig. 16.1 British National Party membership list mashup (TechCrunch Europe, © 2008 Google, Map Data © 2008 Tele Atlas)

cluttered which can reduce their effectiveness as communication devices e.g., Fig. 16.1 (Fairbairn 2006; Rosenholtz et al., 2005).

Mashups are relatively simple to construct and publish online but the combination of the data type, screen display, variable viewing scale, live (and changing) data feeds and uncertainty in metadata presents challenges for revealing patterns cartographically. In order to avoid the conflicts that affect some large volume data visualization exercises the authors have resorted to implementing some classical cartographic techniques which are often not applied to map mashups (perhaps as non-traditional cartographers are unfamiliar with the approaches).

16.2.1 Public Participation GIS and Geodeliberation

In the field of public participation GIS (PPGIS) much work has been done on testing and analysing approaches and processes within consensus building efforts to solve geospatial problems as well as the collaborative use of GIS. Rinner and colleagues (2008) idea of (geo)"deliberation", extended by Cai and Yu (2009) underpins the ideas presented later in the methodology. To avoid the potentially biased, local knowledge influences of traditional PPGIS the idea of deliberation proposes cooperation and working towards consensus (or for the application presented in Sect. 16.4, a common data collection ontology). Cai and Yu (2009) suggest a five stage process beginning with an introduction to a problem (1) through exploring issues (2) contributing personal observations (3) establishing a common ground (4) and developing actions (5) and suggest this is a linear, bottom up process. In the domain of student fieldwork, observation and experience of an unfamiliar environment will change assumptions. For this case we suggest geodeliberation is more accurately visualized as an iterative process of discovering exceptions to previously defined rules requiring modification of previously held beliefs (Kirschner et al., 2006; Sandoval 2003).

In order to undertake this process of geodeliberation various tools are required to facilitate communication, visualization, discussion tracking and analysis and sharing of data between participants. These processes need to exist somewhere such as a *collaboratory*, a virtual space which enables geographically dispersed individuals to access data repositories, conversation spaces, and even instrumentation (Cerf 1993; Finholt 2002). Examples exist for research in physical sciences (Kouzes et al., 1996; Olson et al., 1998; Russell et al., 2001; Keahey et al., 2002; Schissel et al., 2002), health sciences (Craver and Gold 2002; Olson et al., 2002), computational science (Kaur et al., 2001) and interdisciplinary research (MacEachren et al., 2006). Collaboratories have not been explored as a support mechanism for geographical fieldwork. In Sect. 16.4 we present an infrastructure that attempts to develop a web-enabled fieldwork collaboratory utilizing Web 2.0 tools and the Twitter social network.

16.2.2 Delphi

We have built upon the principles of the Delphi approach and subsequent web-based systems such as e-Delphi (Pike et al., 2009) in using online map-based social networking as a collaboratory. The goal of Delphi is to supporting asynchronous discussion (Dalkey 1969; Linstone and Turoff 1975; Turoff and Hiltz 1996) by eliciting structured, iterative input from diverse groups to not necessarily reach consensus, but identify key elements of a problem or points of agreement and disagreement. A moderator (an academic staff member in our application) poses questions, prompts interaction, synthesizes feedback, and guides the group (students) towards a goal (standardized data collection schema).

By providing moderation and "expert" oversight to balance the deliberation approaches of Rinner et al., (2008) and Cai and Yu (2009) we impose a top-down structure to ensure that the bottom-up discovery remains within the bounds of the problem to be solved and provide guidance to ensure consistency across iterations of the field course so that data from year to year are comparable. This approach also has pedagogic benefits of blending discovery, experiential and guided instruction techniques.

16.2.3 Web 2.0 and GIS Infrastructure and Services

A significant number of options exist for building interactive mapping applications (Boroushaki and Malczewski 2010). Due to the choice of maps APIs that exist it is possible to create environments which permit PPGIS, geodeliberation and collaboratories and therefore the collection of volunteered geographic information. A critical element of the geodeliberation framework is interpersonal communication which is a central tenet of Web 2.0, along with interoperable datasets, freely available spatial data access tools and application programming interfaces for bringing them all together (Hall et al., 2010).

A spatial framework is valuable for underpinning discussion both for Web 2.0 applications and for geocollaboration and geodeliberation. Whether that is to provide spatial context for the users' comments or to spatially reference the users themselves, many freely available APIs exist. Chow (2008) provides a comparison of the most commonly used maps APIs as well as explaining their conceptual architecture. The Web 2.0 maps APIs have been widely used to display and augment GIS data on the internet for PPGIS tasks, interactive environmental planning (Ghaemi et al., 2009) and geodeliberation (Cai and Yu 2009).

The typical architecture for web mapping applications for either PPGIS or geodeliberation is a database server (for maintaining discussions), a map server and/or spatial database server for providing spatial content (e.g., ArcGIS geodatabases, map tiles or Google Maps), a web server for hosting the application, and a series of clients (e.g., web browsers) for interacting with the application (Rinner et al., 2008; Cai and Yu 2009; Ghaemi et al., 2009). Each of these components provides their own APIs which the developer uses to access the required functionality using prescribed data formats/data access protocols (e.g., PHP, XML, GML, Ajax).

16.2.4 Accessing and Mapping Social Networking Data

Since its creation in 2006, Twitter[1] has become a hugely popular online social networking tool where users post tweets of up to 140 characters in a similar fashion to SMS messaging (Honeycutt and Herring 2009). It was originally based around a simple concept asking users to post tweets to their profile with "What are they

---

[1] The justification for the selection of Twitter as a social networking client is discussed later in this section and in the Sect. 16.4.

doing?"; later re-phrased as "What's happening?". This change reflected a change in usage of Twitter as a tool to comment on issues instead of just discussing the individual's actions (Mischaud 2007; Java et al., 2007). The spatial data generated via Twitter feeds has received little attention. Until the implementation of automated geotagging in late 2009, the spatial representation of tweets had been largely overlooked and the only visible spatial expression was in an individual's profile (Honeycutt and Herring 2009; Field and O'Brien 2010a). The geotagging functionality in the Twitter API enables a tweet to be automatically georeferenced from an onboard GPS receiver and then the tweet can be spatially represented on a map. Still, in addition to using the location of tweets as a proxy for a tweeter's location, there remains a range of potential ways in which the spatial expression can be enhanced and utilized beyond simple depiction on a map.

Many non-map examples illustrate visualizations which give an overview of the interconnectedness amongst a collection of tweets. Twitarcs, for example, provides a graphical view of related tweets with arcs being used to relate common, repeated terms or Twitter IDs. By extending the linear timeline used in Twitter a richer information source giving the ability to explore relationships amongst ideas, themes and comments is presented.[2] Similarly, Social Collider organizes tweets based on common terms, Twitter IDs or hashtags using a scattergraph with the horizontal axis representing the common terms and the vertical axis time. The emerging pattern illustrates the pattern of tweets over time, identifying discussion threads, key moments in time where clusters of tweets exist and the way in which the pattern changes temporally.[3]

The use of embedded information in a spatial context is best exhibited by a map developed during the snowfall on 1st February 2009 in the United Kingdom. As snow began to fall a #uksnow tag emerged on Twitter providing real time reports of snowfall where the tweet had "#uksnow", a UK postcode (first part only) and optionally a subjective score out of 10 indicating how hard it was snowing. Darbyshire (2009) reverse geocodes the postcode component of the tweet to generate a place marker on Google Maps base data symbolized to represent snow intensity.

More innovative approaches determine tweet locations from GPS derived spatial coordinates, coordinates embedded in a tweet, by geocoding locations mentioned in the tweet or leveraging geolocation controls in the Twitter API (Fig. 16.2).

16.2.5 Summary

While infrastructures currently exist for collecting volunteered geographic information its use is questionable without imposing a top-down framework to ensure consistency (Goodchild 2008). Web 2.0 provides a set of tools for enabling data

---

[2] Twitarcs: http://www.neoformix.com/Projects/TwitArcs/TwitArcs.html.
[3] Social Collider: http://socialcollider.net/.

Fig. 16.2 UKsnow Twitter Map (Darbyshire, ©2009 Google, Map Data ©2009 Tele Atlas)

contribution, information exchange and negotiation for GIS and the process of data collection from users. The challenge is to link these tools to manage spatial and aspatial data collection in a structured manner while facilitating collaboration.

16.3 Context

While the development of the application is explored in the next section it is important to place it in context before discussing the objectives of this research. As part of a 3 year Bachelor's degree in GIS, second year students (and GIS MSc students) take a class in Mobile GIS comprising a 1 week field course in Malta for data collection and post field course analysis tasks. The field course introduces a range of mobile data capture technologies to satisfy a set of pedagogic objectives. The students operate under the simple premise that they arrive on the island with no data and by the end of the week will have acquired a range of datasets to support post-fieldcourse analytical work. A sequence of exercises begins with gaining familiarisation with consumer-grade GPS devices for navigation and the capture of waypoints, tracklogs photos, video and audio recordings for a virtual tourism experience. As the week progresses the use of mobile GIS solutions for land use surveying using hand-held PDAs running Esri® ArcPad® integrated with ArcGIS Server, differential GPS, Real-time Kinematic GPS and other surveying techniques are introduced.

The dominant learning paradigm is data gathering with mobile devices in small-groups in Malta and then collaboration and data analysis at a later stage in a desktop environment at Kingston University London. Historically, this data gathering process has created significant difficulties in the field and led to problems in the analysis stage if students have not collected data within a common ontology. The tools demonstrated here are used heavily in the land use surveying exercises after being introduced to students in the navigation exercise.

The fostering of a collaborative approach to data gathering is only partially served by this small group approach (Drummond et al., 2006) and prior experience has demonstrated limitations in terms of what different groups of students achieve both in terms of productivity and category agreement. Student groups are spatially dispersed across a 2 × 2km (1.6 × 1.6mi) study area and without means of communication & collaboration. Subsequently they operate autonomously, gathering data with different sampling strategies (e.g., classifying features using different object types, capture resolutions, attribution and detail). These different collection and classification strategies (their ontology of data collection) and/or different epistemologies about category types lead to inconsistencies when data are combined as demonstrated by Gahegan and Brodaric (2001). Students' results are often inaccurate due to poor quality analysis resulting from poor data capture techniques.

To meet our ultimate goal of improving collaborative data gathering our objectives were:

- To create mechanisms for student–student group interaction when small groups are dispersed across the study area;
- To provide a means to collaborate remotely with staff in order to bring together interim results or develop methodologies as part of the exercise itself;
- To provide a means of enhancing remote student–staff interaction for the purposes of support; this could be to assess interim work for formative feedback, the modification of data by students and/or staff changes to data collection requirements by monitoring data streamed to the ArcGIS server; and
- To develop content that can be delivered remotely in the field to enhance student experiential learning (e.g., to provide challenges during an exercise or to introduce changes to data requirements to assess adaptability and ingenuity).

As part of a wider research project (Linsey et al., 2010), our approach was to develop a mobile learning collaboratory where students could interact through their personal devices. The use of the internet was critical to the development of this collaborative learning environment and provided a suitable mechanism to visualize the progress and quality of student work during the exercises. This visualization was via an online map that underpinned the collaboration. It was intended to act as a spatial reference to improve student familiarity with an unfamiliar environment as well as mapping the locations of student groups, and visualizing productivity of student groups within their study areas. A communication method was required and the project made use of two, TxtTools (a University implemented bulk SMS sending & receiving application) and the Twitter microblogging service. Twitter enables us to demonstrate how spatially referenced asynchronous and spatially distant collaboration can take place through this medium. The underlying map is used as a basis for discussion and consensus building in addition to being a visualization aid as noted above.

We develop a twitter map as a collaborative application medium (but with two application dependent views). The first "collaboration" application explores how they might be used in real-time collaboration to develop a common conceptual understanding in support of data capture. The fundamental research question here focuses on whether such mashups have any practical value other than visualizing tweets. The second "race" application focuses on the architecture of the information visualization itself taking the problem of multiple coincident points as a case study. The purpose of this application is to explore whether there are ways of visualizing tweets that give added value through alternative visualizations in addition to the objectives described earlier. The purpose of organizing the aspatial data over a map is to attempt to reveal semantic elements and to visualize spatio-temporal patterns.

16.4 Methodology

To meet the objectives outlined earlier we built a geocollaboratory linking together multiple web map services (using Google Maps API and Esri's ArcGIS Server Javascript API), a communication medium (Twitter) and other services to visualize the content of tweets and interrelationships between them. We make use of a standard client and server architecture model (Fig. 16.1) with multiple servers storing map data (a Google Maps server and our own ArcGIS Server and ArcSDE servers). We make use of a Word Cloud to visualize tweets contextually).

The outputs of this geocollaboratory centered on the "TweetMap" views which spatially reference tweets extracted via JSON through the Twitter API overlaid on data from the map services.

The remainder of this section justifies the selection of the tools, environments used and discusses the implementation of the two TweetMap views by demonstrating their use by staff and students.

In our implementation we utilize Twitter as our primary communication medium in preference to TxtTools as Twitter has an open and accessible API and data are more easily exported from Twitter for later analysis. TxtTools requires a web client to broadcast messages and access is restricted to staff only. Twitter was chosen in preference to Facebook as earlier surveys of students (Linsey et al., 2010) indicated a desire for students to keep personal and "professional" social networks separate (most students did not have Twitter accounts) and the 140 character limit of Twitter forced students to have focussed discussions.

We have selected the Google Maps API and the ArcGIS Server Google Maps Javascript API in order to include both Google Maps data as well as data held within our ArcGIS Server (which contains the background data students have on their PDAs assuring consistent spatial context). The use of ArcGIS Server and ArcSDE also enable the periodic collection of data for students to permit later analysis of changing data collection methodologies (through database versioning).

While we could have made use of other map APIs Google's was selected due to an overwhelming familiarity to students (100% of the class had experience with Google's implementation to the exclusion of alternative products), it is a mature API with large developer community and has good integration with ArcGIS Server (which Bing Maps also exhibits).

16.4.1 Supporting Collaborative Investigation Using Twitter Map Mashups

Student fieldwork often reflects the way in which geographically dispersed teams or individuals work (similar to the process of geodeliberation and the applications for collaboratories). The first TweetMap was built as a mechanism to explore the value of using social networking as a mechanism for collaborative learning using the #uksnow map as a framework (Marsh 2009).

The 2009 iteration of the fieldcourse (Field and O'Brien 2010b) made use of a beta release of the map (Fig. 16.3) with student and staff feedback used to enhance the design of the interface and types of data displayed on the map. Issues identified through user feedback included how to deal with coincident tweets, how to manage threaded discussions and how to link tweets discussing similar concepts. These improvements were incorporated into the 2010 version of the map.

Further testing at the AGI Geocommunity 2009 conference informed development work for the 2010 Malta fieldcourse improving the spatial and temporal representation. User feedback from this event confirmed psychology and visualization theories that a Twitter avatar provided a good marker symbol and that a visual

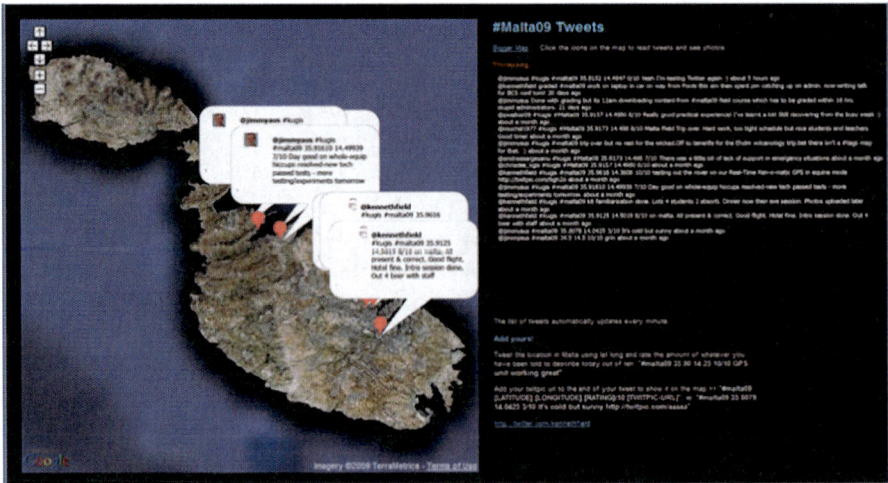

Fig. 16.3 Twitter as a collaborative mapping tool on Malta in 2009

timeline emphasized the temporal dimension of tweets. Further design work was undertaken to specifically tackle the figure-ground relationship and to provide a more appropriate base map. The Google Maps API v3.0 (Google 2010) was used which allowed some modification to the map style. In order to have full control over the underlying map service, ArcServer™ and the ArcGIS® API for Javascript were used to construct the map so that our base layer of choice could be served in conjunction with Google Maps data. The use of map services from our own ArcGIS server added value by providing students with historic imagery for use in the field as additional context. We also provided the option of delivering OpenStreetmap data (instead of Google Maps) as a tiled map service using styles designed using Cloudmade.

The 2010 collaborative map used the Twitter API search function to find the #malta10 (an abbreviation for the Kingston University 2010 Malta field course) hashtag to denote content related to the fieldwork activities. The map illustrates tweets spatially arranged using GPS coordinates supplied within tweets that took the following form:

#malta10 [latitude][longitude] [rating] text [twitpic URL]
e.g., #malta10 35.8079 14.0425 3/10 mostly scrubland with some agricultural use http://twitpic.com/aabc

---

Coordinate data are supplied by students using the GPS equipment provided for the exercise or GPS functionality built into their phones. The students profile or Twitter's geolocation API were not used to ensure accurate locations were recorded – only 22.5% of tweets (166 out of 738) had correctly formed automatically provided geolocations.

The Twitter search API returns a JSON formatted data stream of all tweets containing the #malta10 string. The tweets are parsed for textual content, the latitude and longitude, rating and any images or links present. The latitude and longitude values are passed to the standard Google Maps API function GMarker along with the appropriately scaled icon (based on the student supplied rating allowing rapid visualization of student response to various questions set) and added to the map.

The textual content of the tweet is processed through a Javascript Word Cloud library to determine the most popular topics of discussion among the group which are displayed both at the bottom of the map (see Fig. 16.3) as well as graphically around the outside of the mapped tweets. The content of the tweet and any HTML links or image data are copied into an HTML popup linked to the marker enabling mouseOver and mouseClick events on the marker (Fig. 16.5).

The students used the map as part of their data collection workflow during 2 days in the study area. The first day was a largely unstructured exercise where the students were given brief instructions about the need to map the landuse of the study area. The following day after the students had gained familiarity with the area, small groups were assigned "service areas" (Fig. 16.6) and given the same instructions. The instructions were deliberately vague to encourage the students to use the Twitter map to collaborate.

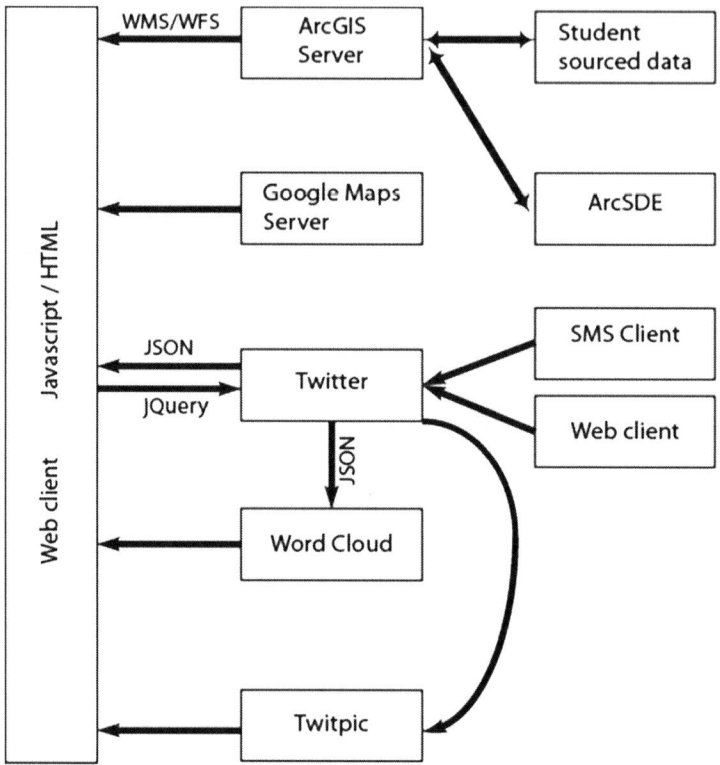

Fig. 16.4 Fieldwork geocollaboratory architecture

Figure 16.7 illustrates the various elements discussed within the methodology. The greyscale photo overlaying the Google base map data is supplied from our ArcGIS Server. The tweets are reorganized around a central point to depict a discussion theme of tourism with common terms linked to the tweets which make reference to them. Only the most commonly used terms are mapped. The line thickness denotes the position of the term in the tweet as a primitive indicator of significance. The search bar for themes is shown in the bottom right corner and supports logical operators such as AND, OR and NOT.

The students used Trimble Juno PDAs and ArcPad for data collection and made supplementary use of Garmin eTrex devices as a backup and for providing tweet coordinates. Students sent tweets via SMS from their mobile phones and had the ability to send and receive data via 3G cellular network to an ArcGIS Server. The high cost of 3G access in Malta (approx. $8USD/Mb) limited students from sharing data in this manner but they did upload data to the server via local WiFi hotspots and/or via the staff's portable WiFi hotspot located at their initial starting location.

As students surveyed their allocated areas they utilized Twitter and twitpic (via multimedia message service) to communicate with staff about technical issues often with advice from other students, deliberated about data collection and sampling

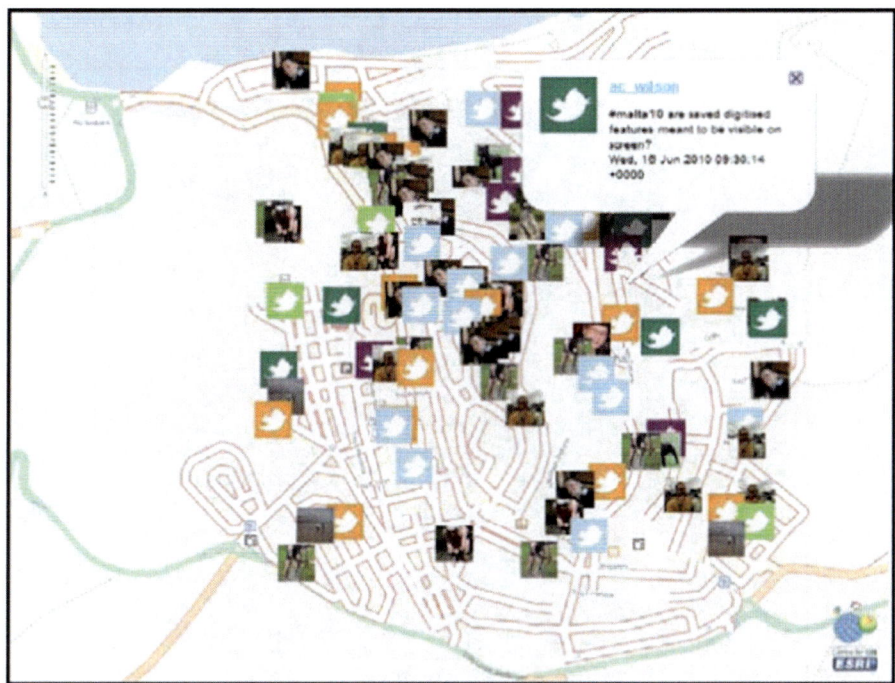

Fig. 16.5 #malta10 TweetMap used for the land use mapping exercise in Mellieha and utilizing alternate OpenStreetMap data (base map ©2010 CloudMade -Map data CCBYSA 2010 OpenStreetMap.org)

strategies before settling on a common approach which was disseminated using Twitter. A review of the students' discussion is presented in the Results below (Sect. 16.5).

16.4.2 Visualizing Spatio-temporal Data Using Twitter Map Mashups

The TweetMap was introduced to students as part of an equipment and environment familiarization exercise comprising a navigation challenge using either paper maps or GPS receivers. Navigating via 5 waypoints across the cities of Sliema and Valletta in Malta, at each waypoint the students were to tweet the number of the waypoint and a series of keywords describing the immediate local environment from their perspective.

Tweets were in the format:

#malta10 #wp4 time 4 a cup of tea. Fishing boats, cathedral, spire, cockneys restaurant, police station bar?

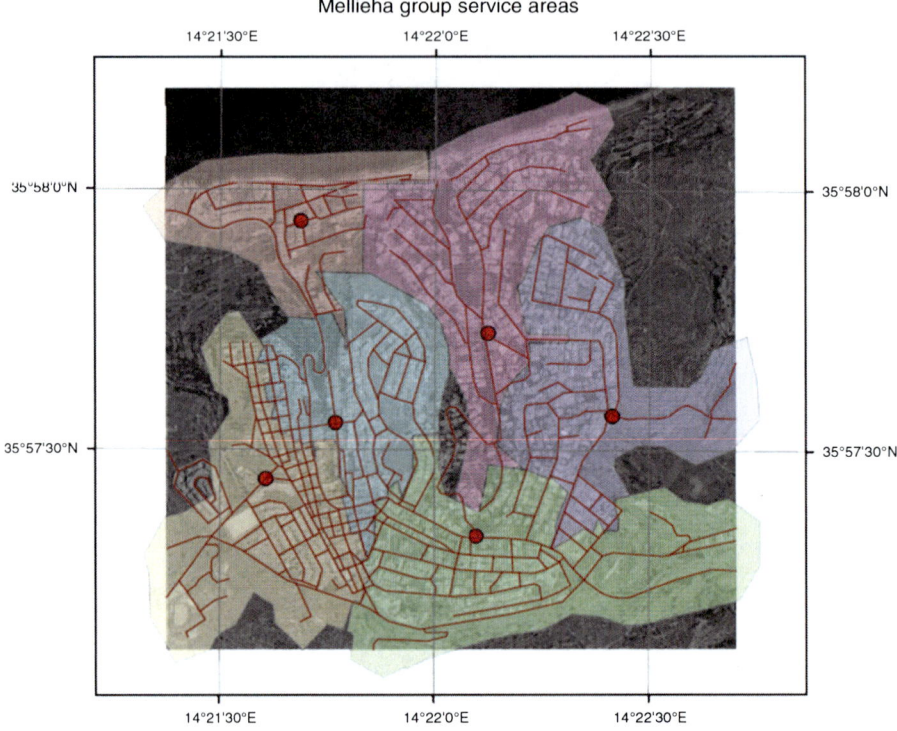

Fig. 16.6 Different *colored* polygons representing work areas assigned to student groups

As with the geocollaboration example, tweets were extracted from the Twitter search API by searching for the #malta10 hashtag. The students did not have to supply the latitude and longitude as the location of the waypoints was known and used to locate the tweets. As each student would be tweeting from the same locations a Fibonacci spiral was used to arrange the tweets around each waypoint. The latest tweets would be located at the centre of the spiral and displayed using a smaller avatar while the earliest tweets (signifying leading the "race") would be larger and at the outside of the spiral (Fig. 16.8).

The main purpose of this map was to allow staff to monitor student progress throughout the race, to record student commentary about the relative effectiveness of paper vs. map based navigation, document student navigation issues as soon as they arrived to inform a reflective discussion after the event and as a familiarization exercise with Twitter prior to the land use collaboration exercise. As no spatial data were needed this version of the map does not make use of the ArcGIS Server functionality. The circle of terms around the outside of each waypoint records the students' commentary about the waypoint. Larger terms with darker backgrounds denote the higher frequency of those terms with smaller, paler backgrounds denoting less frequently used terms.

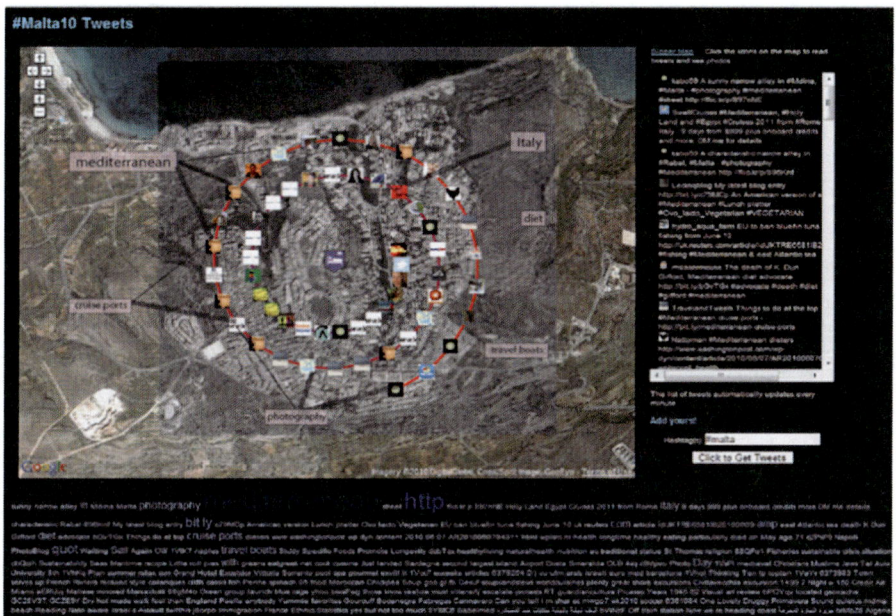

Fig. 16.7 Tweet visualization environment demonstrating a discussion theme

16.5 Results

The #malta10 TweetMap provides a number of ways of supporting collaborative learning. Firstly, student groups were working in disparate locations (see Fig. 16.4) and the Twitter map allowed a common communication framework in which students could post messages, work in collaboration with other groups and communicate with staff. Students registered for Twitter accounts prior to the trip after a preliminary survey of personal technologies found that all students on the trip had a mobile device capable sending Twitter messages (either posting messages directly to their Twitter account through a web interface or indirectly as an SMS message through a third party application).

The #malta10 TweetMap therefore provided an online social network for students on the fieldcourse that pulled information from individual Twitter accounts into a common, shared platform that added value by defining the spatial context and enabled collaborative learning.

The loosely defined task of the landuse mapping exercise forced students to interact as demonstrated by them sharing images of unfamiliar landuse types via Twitter and asking questions such as "*Is this garrigue?*" and "*#malta10 35.95979 14.35880 establishing whether farm land is derelict or not is difficult twitpic.com/ atj12*". Responses from classmates who provided alternate views (bottom-up) and staff who could share textbook examples, images and definitions (top-down)

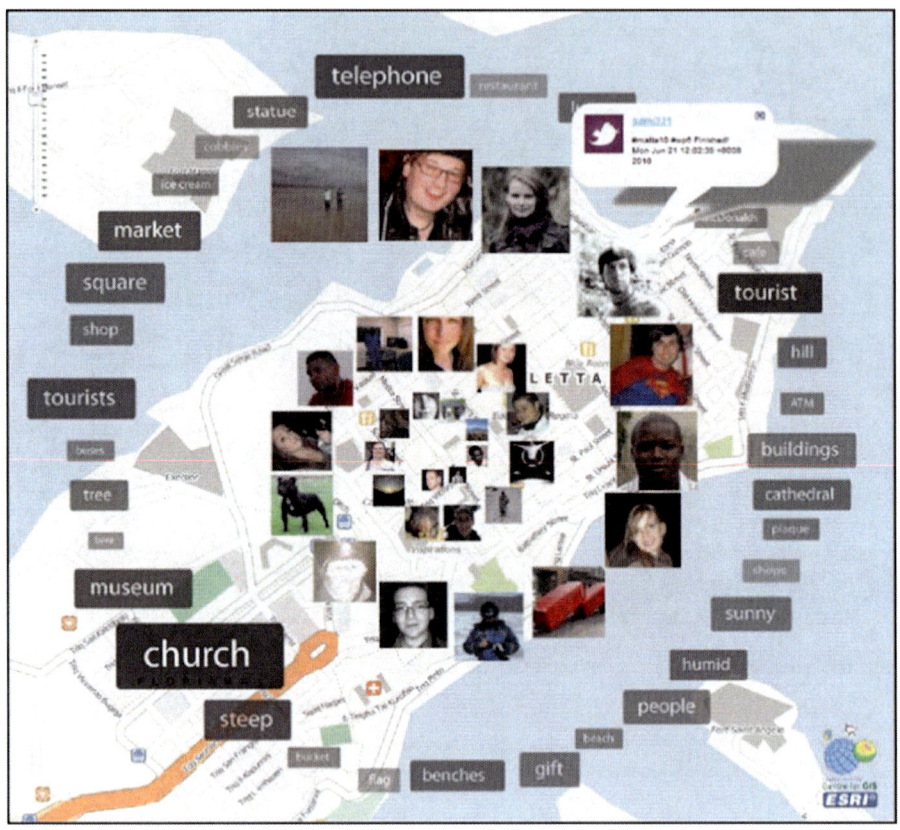

Fig. 16.8 #malta10 TweetMap at medium scale zoom level used for the land use mapping exercise in Mellieha (base map ©2010 CloudMade -Map data CCBYSA 2010 OpenStreetMap.org)

informed and guided the implementation of a data collection schema (ontology) based on the students sharing their epistemologies.

Students debated this need for consistent data collection via Twitter (e.g., "*we need to sort out standardisation*") and developed an ad-hoc schema after deliberation (e.g., "*@rafchris #malta10 map these cats: ftpth, main rd, resid rd st furn aband agri, agri, agri bld, beach, com bldg, garrigue, maqu/@wedwobin*") and built on experience of their environment the previous day which was then disseminated via Twitter.

The students also shared technical tips for mapping different land use//land cover types using different technology (e.g., "*#malta10 students who are complaining about poor signal quality are you using the external antenna?*" and "*great experience today, how a combination of diff devices to capture data and get one total output*") and students reported a gain in efficiency of data collection over the previous day (e.g., "*Efficiency much better – approx. over 1/3 of area done*" and "*productivity down due to heat and hills but much better than yesterday*").

The threaded nature of discussions between students could also be visualized on the map (Fig. 16.4) using the spiral technique of Fig. 16.6 but centred over the study area in Mellieha or the first tweeter to use a term, concept or ask a question (using the search box on the map page). The Twitter "reply" field was utilized in conjunction with jQuery to find all of the tweets which referenced the earlier tweet of interest. The IDs of these tweet replies were also searched to capture any subsequent replies. An on-the-fly list of replies to tweets was constructed and ordered chronological and the Fibonacci spiral of tweets created and added to the map. This minor functional extension allowed staff to follow and contribute to a threaded discussion and to monitor the contributions of students for summative assessment.

16.6 Discussion

As illustrated above the 2010 #malta10 TweetMap was used in a number of ways during the field course. Students were required to use it as a general collaboratory during the field course which enabled the collection of a chronological field log (Fig. 16.9). 738 tweets were exchanged during the week on all aspects of the tasks set with approximately half of these occurring during the land use mapping exercise.

The #malta10 TweetMap was used to support the same land use mapping exercise in Mellieha as the #malta09 TweetMap supported. The architecture and interface improvements in addition to pedagogic improvements and increased student use of the environment proved valuable. The students were able to exchange ideas, interpretations of landscapes and the search for common frameworks for data collection. The students reported (via Twitter and daily reflective blog & video diaries) greater productivity on the second day of the land use exercise and noted that they felt that the collaboration and deliberation enabled by the Twitter map had improved their productivity:

> "I have forgotten to mention the use of twitter so far. Today this has been very useful in resolving issues that have arisen. I took a while to follow everyone and set up text forwarding but it was worth it. If it wasn't for the ridiculous roaming charges, it got me thinking about the use of Skype to contact students and tutors either over video, audio or text. Also you could use the conference calling feature to contact many people at once. I did attempt to geo tag my texts automatically using the phones gps receiver however this did not work well as it did not save the setting and would have to set it up every time I wanted to text twitter."

and

> "As well as this twitter was used between the six groups to work out a specification of what features to map so that data collected between the six groups would have similar characteristics when joined together to create the land cover(not land use!) map."

The use of Twitter also enabled current and former students who had attended the University and other interested followers of the field course to contribute from a

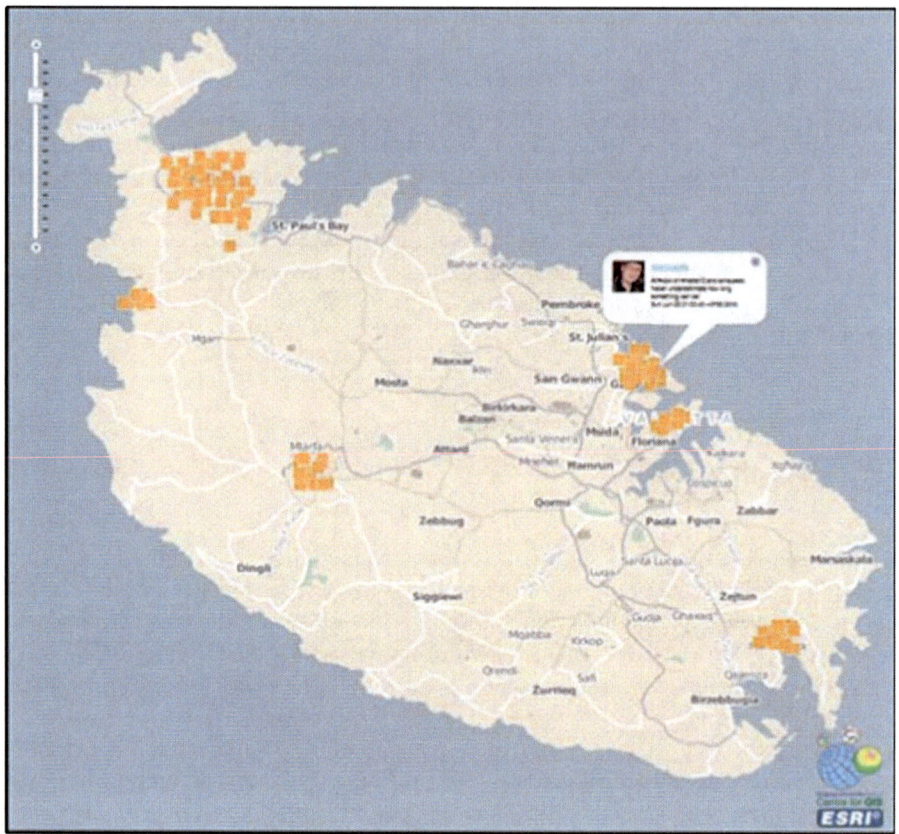

Fig. 16.9 #malta10 TweetMap at small scale zoom level used as a general fieldcourse log (base map ©2010 CloudMade -Map data CCBYSA 2010 OpenStreetMap.org)

distance widening the participation of the trip (*"@Muchel1977: nice sitting in Germany reading #malta10 tweets. Some nice memories. Enjoy your trip"*). None of these external observers contributed to the discussion perhaps highlighting that local observation was important for category formation.

Figure 16.10 depicts data from the 2008 version of the Malta field course where the TweetMap was not used for category agreement. The area highlighted by the dashed circle shows two problem areas. Firstly the small trapezoidal polygon overlapping the larger brown polygon demonstrated category disagreement. The trapezoidal polygon is classified as "agricultural terraces" while the brown polygon is "abandoned agriculture". The land here is not abandoned agriculture but is partly terraced. The larger polygon also serves to illustrate misunderstanding about the "abandoned agriculture" category. In this year these areas were fallow but did show evidence of regular maintenance and should have been classified as "agriculture."

Figure 16.11 (from the 2010 Malta field course) does not demonstrate this type of inconsistency in data collection as students were able to confer and reach agreement

16 Mapping Social-Network Interactions

Fig. 16.10 2008 Data/category disagreement within area highlighted by the *dashed circle* (see the explanation in text)

about the categories. The dashed boundary polygon was captured in 2008 while the solid boundary polygons were captured in 2010. Visual comparisons of the 2008 with 2010 data indicate that students produced more consistent data with the aid of the geocollaborative framework. Figure 16.11 also illustrates higher resolution data capture as the students were able to more accurately identify different landuse types. The dashed line polygon demonstrates the large, low resolution capture strategy some groups undertook in 2008 while the smaller underlying polygons demonstrate the higher resolution 2010 capture strategy implemented across the study area. The students also uniformly identified individual agricultural boundaries derived from their discussions about consistent data collection (Fig. 16.11).

Student / staff collaboration was enhanced through the use of Twitter and the monitoring infrastructure of the TweetMap geocollaboratory. Determining student location based on tweets allowed progress to be coarsely estimated. Data updates from the field provided a more accurate measure of student progress while also allowing potential data capture issues to be noted early enough to allow recapture. The bottom-up student discovery and top-down staff led guidance learning paradigms were also enabled by Twitter and the TweetMap.

Students were able to liaise with staff to gain technical support from remote locations within the study area. Previous experience had demonstrated that students experiencing technical difficulties would cease data collection thereby generating

Fig. 16.11 2008 and 2010 student data collection comparison (*dashed line* shows low resolution/accuracy 2008 data, underlying *smaller polygons* collected by 2010 students)

an incomplete data set. Twitter enabled student issues to be resolved and other students could take preventative or adaptive measures if necessary (e.g., fitting an external aerial to their GPS).

The interaction and deliberation of students via Twitter supported collaborative ontology development while the visualization of threaded discussion with basic semantic analysis allowed staff to monitor, guide and influence the ontology to ensure that it was fit for the purposes of the exercise.

16.7 Conclusions

The application presented here has built a collaborative infrastructure that is implemented using cartographic and visualization techniques to not only be effective but also deliver a positive user reaction (Hodza 2009). We have drawn together the concepts of geodeliberation and geocollaboratories through Web 2.0 and GIS server and web service technology to extract and visualize content derived from Twitter. We impose order on this volunteered geographic information through the e-Delphi approach of structured consensus building to build a common framework and schema for data collection. The individual methods are not necessarily new but by joining them in a unified framework for fieldwork support we link the individual elements and place a greater focus on how data is conceptualized rather than emphasizing the process of collection. The integration of social network tools with the existing techniques is novel.

As noted in the discussion each of the objectives of providing methods of interaction for dispersed student groups, enabling monitored staff–student collaboration, enabling remote technical support and providing methods of managing and communicating adaptations to the requirements for student work were met.

A number of further developments are planned for the TweetMap to enhance future fieldtrips. The use of hashtags for the trips could be supplemented with a Twitter fieldtrip username for all students and staff to follow (e.g., @KUMaltaTrip). This would enable potentially greater engagement with trip alumni who can follow that user after the trip and interact with subsequent iterations of the fieldtrip. Using SMS notifications to deliver new tweets to followers of the fieldtrip username would also remove the need to use txttools for mass communication with students, streamlining the communication methods.

A greater level of semantic analysis will be implemented. The semantic analysis of tweet content is intended to allow a greater linking between threaded discussions For example if multiple groups of students are having concurrent but separate discussions of the same issue these discussions could be linked by content rather than by who is replying to whom.

A MySQL database containing tweets could be integrated into the architecture to allow a greater degree of data analysis as well as removing the reliance on Twitter for archiving tweets (the current search limit is 5–7 days). The application could also be more tightly coupled with ArcGIS Server Geoprocessing services to allow automated analysis of student productivity and mapping of uncertainty (e.g., where were students asking most questions about a data type).

Acknowledgements This work was part funded by the transforming curriculum delivery through technology JISC e-learning programme Mobilising Remote Student Engagement (MoRSE), a collaborative project between researchers at Kingston University London and De Montfort University, Leicester. It is also part-funded by a SPLINT-CETL Honorary Visiting Fellowship award at University of Leicester and part-funded by the Geography, Earth and Environmental Sciences (GEES) Subject Centre Learning and Teaching Development fund programme Mobile Decision-making in the Cloud (MobiDIC) project.

References

Boroushaki S, Malczewski J (2010) Measuring consensus for collaborative decision-making: a GIS-based approach. Comput Environ Urban 34(4):322–332
Cai G, Yu B (2009) Spatial annotation technology for public deliberation. Trans GIS 13:123–146
Cerf V (1993) National collaboratories: applying information technology for scientific research. National Academy Press, Washington, DC
Chow TE (2008) The potential of maps APIs for internet GIS applications. Trans GIS 12 (2):179–191
Craver JM, Gold RS (2002) Research collaboratories: their potential for health behavior researchers. Am J Health Behav 26(6):504–509
Dalkey NC (1969) The delphi method: an experimental study on group opinion. The RAND Corporation, Santa Monica
Darbyshire K (2009)UK snow map. [cited] http://nocto.com/uksnow/
Drummond J, Billen R, Joao E, Forrest D (2006) Dynamic and mobile GIS: investigating changes in space and time, 1st edn. CRC Press, Boca Raton

Fairbairn D (2006) Measuring map complexity. Cartogr J 43:224–238
Field KS, O'Brien JA (2010a) Cartoblography: exploring the spatial context of micro-blogging. Paper read at GISRUK 2010, London, UK
Field KS, O'Brien JA (2010b) Cartoblography: experiments in using and organising the spatial context of micro-blogging. Trans GIS 14:5–23
Finholt T (ed) (2002) Collaboratories. American Society for Information Science and Technology, Washington, DC
Gahegan M, Brodaric B (2001) Learning geoscience categories in situ: implications for geographic knowledge representation. Paper read at ACM-GIS 2001 the ninth ACM international symposium on advances in geographic information systems, Atlanta
Ghaemi P, Swift J, Sister C, Wilson JP, Wolch J (2009) Design and implementation of a web-based platform to support interactive environmental planning. Comput Environ Urban 33(6):482–491
Gibin M, Singleton A, Milton R, Mateos P, Longley P (2008) An exploratory cartographic visualisation of London through the google maps API. Appl Spat Anal Policy 1(2):85–97
Goodchild MF (2008) Commentary: wither VGI? GeoJournal 72(3–4):239–244
Google (2010) Google Maps API [cited 1 April 2010] http://code.google.com/apis/maps
Graham M (2009) Neogeography and the palimpsests of place: web 2.0 and the construction of a virual earth. Tijdschrift vood economicshe en sociale geografie 9999
Haklay M, Singleton A, Parker C (2008) Web mapping 2.0: the neogeography of the geoweb. Geography Compass 26:2011–2039
Hall GB, Chipeniuk R, Feick RD, Leahy MG, Deparday V (2010) Community-based production of geographic information using open source software and web 2.0. Int J Geogr Inf Sci 24(5):761–781
Hodza P (2009) Evaluating user experience of experiential GIS. Trans GIS 13(5–6):503–525
Honeycutt C, Herring SC (2009) Beyond microblogging: conversation and collaboration via Twitter. Paper read at forty-second Hawaii international conference on system sciences, Los Alamitos
Java A, Song X, Finin T, Tseng B (2007) Why we Twitter: understanding microblogging usage and communities. Paper read at KDD, San Jose, California
Kaur S, Mann V, Matossian V, Muralidhar R, Parashar M (2001) Engineering a distributed computational collaboratory. In: 34th annual Hawaii international conference on system sciences, vol 9, p 9026
Keahey K, Fredian T, Peng Q, Schissel DP, Thompson M, Foster I, Greenwald M, McCune D (2002) Computational grids in action: the national fusion collaboratory. Future Gener Comp Sy 18(8):1005–1015
Kirschner PA, Sweller J, Clark RE (2006) Why minimal guidance during instruction does not work: an analysis of the failure of constructivist, discovery, problem-based, experiential, and inquiry based teaching. Educ Psychol 41:75–86
Kouzes R, Myers JD, Wulf WA (1996) Collaboratories: doing science on the internet. Computer 29(8):40–46
Linsey T, Ooms A, Downward S, Field KS, O'Brien J (2010) Mobilising remote student engagement on field trips. Paper read at association for learning technology annual conference, September 2010, Nottingham
Linstone H, Turoff M (1975) The delphi method. Addison-Wesley, Reading
MacEachren AM, Pike W, Yu C, Brewer I, Gahegan M, Weaver SD, Yarnal B (2006) Building a geocollaboratory: supporting human-environment regional observatory (HERO) collaborative science activities. Comput Environ Urban 30(2):201–225
Marsh B (2009) #uksnow Tweets. [cited 20 Feb 2009]. http://benmarsh.co.uk/snow/
Mischaud E (2007) Twitter: expressions of the whole self. An investigation into user appropriation of a web-based communications platform. Department of Media and Communications, London School of Economics, London

Olson GM, Atkins DE, Clauer R, Finholt TA (1998) The upper atmosphere research collaboratory. Interactions 5(3):48–55

Olson GM, Teasley S, Bietz MJ, Cogburn DL (2002) Collaboratories to support distributed science: the example of international HIV/AIDS research. Paper read at 2002 annual research conference of the South African institute of computer scientists and information technologists on enablement through technology, Pretoria

Pike W, MacEachren A, Yarnal B (2009) Infrastructure for collaboration. In: Yarnal B, Polsky C, O'Brien J (eds) Sustainable communities on a sustainable planet – the human-environment regional observatory project. Cambridge University Press, New York, pp 34–58

Rinner C, Keßler C, Andrulis S (2008) The use of web 2.0 concepts to support deliberation in spatial decision-making. Comput Environ Urban 32(5):386–395

Rosenholtz R, Li Y, Mansfield J, Jin Z (2005) Feature congestion: a measure of display clutter. Paper read at SIGCHI conference on human factors in computing systems, Portland, Oregon

Russell M, Allen G, Daues G, Foster I, Seidel E, Novotny J, Shalf J, Laszewski GV (2001) The astrophysics simulation collaboratory: a science portal enabling community software development. Paper read at 10th IEEE international symposium on high performance distributed computing, San Francicso, California

Sandoval WA (2003) Conceptual and epistemic aspects of students' scientific explanations'. J Learn Sci 12(5–51)

Schissel DP, Finkelstein A, Foster IT, Fredian TW, Greenwald MJ, Hansen CD, Johnson CR, Keahey K, Klasky SA, Li K, McCune DC, Peng Q, Stevens R, Thompson MR (2002) Data management, code deployment, and scientific visualization to enhance scientific discovery in fusion research through advanced computing. Fusion Eng Des 60(3):481–486

Simon R, Fröhlich P, Anegg H (2007) Enabling spatially aware mobile applications. Trans GIS 11(5):783–794

Turoff M, Hiltz S (1996) Computer based delphi processes. In: Adler M, Ziglio E (eds) In gazing into the oracle: the delphi method and its application to social policy and public health. Kingsley, London

Chapter 17
Online Map Service Using Google Maps API and Other JavaScript Libraries: An Open Source Method

Shunfu Hu

Abstract There has seen increasing interest in developing online map services using Google Maps Application Programming Interface (API), Yahoo! Maps API, Microsoft Bing Maps API, Nokia Ovi Maps API, and ESRI ArcGIS API. However, such online map services are mainly "mashups" in nature, meaning that they utilize Maps API as a platform and combine other spatial data from multiple sources to create new services. The objective of this chapter is to demonstrate an online mapping application that focuses not only on the functionality to display points of interest with customized icons and the information associated with them, but also on the sophisticated functionalities for marker clustering, searching, filtering, and tabbed interface that offer the user the capability to manipulate the data, which is lacking in most documented web mapping services. A case study of developing an online map service to display the locations of hundreds of gardens on the Internet for the United States Department of Agriculture (USDA) People's Garden initiative is presented. Google Maps API, Google Geocoder and other JavaScript libraries such as jQuery, XML, MarkerClusterer, Spry Framework for Ajax, all free and open source, are employed to develop this online map service. It is anticipated that the online map service demonstrated here can be used in most of the web browsers such as Microsoft Internet Explorer (IE) 7.0+, Google Chrome, Mozilla Firefox, and Apple Safari.

17.1 Introduction

The launch of Google Maps in 2005 has revolutionized online mapping service applications on the Internet. Based on Asynchronous JavaScript and XML (AJAX), Google Maps introduces a new type of server/client interaction that maintains a

S. Hu (✉)
Department of Geography, Southern Illinois University, Edwardsville, IL, USA
e-mail: shu@siue.edu

constant connection to the server for more immediate downloading of additional map information (Peterson 2007). In addition, Google also provides programmers free access to its code in the form of an Application Programming Interface (API). The API consists of a set of routines or functions that can be called by a programmer using JavaScript, php, or similar scripting language (Udell 2009). With the current version 3, it is not necessary to acquire an API key to use the Google Maps API. The new version supports both traditional desktop browser applications as well as mobile devices such as the Apple iPad and iPhone. It also supports multiple web browsers such as Internet Explorer 7.0+, Firefox 3.0+, Safari 4+, Chrome, Android, BlackBerry, and Dolfin, all of which having a full JavaScript implementation. These features make Google Maps JavaScript API the most commonly used mapping API for web-based mapping. Other mapping APIs are also available for online mapping, including Yahoo! Maps API, Microsoft Bing Maps API, Nokia Ovi Maps API, and ESRI ArcGIS API.

Recently there has seen increasing interest in utilizing Google Maps API to implement web-based mapping services, from simple applications that display just a few points of interest with pop-up windows to sophisticated map mashups (Haubrock et al., 2007; Johnston and Jensen 2009; Peng and Wu 2010; Roth and Ross 2009; Chow 2008; Niccolai 2008; Pan et al., 2010). Scholefield (2008) developed a web-based map service for tourism of eighteenth and nineteenth century Edinburgh using Google Map API, Oracle Relational Database Management System (RDMS), Structured Query Language (SQL), Perl, eXtensible Markup Language (XML), JavaScript, Hypertext Markup Language (HTML), eXtensible HTML (XHML), and Cascade Style Sheet (CSS). Similarly, Pejic et al., (2009) developed an eTourism application using Google Map API to present prominent points of tourist destinations. Bildirici and Ulugtekin (2010) demonstrates a web mapping service with Google Maps (API V2) mashups in which points, polylines and polygons from the data stored in Keyhole Markup Language (KML), XML, and Geodatabase format are overlaid with Google Maps through JavaScript code. Liu and Palen (2010) examine the use of Google Maps mashups in the crisis management for nine natural disasters such as earthquakes, fires, sea level rise, using near real-time and publicly available data feeds. Hu (2012) discusses a new approach of mashups in multimedia mapping that utilizes Google Maps API, Yahoo! Flickr API, and the YouTube API to combine spatial data, multimedia information and functionality from multiple sources to create the online visitor guide for the Southern Illinois University Edwardsville campus.

The objective of this chapter is to demonstrate an online mapping application that focuses on not only the simple functions of displaying points of interest and the information associated with them, but also the search functions that offer the user the capability to manipulate the data, which is lacking in most documented web mapping services. Google Maps JavaScript API, and other JavaScript libraries such as JQuery, XML, and MarkerClusterer are employed. A case study of an online mapping service to display and search over 600 locations of the gardens from the United States Department of Agriculture (USDA) People's Garden Initiative is demonstrated.

17.2 Methodology

17.2.1 Data Set

The USDA People's Garden Initiative is an effort that challenges its employees to establish People's Gardens at USDA facilities worldwide or help communities create gardens. People's Gardens vary in size and type, but all have a common purpose – to help the surrounding community and the environment (USDA 2011). The data set for this project includes the name of each garden, the city, state, and zip code of the garden, the street address of the garden's owner, the type of the garden (1 – USDA Facilities; 2 – Schools; 3 – Other Places Within the Community; 4 – Faith-based Centers; and 5 – Other Federal Agencies), the geographic location (i.e., latitude and longitude) of each garden in decimal degrees, and more importantly what are planted in each garden. The data set is collected initially through the USDA People's Garden online registration process and transformed into XML file format (see the section below).

17.2.2 Use of XML to Store the Data

There are different ways to store and prepare the data for display in Google Map API v3. One common method is to use PHP (or Hypertext Preprocessor) and MySQL. PHP is a widely used general-purpose scripting language for web development and can be embedded into HTML (PHP 2011). MySQL is an open source relational database management system (RDBMS) that offers multi-user access to a number of databases (MySQL 2011). The use of PHP and MySQL involves three steps: (1) creating a table in MySQL; (2) populating the table from an external Excel or spreadsheet into MySQL database; and (3) outputting XML or HTML with PHP. The only limit in this method is that it performs best with more than 10 markers but less than 100 markers at the same time (Jhnidk 2010). This method is not useful to handle the over 600 markers for this project. Therefore, the author decides to use XML directly.

With XML, data can be stored in a single or separate XML files in plain text format, which provides a software- and hardware-independent way of storing data, sharing data, and transporting data among multiple platforms (i.e., computer operating systems as well as web browsers) (W3Schools 2011). With a few lines of JavaScript code, a programmer can read an external XML file and update the data content of a web page. Below is an example of the XML file (e.g., gardens.xml) that contains the information for one garden named Amherst MA USDA People's Garden. For each garden, all relevant information is placed in a pair of <pt> and </pt>tags. There are over 600 pairs of such tags in the XML file.

```
<gardens>

<pt name="Amherst MA USDA People's Garden" city="Amherst" state="MA"

zipcode="01002" address=" 445 West Street" CIndex="1" longitude="-72.50323"

latitude="42.377651" html= "Vegetables and Herbs"></pt>

</gardens>
```

Notepad++, a free source code editor that supports several programming languages running under the Microsoft Windows environment (http://notepad-plus-plus.org/), was used to prepare the XML data file for all of the gardens. It was also used to edit all of the JavaScript code for the project.

17.2.3 Use of jQuery JavaScript Library to Load the XML File onto the Google Map

jQuery is a cross-browser JavaScript library designed to simplify the client-side scripting of HTML. It is free and open source software (FOSS) designed to create dynamic web pages (jQuery 2011). To use jQuery JavaScript library, the programmer has to attach the external jQuery JavaScript file to an XHTML page as follows:

```
<script type="text/javascript"
  src="http://ajax.googleapis.com/ajax/libs/jquery/1.5.2/jquery.js"></script>
```

This script tag is placed in the head section of the page. Note that src links to Google's content distribution network (CDN) to load the jQuery core file. Then, jQuery.get function is used to load the XML file (e.g., gardens.xml) at the initialization of the Google Maps (i.e., *function initialize()*) and jQuery(data).find("pt") is used to retrieve the information for each garden (notice the <pt> and </pt>tags mentioned above). A variable, xmldoc, is declared to withhold the information (e.g., name, city, address, etc.) for each garden. Another variable, latlng, is declared to withhold the locational information (i.e., latitude (y) and longitude(x)).

```
jQuery.get("gardens.xml", {}, function(data) {

jQuery(data).find("pt").each(function() {

var xmldoc = jQuery(this);


var latlng = new google.maps.LatLng(parseFloat(xmldoc.attr("y")),

    parseFloat(xmldoc.attr("x")));

});

});
```

17.2.4 Use of Google Maps JavaScript API V3 to Display the Data

Linking the web page with the Google Maps API is pretty straightforward in version 3. A single XHTML element is just what is needed as below:

<script type="text/javascript" src="http://maps.google.com/maps/api/js?sensor=false">

</script>

This standard XHTML directive to include an external JavaScript file, served by maps.google.com. This element is added to the head section of the page. The next key step is to initialize the API and load the map onto the page as follows.

function initialize( ) {

    // Set the starting map viewport, based on center coordinates and zoom level

    var myLatlng = new google.maps.LatLng(38.1108515967, -95.7223081589);

    var myOptions = {

        zoom: 4,

        center: myLatlng,

        mapTypeId: google.maps.MapTypeId.ROADMAP

        }

    // initialize the core map object

    var map = new google.maps.Map(document.getElementById("map_canvas"),

            myOptions);

    // add markers to the map and notice the latllng variable declared in jQuery

    var marker = new google.maps.Marker(latlng);

}

17.2.5 Dealing with Marker Clustering on the Map Using MarkerClusterer JavaScript Library

Displaying a large number of locations or markers often causes both visual overload and sluggish interaction with the map. Clustering techniques simplify the data visualization by consolidating markers that are very close each other on the map in an aggregate form. There are a few approaches to dealing with marker clustering on the map, including Grid-based Clustering, Distance-based Clustering, Viewport Marker Management, MarkerClusterer, Fusion Tables, and MarkerManager (Mahe and Broadfoot 2010). There are over 600 gardens for the People's Garden project and the use of any clustering technique is thought be beneficial to simplify the data visualization. The author decided to use the MarkerClusterer, a client-side utility library that applies grid-based clustering to a collection of markers. To use it, the programmer has to first make a declaration in the head section to link the MarkerClusterer utility library to the web page as follows:

```
<script type="text/javascript" src="markerclusterer.js"></script>
```

In the markerclusterer.js, the programmer can specify how many markers (e.g., 10) to become one cluster and the size of the grid (e.g., 10 miles × 10 miles). The programmer then has to create a new markerClusterer object and pass the map to it as follows:

```
var markerclusterer = new MarkerClusterer(map, markers);
```

After the MarkerClusterer finishes going through the collection of markers, clusters will be defined.

17.2.6 Development of Searching and Filtering Functions

It is required for this project to provide the user with search functions such as Find Gardens by Address or Zip Code and Find Gardens by State so the user can have the option to see only selected gardens around that address or zipcode, or within that state. Furthermore, search functions need to be accompanied by filtering the type of gardens, a function called Filter by Type. The first two functions, Find Gardens by Address or Zip Code and Find Gardens by State, were accomplished by using Google Maps API's geocoding process, which converts addresses (e.g., "1,600 Amphitheatre Parkway, Mountain View, CA") into geographic coordinates (e.g., latitude 37.423021 and longitude −122.083739), that the programmer can use to place a marker or to position the map. The Google Maps API provides a geocoder class for geocoding addresses dynamically from user input. The programmer has to create a new geocoder object within the Google Maps *function initialize()* as follows:var geocoder = new google.maps.Geocoder(); and then create a new *codeaddress ()* function as follows:

17 Online Map Service Using Google Maps API and Other JavaScript Libraries

```javascript
var geocoder = new google.maps.Geocoder();
```

and then create a new codeaddress () function as follows:

```javascript
function codeAddress() {
    // Get the user's address input and the user's selection of garden type
    var address = document.getElementById("address").value;
    var selType = document.getElementById("select_type").value;
    // Google standard geocoding function
    geocoder.geocode( { 'address': address}, function(results, status) {
        if (status == google.maps.GeocoderStatus.OK) {
            map.setCenter(results[0].geometry.location);
            map.setZoom(11);
            var marker = new google.maps.Marker({
                map: map,
                position: results[0].geometry.location
            });
        } else {
            alert("Geocode was not successful for the following reason: " + status);
        }
    });

    // Function for Filter by Type
    for (var i=0; i<markers.length; i++) {
        if (selType == "0"){
            markers[i].setVisible(true);
        }
        else if (selType != "0" & markers[i].mycategory == selType) {
            markers[i].setVisible(true);
        }
        else {
            markers[i].setVisible(false);
        }
    }
}
```

17.2.7 Use of JavaScript and CSS to Design the Layout of the Web Site

In the design of the web application, a two-column layout design was adopted, including a column for a map container and the other column for a sidebar. In the map container, the Google map or satellite imagery is displayed, and points of interest (e.g., gardens) are marked with the customized marker icons – the green balloon for gardens at USDA facilities, the yellow balloon for gardens at schools, and so on.

The sidebar contains a list of the same gardens marked on the map. Such content is handy to the page visitor because it enables him/her to the points of interest at a glance. If the user is interested in a particular garden, for example, but does not know where it is, the sidebar list lets them go straight there without having to hunt around the map for it. Therefore, it is beneficial to establish a real linkage between the sidebar and the map.

In order to provide the user with interaction with the map, a few standard Google Maps controls are added, such as Pan and Zoom controls; Map Scale control; and Map Type control: Roadmap and Satellite. In addition, tooltips (e.g., garden name) to the markers are provided, along with clickable marker icons with Google Maps API's standard Infowindow, that provides the information about each garden (i.e., garden name, address, city, state, zip code, etc.). Unlike the Google standard infowindow, which is displayed as a pop-up on the map when an icon is clicked, the infowindow here is displayed in the sidebar to leave the entire map visible.

In addition, in order to allow the user to link the search functions (e.g., Find Gardens by Location or Find Gardens by State) with the map display, the author decided to provide tab panels as a user interface. This was developed using the Adobe's Spry 1.6 framework for Ajax, which is a JavaScript library that provides web designers with the ability to build more interactive web pages (Adobe 2011). To do so, the programmer has to download SpryTabbedPanels.css and SpryTabbedPanels.js (all open source) from the Adobe Labs. The former links the CSS style sheet that styles the tabbed panel; the latter links the Spry TabbedPanels JavaScript library with the search functions. Both of these need to be placed in the head section as follows:

```
<link href="SpryTabbedPanels.css" rel="stylesheet" type="text/css" />

<script src="SpryTabbedPanels.js" type="text/javascript"></script>
```

Working in conjunction with the tab Find Gardens by Location is the HTML <input> tag and input field that allow the user to type in address or zip code from the keyboard. The HTML code looks as below:

```
<input id="address" size="40" type="textbox"

value="Please type your address or zipcode here">
```

17 Online Map Service Using Google Maps API and Other JavaScript Libraries 273

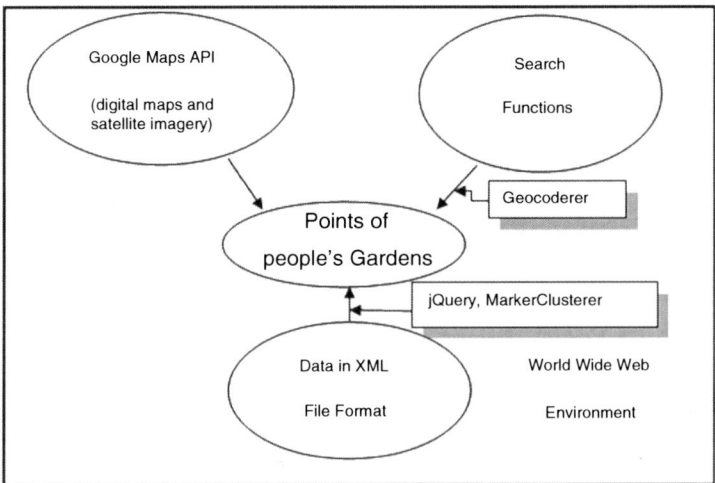

Fig. 17.1 A conceptual framework of mapping a large number of points of interest using Google Maps API that includes the functions of data uploading, display and search

Then, the <select> tag is used to create a select list (drop-down list) and the <option> tags inside the select element define the available options in the list. For instance, Filter by Type allows the user to choose one of the five garden types with 0 as a default value for no filtering. It was implemented with the following HTML code:

Similarly, Find Gardens by State allows the user to choose one state from the list with "All States" as the default value to show all the gardens in the entire country.

Finally, all the pieces have to be put together. JavaScript is the native language of Google Maps. In addition, Google Maps is based on XHTML, and can be formatted with CSS (Cascading Style Sheet) (Udell 2009). Therefore, JavaScript, XHTML, and CSS are used to develop all the functionalities, including uploading XML data files via jQuery, displaying points of gardens using customized icons via Google Maps API, and the search functions via Google Geocoder. Figure 17.1 illustrates a conceptual framework for the integration of the Google Maps API and other JavaScript libraries in the World Wide Web environment.

17.3 Results

The use of Google Maps API V3 provides a very efficient mechanism to deliver digital cartographic information to the Internet user with fast response time and user-friendly interaction. With Google Maps standard Map Type control, the user is able to choose one of the two map types: Roadmap and satellite imagery. Figure 17.2 shows an outlook of the online map service for the People's Garden project

Fig. 17.2 The use of MarkerClusterer JavaScript library to deal with marker clustering to avoid the cluster of the gardens points. The sidebar displays a Flickr slideshow (*top*) and a YouTube movie (*bottom*)

in Google Chrome. At the initial launch of the web page, points of gardens are displayed within the map container as clusters. This gives the user a clear idea where most of the gardens are concentrated. The sidebar on the right side displays a Flickr slideshow of digital photos of vegetable gardens and their produces such as tomatoes, cauliflowers, cucumber, and so on as well as a YouTube movie about how to start an organic garden. This same sidebar can display garden information if the user clicks an icon on the map. Notice that in Fig. 17.2 there are two tabs, Find Gardens by Location and Find Gardens by State. The user can click one of them to perform the search functions. Also, notice the rounded corners for the tabs, the map container, and sidebar.

Figure 17.3 shows the result of a search function, Find Gardens by State. In this case, the user selects the state of Illinois, all of the gardens in the state are displayed on the map with the customized marker icons. The map legend at the bottom of the screen provides the five types of gardens shown on the map. This unique feature provides the user with a clear understanding about the gardens he/she is looking at. Then, the user can click on any icon to get specific information about that garden, which is displayed in the sidebar on the right side of the screen. Here, the user can also filter the gardens by specifying the garden type to display only the gardens of that type (e.g., At USDA Faculties).

Figure 17.4 shows the result a Filter by Type function. In this case, the user first select all the gardens in the entire United States (All States as default) and then filters the gardens by a type (e.g., At USDA Faculties). Figure 17.5 shows the result of another search function, Find Gardens by Location, using addresses or zip codes. The user can first type in the street address, city, and state, or a zip code, and the

17 Online Map Service Using Google Maps API and Other JavaScript Libraries 275

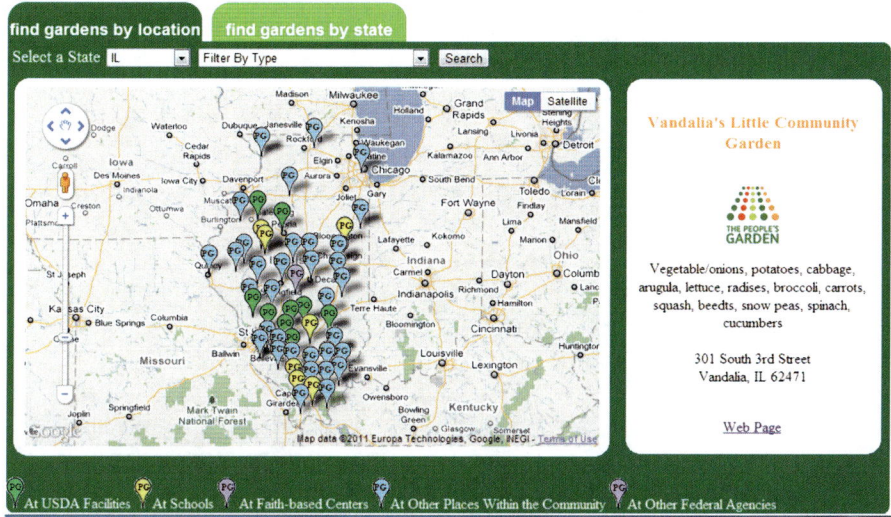

Fig. 17.3 Find Gardens by State displays only the gardens in a selected state. The user can also filter the gardens by one of the five types

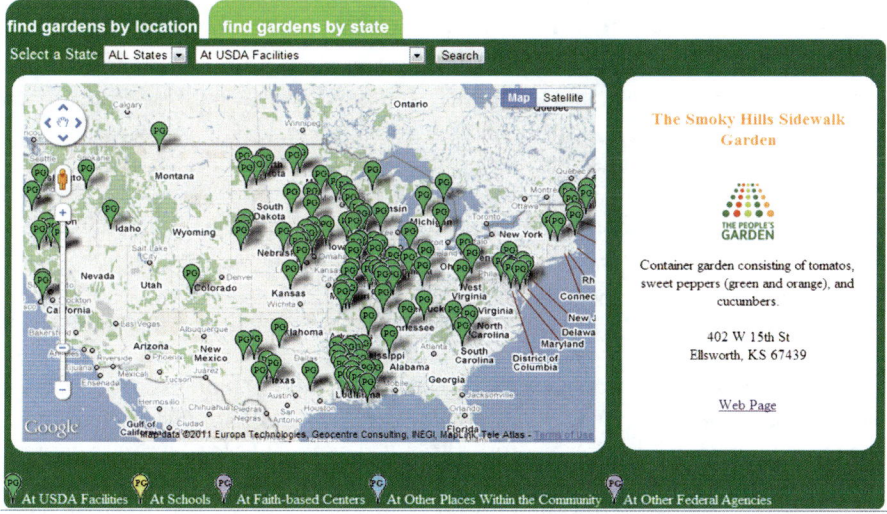

Fig. 17.4 Filter by Type (in this case, at USDA facilities) displays the gardens in that category. If the user clicks an icon on the map, the information about that garden is displayed in the sidebar

map will zoom into that address or zip code and display all of the gardens around it. The user can then filters the gardens by type (e.g., at USDA Facilities). Again, the user can click on an icon at any time to display the garden information in the sidebar.

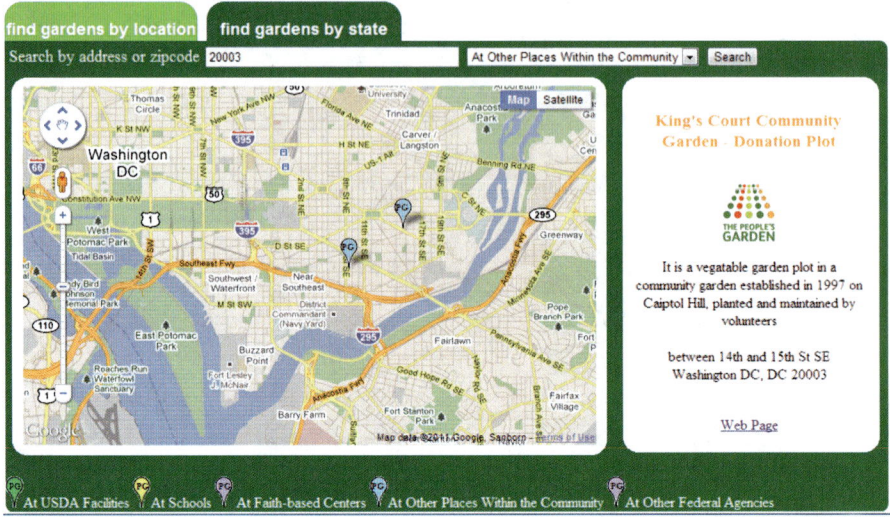

Fig. 17.5 Find Gardens by Location (address or zipcode) will zoom to that location. If the user clicks an icon on the map, the information about that garden is displayed in the sidebar

17.4 Conclusion and Discussion

This chapter has demonstrated an online mapping service that was successfully carried out using the open source method, which includes Google Maps API v3, Google Geocoding and other JavaScript libraries such as jQuery, MarkerClusterer, and Spry Framework for Ajax.

The case study presented in this chapter provides the advanced functionality to display the hundreds of USDA gardens with customized icons and map legend. It also provides the sophisticated functionalities for marker clustering, searching, filtering, and tabbed interface that offer the user the capability to manipulate the data, which is lacking in most documented web mapping services.

Although it was intended to develop such an online mapping service for use under different web browsers, there were, however, compatibility issues in some of the web features. For instance, Microsoft IE (7, 8 and 9) and Mozilla Firefox did not support rounded corners for tabbed panels. Apple Safari for iPad and iPhone did not support Flash movies. All features of the map service worked under Google Chrome. The author suggests checking the official documentations for different web browsers before implementing JavaScript code.

And finally, XML files can be easily uploaded onto a web page, the preparation of such a file is very time consuming, especially when there are hundreds or even thousands of markers to be displayed on the map. A better solution is to use the Google Fusion Tables API to insert, update, and query data after uploading data tables directly from spreadsheets, CSV, or KML files through JavaScript code. The author will implement this procedure in the next phase of the project.

Acknowledgement The author would like to thank three graduate students, Teresa Kysor, Bradley Craddick, and Rachel Byrd, at the Department of Geography, Southern Illinois University Edwardsville for their assistance in compiling the information for over 600 gardens. Their hard work made this project possible.

References

Adobe (2011) Working with Spry 1.6. http://livedocs.adobe.com/en_US/Spry/SDG/help.html. Accessed 10 May 2011

Bildirici IO, Ulugtekin NN (2010) Web mapping with Google maps mashups: overlaying geodata. A special joint symposium of ISPRS technical commission IV & AutoCarto in conjunction with ASPRS/CaGIS 2010 fall specialty conference, Orlando, FL, 15–19 November

Chow TE (2008) The potential of maps APIs for internet GIS applications. Trans GIS 12(2):179–191

Haubrock S, Wittkopf T, Grünthal G, Dransch D (2007) Community-made earthquake intensity maps using Google's API. In: Proceedings of the 10th AGILE international conference on geographic information science, Aalborg, Denmark, CD, 8

Hu S (2012) Multimedia mapping on the Internet using commercial APIs. In: Peterson MP (ed) Online Maps with APIs and WebServices. Springer, Berlin, pp (need page number for this chapter in this book)

Jhnidk (2010) Use PHP, MySql and Google Map API v3 for displaying data on map. http://tips4php.net/2010/10/use-php-mysql-and-google-map-api-v3-for-displaying-data-on-map/. Accessed 10 May 2011

Johnston LR, Jensen KL (2009) Maphappy: a user-centered interface to library map collections via a Google maps "Mashup". J Map Geogr Libr 5(2):114–130

jQuery (2011) The write less, do more, JavaScript library. http://jquery.com/. Accessed 9 May 2011

Liu SB, Palen L (2010) The new cartographers: crisis map mashups and the emergence of neogeographic practice. Cartogr Geogr Inf 37(1):69–90

Mahe L, Broadfoo C (2010) Too many markers! http://code.google.com/apis/maps/articles/toomanymarkers.html#gridbasedclustering. Accessed 28 Apr 2011

MySQL (2011) The world's most popular open source database. http://www.mysql.com/. Accessed 12 May 2011

Niccolai J (2008) So what is an enterprise mashup, anyway? http://www.pcworld.com/businesscenter/article/145039/so_what_is_an_enterprise_mashup_anyway.html. Accessed 12 Dec 2010

Pan B, Crottsa JC, Mullerb B (2010) Developing web-based tourist information tools using Google Map. http://www.ota.cofc.edu/pan/PanCrottsMullerDevelopingGoogleMap.pdf. Accessed 7 May 2011

Pejic A, Pletl S, Pejic B (2009) An expert system for tourists using Google Maps API. In: 7th international symposium on intelligent systems and informatics, SISY '09, Subotica

Peng X, Wu X (2010) Digital campus map publishing based on Google Map API. J Geomatics 35(1):25–27

Peterson MP (2007) International perspectives on maps and the internet: an introduction. In: Peterson MP (ed) International perspectives on maps and the internet. Springer, Berlin, pp 3–10

PHP (2011) What is PHP? http://www.php.net/. Accessed 9 Feb 2011

Roth RE, Ross KS (2009) Extending the Google Maps API for event animation mashups. Cartogr Perspect 64:21–31

Scholefield K (2008) Web based map services for scientific tourism: a case study of eighteenth and nineteenth century Edinburgh. Master of Science Thesis. http://hdl.handle.net/1842/2475

Udell S (2009) Beginning Google maps mashups with mapplets, KML, and GeoRSS. Apress, New York

United States Department of Agriculture (USDA) (2011) People's garden initiative. http://www.usda.gov/wps/portal/usda/usdahome?navid=PEOPLES_GARDEN. Accessed 9 Feb 2011

W3Schools (2011) XML basics and XML JavaScript. http://www.w3schools.com/xml/xml_whatis.asp. Accessed 9 May 2011

Chapter 18
Online Information Dissemination at the Wisconsin State Cartographer's Office Using Map Services and APIs

Howard Veregin and Timothy Kennedy

Abstract Since its creation in 1974, the Wisconsin State Cartographer's Office has been responsible for disseminating information about the availability of geospatial resources for the state. Over time this activity has evolved from the production of paper catalogs to custom programming of Web-based mapping applications. From 1974 to 1993, the office published paper catalogs detailing the availability of aerial photography, maps and geospatial data, and geodetic control data for Wisconsin. With the 2011 release of the Wisconsin Historic Aerial Image Finder, and new versions of its ControlFinder and PLSSFinder applications, the State Cartographer's Office has completed another chapter in its transition from paper maps to interactive Web-based mapping. These new "Finder" applications incorporate map services and commercial map APIs in an open source geospatial software framework for map-based information delivery over the Web. This chapter provides an overview of the SCO's Web-based mapping applications, to demonstrate how Web-mapping technology has impacted service delivery to SCO customers, and affected data and software maintenance workflows and activities in the office.

18.1 Introduction

The Wisconsin State Cartographer's Office (SCO) was created in 1973 with the passage of Assembly Bill 300. The statute tasked the new office with helping coordinate mapping programs at all levels of government in the state, providing cartographic consulting services, and collecting and distributing information on maps and mapping programs throughout Wisconsin. The SCO's current areas of

---

H. Veregin (✉)
Wisconsin State Cartographer's Office, University of Wisconsin-Madison, Rm. 384 Science Hall, 550 N. Park St., Madison, WI 53706, USA
e-mail: veregin@wisc.edu

responsibility are detailed in Chapter 36 of Wisconsin State Statutes (http://legis.wisconsin.gov/statutes/stat0036.pdf). For the purposes of this paper the most important of these responsibilities are:

- Cataloging current and historic maps relating to Wisconsin.
- Disseminating information about mapping methods, map and air photo indexes, control data, and other information to help facilitate an effective mapping program for the state.
- Distributing special maps and map information to promote mapping and map use in the state.

The SCO has long been on the forefront of using technology to help make these activities more efficient and provide information that is more current and more easily accessible. Although the first catalogs were necessarily manually-produced paper products, computers were quickly introduced to help automate catalog production. Later innovations included distribution of data on floppy disks and CDs, an electronic Bulletin Board System to allow remote access to data, and a Web site with data download capabilities.

For more than a decade the SCO's efforts in this domain have focused on interactive Web-based applications that allow users to obtain information about different types of geospatial data, including aerial photography and GIS data layers. These applications feature a map interface permitting users to perform spatial queries, visualize results cartographically, and explore and interact with the data. Early versions of these applications involved labor-intensive, highly customized coding due to a lack of prototypes, software libraries and applications frameworks. As Web-mapping technology has evolved over time, these applications have been reengineered with major changes to their code base, functionality, and interface design.

This chapter discusses the evolution of the SCO's efforts to employ Web-mapping technology to support its cataloging and information dissemination mission. We pay particular attention to applications that serve information about aerial imagery, geodetic control, and PLSS (Public Land Survey System) data – thematic categories that have been part of the SCO's efforts since its inception. Currently these applications are built around open source Web-mapping software and standards, and commercial APIs providing access to base maps and geospatial functions. Each of these mapping applications has evolved significantly over time with changes in user needs and mapping technology. In this chapter we highlight some of these changes, with the goal of explaining how software design decisions have been impacted by user needs and technological capabilities. We attempt to demonstrate how advances in Web-mapping technology have impacted service delivery to SCO customers and affected behind-the-scenes data and software maintenance workflows and cycles. We emphasize the importance of user-based evaluation and feedback as a key to the success of the SCO's online applications, and provide specific examples where such feedback was incorporated to better align system functionality with user needs.

18.2 From Paper Maps to Online Applications

18.2.1 Early Paper and Digital Catalogs

The SCO's earliest paper and digital catalogs contained information about aerial imagery, geodetic control, and maps and datasets produced at the state and local level. After more than 35 years these categories remain at the forefront of the SCO's information dissemination program. While the technology used to disseminate information has gone through several significant shifts, the data categories valued by SCO customers have remained remarkably consistent over the years (Fig. 18.1).

The SCO introduced its first paper catalogs in the mid-1970s. These catalogs documented Federal geodetic control data available for the state, and aerial imagery publically available from a variety of sources, including the USDA (US Department of Agriculture), USGS (US Geological Survey), NASA (National Aeronautics and Space Administration), regional planning commissions, counties, and state agencies. Soon after, the SCO began to publish paper county cartographic catalogs of publically available geospatial data at the county level. The county catalog series was published after a 4 year period of testing, evaluation, and recommendations from potential users – the first large-scale effort on the part of the SCO to gather and synthesize requirements from its user base. Ultimately the catalogs contained sample available maps and repository locations for such diverse data as location

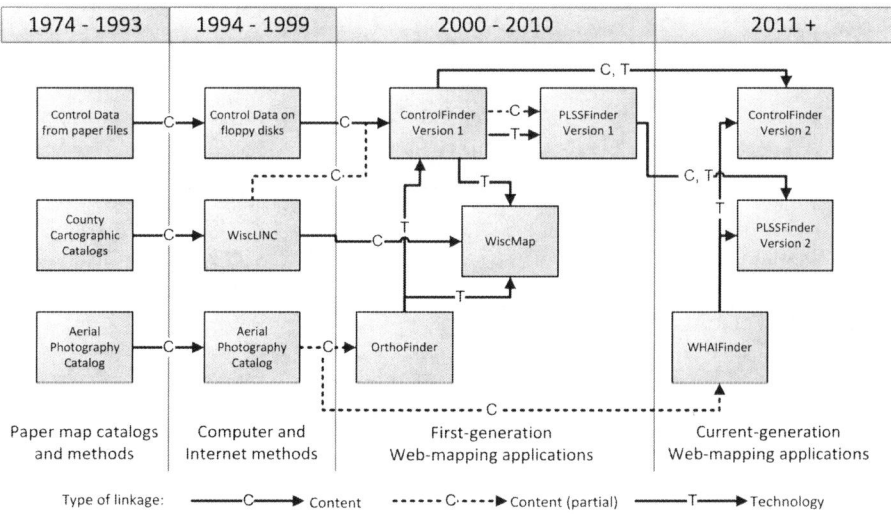

Fig. 18.1 Evolution of the SCO's information delivery platforms. Content linkages join similar types of information (e.g., aerial photography) delivered using different technology. Partial content linkages are indicative of a weaker link (e.g., aerial photography and orthophotography are related but not identical). Technology linkages join applications based on similar technology, regardless of content

and index maps, topographic maps, hydrographic maps, aerial photography and satellite imagery, geodetic control, geologic maps, soil maps, and land use maps (Figs. 18.2, 18.3). The SCO continued to publish the county cartographic catalog series at an average rate of 5.5 counties per year. Further evaluation and testing indicated high user interest in the program and the series remained a top office priority into the 1990s.

Computing technology found its way into SCO operations in the early 1980s, when the office tested the ability of micro-computers to automate catalog production. While the catalogs remained a paper-based product, computers were used for word processing and graphics production. The testing of digital production methods also allowed the SCO to advise local governments on the procurement of digital systems.

The SCO's initial foray into the creation of a digital product came in 1993 with the distribution of NGS (National Geodetic Survey) control data on 3.5″ floppy disk. Several years later, USGS third-order control points, digitized by the SCO from paper USGS topographic maps, were also packaged and distributed on disk. With the advent of SCO's first Web site in 1996 came the publication of the aerial photography catalog on the Internet. Data that had previously been available in paper map format was reformatted as HTML pages; database technology was not yet employed for data storage. Maintenance was cumbersome, requiring tedious update procedures and manual extraction of data as the system later evolved to incorporate true database components. In the late 1990s, a simple map interface was added allowing users to click on a county outline to extract information. At the same time the catalog was ported to a database. The resulting online Aerial

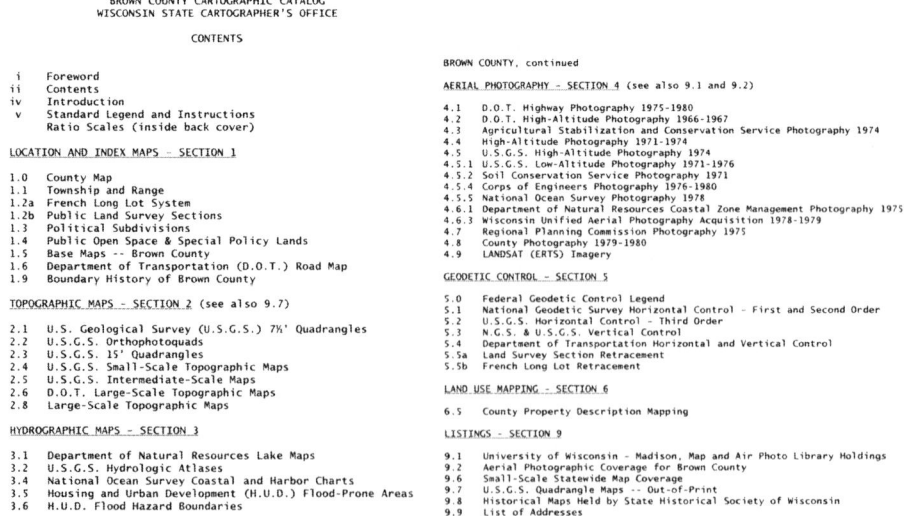

Fig. 18.2 Table of contents from a county cartographic catalog

18 Online Information Dissemination at the Wisconsin State Cartographer's Office

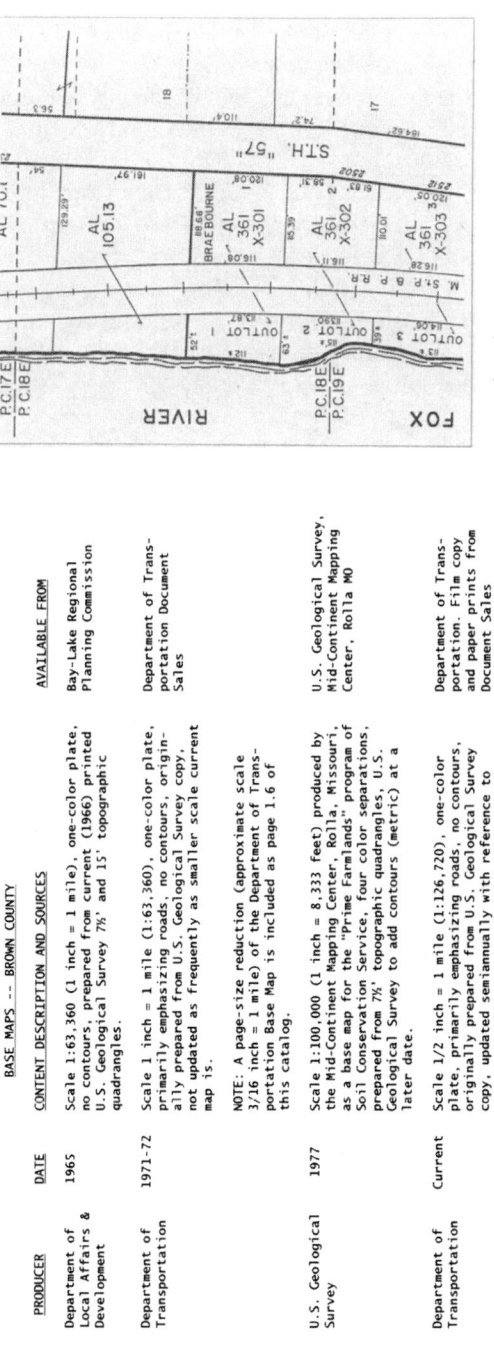

Fig. 18.3 Sample page from a county cartographic catalog. The *left side* is a sample listing of available maps. The *right side* is a sample listing of available tax maps

Photograph Catalog (or APCatalog) thus became a stand-alone application on the SCO Web site (Fig. 18.4).

Efforts at this time also focused on the development of WiscLINC (Wisconsin Land Information Clearinghouse), which provided online access to metadata information previously available through the SCO's county cartographic catalogs. WiscLINC was developed as Wisconsin's NSDI (National Spatial Data Infrastructure) clearinghouse node using the ANSI/NISO Z39.50 standard for catalog services according to FGDC (Federal Geographic Data Committee) specifications (FGDC 1999). An in-house WiscLINC prototype was created and minimally populated under a 1994 FGDC CAP (Cooperative Agreement Program) grant awarded to the SCO. The application was subsequently expanded and enhanced under contract to the Wisconsin Department of Administration in the late 1990s. WiscLINC was based on a map interface similar to the APCatalog to allow users to search and retrieve geospatial metadata by county.

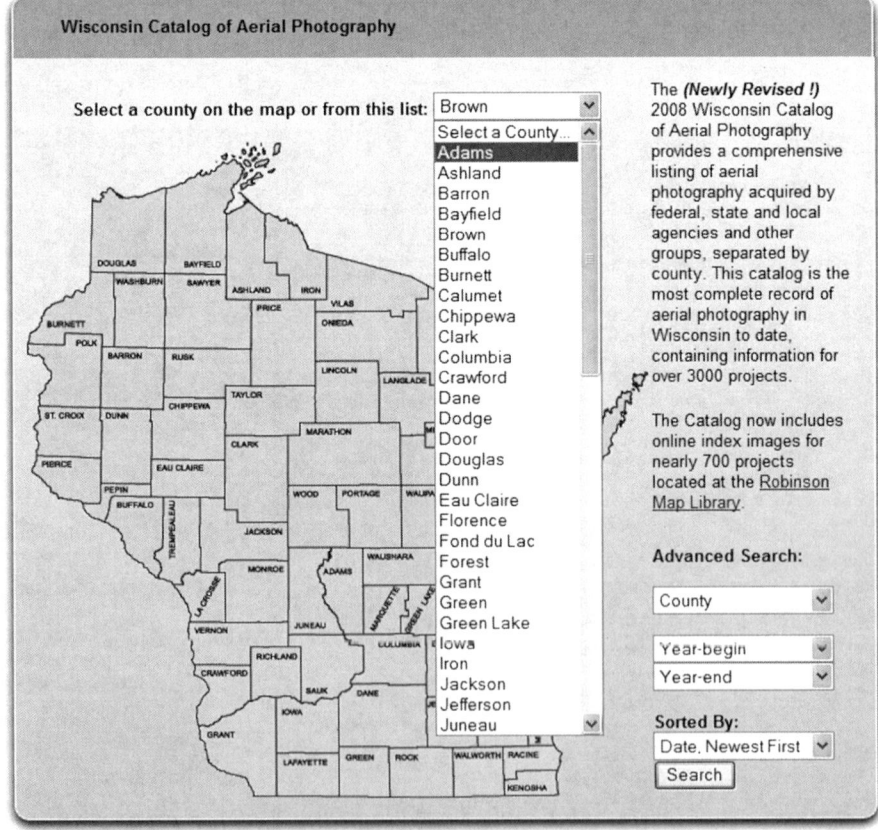

Fig. 18.4 APCatalog interface. Users can click on a county or choose a name from the pull-down list in order to get a listing of aerial photographic projects for that county. http://www.sco.wisc.edu/aerial-photography-catalog.html

18.2.2 First Web-Mapping Applications

The SCO's first truly interactive Web-mapping application, OrthoFinder, was conceived as an online catalog of orthophotography for the state. This catalog was designed to solve concerns about publishing only text-based information, by including cartographic representations of the spatial footprints of orthophotography project areas. OrthoFinder's development was assisted by an FGDC CAP grant awarded to the SCO in 2000. Goals for the grant included providing public access to orthophotography and associated metadata through a Web-based interface built around open source software, specifically MapServer (Table 18.1). Dane County, Wisconsin, was used as a case study to demonstrate proof-of-concept. The Ortho-Finder application was completed in 2001 using custom Java programming on top of MapServer. Issues with data availability limited the ability of the SCO to offer actual imagery and spatial footprints of project areas for the entire state. Still, OrthoFinder provided the SCO with valuable training in the development of Web-mapping applications.

Table 18.1 Web-mapping applications components

Component	Description
MapServer	An open source platform for publishing spatial data and interactive mapping applications to the Web. Originally developed in the mid-1990s at the University of Minnesota. http://mapserver.org/
OGC	The OGC (Open Geospatial Consortium) is a non-profit standards organization leading the development of standards for geospatial and location based services. http://www.opengeospatial.org/
OGC WFS	Web Feature Service. An OGC standard providing for sharing of geographic features from distributed databases and their integration into Web-map frameworks and GIS clients
OGC WMS	Web Map Service. An OGC standard providing for sharing of map images and their integration into Web-map frameworks and GIS clients
PostgreSQL	A widely used open source object-relational database system. http://www.postgresql.org/
PostGIS	Adds support for geographic objects to the PostgreSQL object-relational database, thus spatially enabling the PostgreSQL server. http://postgis.refractions.net/
OpenLayers	A JavaScript library for displaying map data in web browsers. http://openlayers.org/
GeoMOOSE	A Web Client JavaScript Framework for displaying distributed cartographic data. http://www.geomoose.org/
Google APIs	Application programming interfaces to request Google basemap tiles and mapping functions. http://code.google.com/apis/maps/index.html
Fedora Commons	Flexible Extensible Digital Object Repository Architecture. Developed by researchers at Cornell University for storing, managing, and accessing digital content. http://fedora-commons.org/
UWDCC Fedora API	Application programming interface authored by UWDCC (University of Wisconsin Digital Collections Center) for access to air photo metadata, thumbnails, and images

Following OrthoFinder the SCO focused its efforts on mapping of geodetic control data. The SCO had long been involved in disseminating geodetic information to the surveying community in the state. Traditionally, users obtained information by calling the office; a staff member would conduct labor-intensive research using paper maps and then relay information back to the user. Later this approach was supplemented with the use of digital media. In the early 2000s the SCO became committed to the idea of finding a way to permit users to perform this research function over the Web. This would allow on-demand access to information and significantly streamline staff operations. Experience with other Web applications, including OrthoFinder, provided the necessary background training needed to develop a new application serving geodetic control data.

The result of these efforts was ControlFinder, first released in April, 2003. ControlFinder was built on open source software including MapServer. The application included control data from the USGS, NGS, and some Wisconsin counties, stored in Esri shapefiles. The system had the ability to link automatically to online tie sheets (official survey documents containing information about control points) maintained by federal agencies and local governments. At the time of its release, ControlFinder was probably the only Web-based application in the country to provide integrated geodetic control information on a statewide level.

Construction of ControlFinder took approximately 1 year. The custom programming was completed, tested, and implemented using in-house staff and graduate assistant resources. At the time, most application functionality had to be custom-coded due to a lack of Web-mapping libraries and application frameworks. Partly as a result, the application had some limitations. For example, maps were limited to a single size and there was no capability to download data. The base map for the product was built in-house using best available statewide data, and was somewhat generalized in content and appearance.

In 2003, following the initial release of ControlFinder, the SCO received another FGDC CAP grant to create an OGC (Open Geospatial Consortium) compliant map service viewer (Table 18.1) built on MapServer software to enable viewing of online map service data for the state. This mapping application, called WiscMap, allowed users to visualize geospatial data served from some of the distributed online sources cataloged in WiscLINC. WiscMap was adapted from the SCO's OrthoFinder and ControlFinder Web-mapping applications but incorporated a custom map tiling scheme using early tile caching technology.

In 2004, ControlFinder received a major upgrade, funded in part by another FGDC CAP grant. The major impetus for this grant was to demo OGC-compliant Web services, specifically Web Feature Services (WFS), as a way to deliver data to users over the Web (Table 18.1). The project also led to numerous enhancements to ControlFinder. Code was refactored to improve efficiency, data was moved from shapefiles to a PostGIS–enabled PostgreSQL database (Table 18.1), the base map was replaced with USGS DRGs (Digital Raster Graphics), and HMP (Height Modernization Program) points were added. Growth of the ControlFinder user community also led to the incorporation of local geodetic control for additional counties and, later, the incorporation of CORS (Continuously Operating Reference System) datasets (Fig. 18.5).

In the late 2000s the SCO added further enhancements to ControlFinder. DRGs were still being used as base maps but these were losing favor as they became increasingly out-of-date. Custom base maps were again created using state and federal datasets and custom cartographic specifications. In addition, the ability to download control point data in shapefile and KML format was added, providing users with additional tools to support their data needs. A custom SDMS format was also added to support Wisconsin Department of Transportation users. These efforts toward providing direct access paths to data (i.e., downloads) rather than just reporting data existence (i.e., data catalogs) reflect a continuing trend in SCO applications.

The success of the ControlFinder application was due, not just to effective design and execution of software, but to the sustained effort to provide a usable, tangible asset to Wisconsin constituents, primarily the surveying community. Staff annually attended surveying conferences and presented new improvements and enhancements, and actively sought feedback from users about functionality, usability, application design, reference data and content. This specific user-feedback generated real-world problems and solutions that were implemented in the product. The success of ControlFinder also led to the development of a sister application to supply access to Public Land Survey System (PLSS) corner records. County surveyors in Wisconsin asked the SCO to develop and host a product similar to ControlFinder that included county PLSS data, including PLSS corners that had

Fig. 18.5 ControlFinder interface (pre-API version)

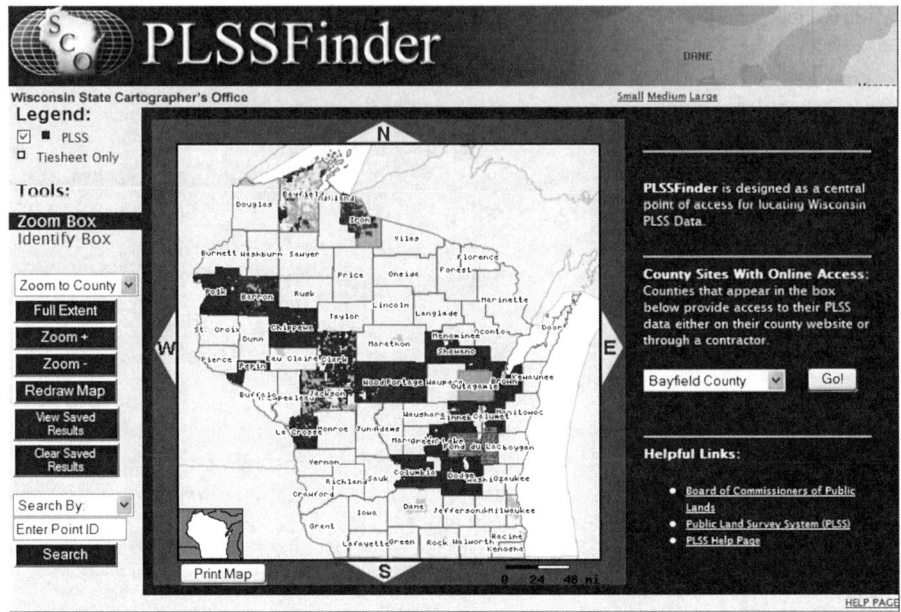

Fig. 18.6 PLSSFinder interface (pre-API version)

been remonumented. In 2007 PLSSFinder was released using the same architecture as ControlFinder (Fig. 18.6).

18.2.3 Incorporation of Commercial Map APIs

From the beginning the SCO's Web-mapping applications were built using open source software and standards. This approach affords maximum flexibility in terms of handling geospatial content in different formats, designing functionality and interface features, creating reusable code, and maximizing interoperability. Starting in the mid-2000s, industry software trends led to further evolution of the SCO's mapping applications, including a new interface design and underlying code base. The stimuli for these developments included the emergence of commercial map APIs offering online access to maps and geospatial services, and the growing availability of open source frameworks like OpenLayers and GeoMOOSE that interface with commercial API services and mapping software (Table 18.1).

From a user perspective, there was also a clear need to upgrade the platform used to disseminate information about aerial photography, which was several years out-of-date given the end of the OrthoFinder project. An opportunity was created in 2008 with the receipt of a grant from the Ira and Ineva Reilly Baldwin Wisconsin Idea Endowment at the University of Wisconsin – Madison, which led to the

18 Online Information Dissemination at the Wisconsin State Cartographer's Office

development of a new application for delivery of historic aerial photographs of the state, called the Wisconsin Historic Aerial Image Finder (WHAIFinder) (Veregin et al., 2010). This project provided the updated code base to reengineer the ControlFinder and PLSSFinder applications to incorporate new technological enhancements and functionality.

The target dataset for the WHAIFinder project was the first comprehensive statewide aerial photography mission flown in Wisconsin, conducted under the auspices of the USDA and USGS from 1937 to 1941. Approximately 38,000 photographs from this collection were scanned and processed for the project. This included tagging of images with metadata to provide search and cataloging capabilities. The photograph collection represents nearly 100% statewide coverage (remarkably, only 85 exposures from the original mission are missing) and is in wide demand as it is the oldest and most complete aerial survey of the state.

The WHAIFinder interface was designed to balance the needs of non-professional users, including map-based navigation, selection and preview functions, with library-based repository storage, delivery and viewing options (Fig. 18.7). Key design goals included emulation of features of paper index maps, use of a current and high-quality base map, incorporation of spatial search functionality (places, addresses, counties, PLSS features, etc.), and the ability for users to download digital image data to their desktops. To achieve these goals, an open source software approach incorporating the Google Maps API was chosen (Table 18.1).

Use of the Google Maps API offered several advantages for system design and maintenance. First, by this time, pre-rendered raster tiled base maps had achieved widespread use in Web-mapping applications, offering a high degree of familiarity

Fig. 18.7 WHAIFinder user interface. http://www.sco.wisc.edu/WHAIFinder/

to even casual users of the Web who were part of the target audience for the WHAIFinder application. This reduced the need to create custom training materials to assist users with basic map reading and navigation functions. Equally important, commercial APIs offer ready access to base maps in a variety of styles, such as street maps, physical maps, and image-based maps. The availability of these maps eliminated the need for the SCO to develop customized base map solutions in-house, as it had done in the past. Also, these base maps were more up-to-date than other ready-made raster solutions that the SCO had employed in the past, such as DRGs. The net effect of these factors was more rapid system development, lower maintenance costs, and higher quality results.

API-based geospatial functions, such as geocoding, also serve to simplify and streamline system development by offering pre-configured access to geospatial functions. For the WHAIFinder application, the Google Geocoding API was adopted to allow access to search functions by address, geographic coordinate or place name. While commercially available base maps and functions provide a foundation for the WHAIFinder application, customization was necessary to meet design requirements. Additional map layers, including counties and PLSS features that do not appear on Google base maps, were added to the WHAIFinder as OGC-compliant Web Map Services (WMS) (Table 18.1). In addition, functionality was developed to allow users to search by county or PLSS feature. The WHAIFinder system is modular in design. Scanned aerial photographs are maintained in a Fedora Commons repository (Table 18.1) that is physically separated from the PostGIS-enabled PostgreSQL database that houses custom map layers and photo center point coordinates. Standard protocols and APIs are used for communication (Fig. 18.8).

As shown in Fig. 18.8, users interact with the system in different ways, including basic navigation, selection and preview of images, and image download. Browsing makes use of Google base map tiles through the Google Maps API, with additional custom layers added as Web Map Services (WMS). Zooming to a location makes use of the Google Geocoding API, while a search based on a county or PLSS feature makes use of custom functions employing base MapServer capabilities. Selection and preview of photos requires calls to both the PostGIS database and the Fedora repository where scanned images are stored. Image download makes use of a custom API allowing access to scanned images in various formats and resolution levels. The full functionality of the WHAIFinder lies in the ability to download actual image data. As with the shapefile and KML download features of ControlFinder and PLSSFinder, the capability to serve actual data out of the interface represents a trend in the SCO's online applications toward direct access paths to data rather than just reporting data existence.

As construction of the WHAIFinder progressed, the SCO began to consider adapting it to the other Finder applications as a way to reduce base map maintenance costs and leverage the efficiencies inherent in the updated code base. The WHAIFinder project in effect paved the way for new versions of ControlFinder and PLSSFinder. Indeed the architecture of all three Finders is very similar, with the major exception being the methods and software used for data storage, access and retrieval (scanned images in one case, versus control points and PLSS corners in

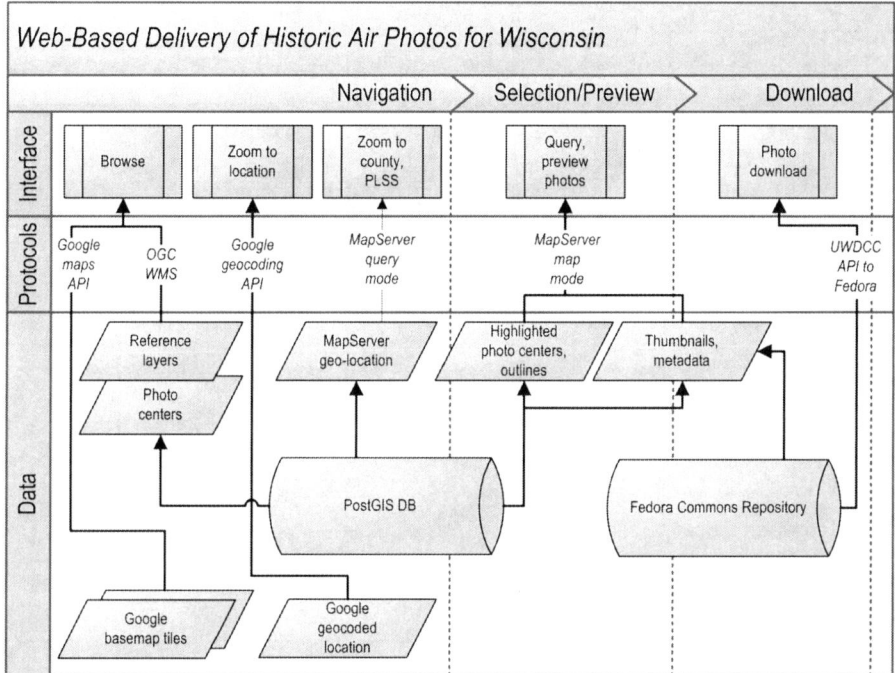

Fig. 18.8 WHAIFinder system functionality. Scanned photographs are stored in the Fedora Commons Repository. Photograph center points and additional spatial layers (counties and PLSS features) are stored in a PostGIS-enabled database. Google base map tiles are combined with these layers in the application interface. Each type of user interaction (navigation, selection/preview, and download) makes use of specific APIs and geospatial services. See Table 18.1 for a more detailed explanation of components

another) (Fig. 18.9). The interface of all three Finders is also quite similar, incorporating Google base maps and a tab structure where functions such as search, map design, identification, and selection are organized (Figs. 18.10, 18.11).

18.3 Effects on Service Delivery

18.3.1 User Interaction

One of the key differences between the SCO's original paper map catalogs and later API-based Web-mapping applications is the level of interactivity permitted and the degree to which users can customize their requests for information. The SCO's Finder applications allow users to interact spatially with the map (e.g., navigate and explore an area of interest), change map scale (e.g., by using the zoom tool), conduct spatial searches (e.g., find a feature or address), perform spatial queries

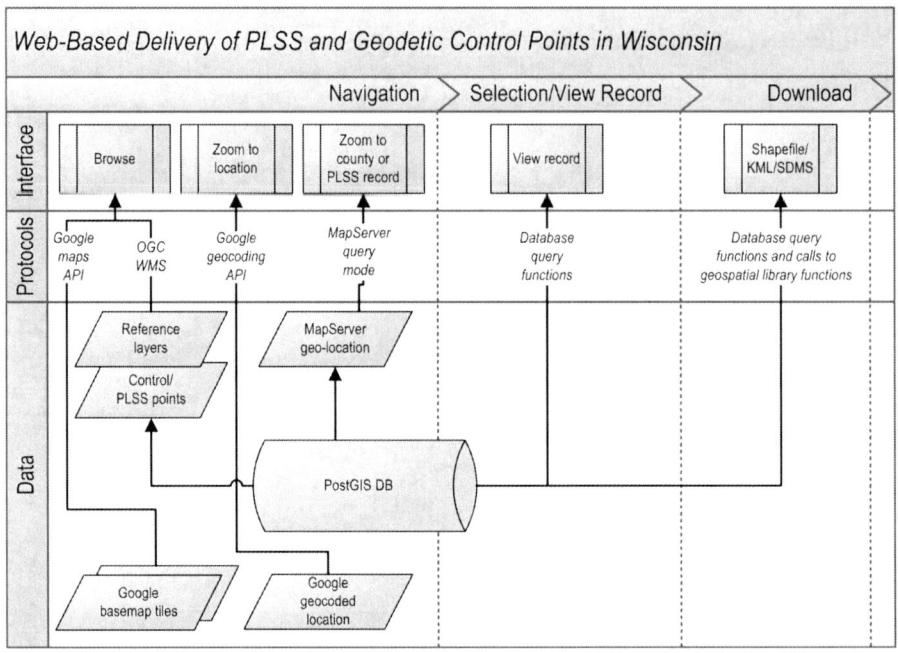

Fig. 18.9 ControlFinder and PLSSFinder functionality. These systems are similar in architecture to WHAIFinder (see Fig. 18.8)

Fig. 18.10 Current ControlFinder interface. http://maps.sco.wisc.edu/controlfinder/

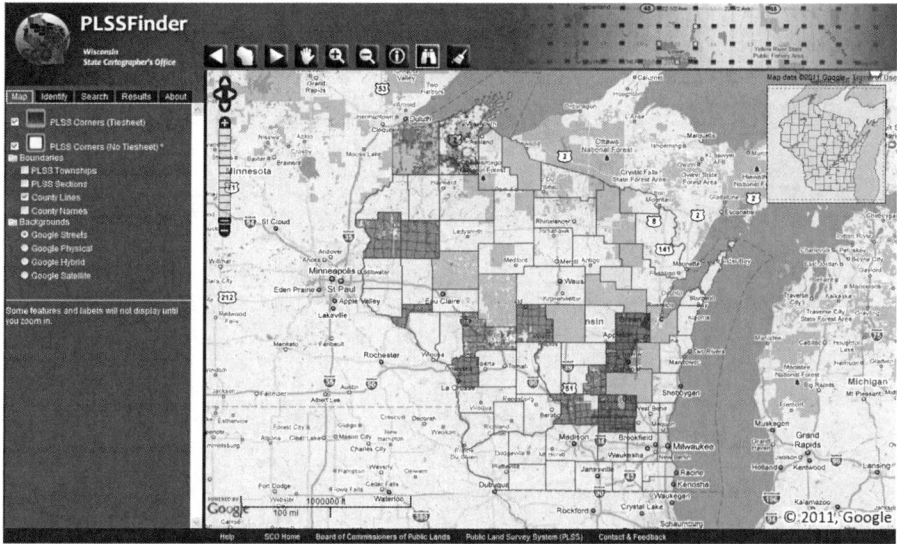

Fig. 18.11 Current PLSSFinder interface. http://maps.sco.wisc.edu/plssfinder/

(e.g., click on a point to select an aerial photograph), visualize results cartographically (e.g., use special map symbols to highlight a selected set of control points), and customize the map display (e.g., turn layers on and off, or change the appearance of the base map).

It should be noted that some of these operations are really analogs of functions that users traditionally performed using paper maps. Indeed, the interface for the WHAIFinder attempts to emulate some of these manual tasks, e.g., by displaying photograph center points and roll-exposure numbers over the base map to simulate the appearance of paper index maps. Other types of interactivity also reflect traditional manual procedures. For example, panning and zooming functions in a digital map interface are analogous to the navigation and exploration functions users perform with paper maps through visual interpretation of map symbols and cartographic standards. Likewise, spatial searches, such as finding a specific populated place, have long been performed with paper maps using an index or gazetteer. In these cases, Web applications offer traditional modes of interaction but in a streamlined, more efficient way, thus heightening user satisfaction and ideally affording greater success in information gathering.

Web-based mapping also offers interactive functionality that is not possible with paper maps. An example is the ability to turn on or off various thematic overlays to customize the cartographic display for specific purposes. Such "on-the-fly" customization is not possible with paper maps due to the obvious limitations of the medium. In the same way, Web-based maps offer much greater flexibility than paper maps in terms of the ability to perform and visualize spatial and attribute queries. In the SCO's paper map catalogs, areas of data availability were symbolized differently, allowing users to conduct basic queries (e.g., "What counties have recently acquired aerial photography?") based on an interpretation of map symbols. In a digital environment, the query can be customized and refined to a much greater degree, and is restricted only by the ability of the underlying data to support the query, not by a priori decisions made by a cartographer.

18.3.2 Information Access

A major advantage of the Web is that it offers improved access to information. There are several dimensions to this, including time-to-publishing or currentness, on-demand access, and the ability to download usable datasets.

The SCO's paper map catalogs published from the 1970s to 1990s required an average production time of over 2 months per county (5.5 counties per year). This implies that the earliest catalogs were many years out-of-date by the time the series was completed. This time lag resulted from the extensive production efforts required to manually compile data into map form, and then print and distribute the map catalogs.

The Web has reduced this time lag by making data available as soon as it has been published to a data repository accessible through an interface. The Web offers instantaneous access to the repository, such that data updates and changes are instantly available to users. This capability requires a modular design in which the interface is separated from the data repository rather than being bundled into a single module (as in the case of some canned Web applications). Standard protocols and APIs can then be used to communicate between the program modules and build page content dynamically. It is noteworthy that this capability exists without the need to "republish" in the traditional sense, whether that means a new edition of a printed map or a new version of a software package distributed on CD. The update and maintenance process is for all intents and purposes transparent to the user.

Instantaneous publishing has a clear advantage in terms of content freshness, which should translate into lower information costs and higher information quality. In addition, these advantages are coupled with service ubiquity whereby access to up-to-date data is possible from any computer with an Internet connection, without the need for special tools or software. Information is available whenever and – with the increasing sophistication of mobile technology – wherever it is needed. These are positive influences, giving users the ability to obtain the information they need for decision-making with minimal constraints. Equally important is the trend toward providing access to actual data, rather than just reporting data existence as in the case of map catalogs and metadata clearinghouses. The Finder applications described in this paper provide access to data in the form of digital images, shapefiles, and KML documents for easy integration into GIS and other software.

These features have much in common with current IT trends, including cloud computing and SaaS (Software as a Service). SaaS characteristics include Web-based access to software on an on-demand basis, no downloads or special software requirements (i.e., access through a browser), centralized updating obviating the need to publish updates and patches to users, and integration into larger frameworks such as mashups ("Software as a service" 2011). Cloud computing refers to on-demand access of third-party computing resources (storage or CPU) with the goal of maximizing the use of the Web to access cost-effective computing power ("Cloud Computing" 2011).

18.3.3 Impacts on Workflows and Skillsets

The technological transformations described above have significantly impacted the work habits of SCO customers and staff. One of the "technological imperatives" of the Web is to offload onto consumers tasks that were once managed or assisted by experts on the supply side. This effect is now ubiquitous, as anyone who has used the Web to book a flight or hotel room can attest to. In a sense it is a natural, almost inevitable, consequence of the Web's capability to automatically and instantaneously deliver customized information to users.

For SCO customers, one implication of these trends is a need to become more fluent, not only with computers and the Web in general, but more specifically with maps, cartography, and geographic information. Users must become more self-sufficient in these areas if they are to use online applications to obtain the information they require, since there is no human expert to assist them. Without adequate training there is the distinct possibility that levels of user satisfaction and success in information gathering will decline despite the power of the new technological means.

Web mapping and the availability of commercial map APIs has also induced a change in office workflows and responsibilities. As noted above, the early Finder applications were built around base maps developed in-house using best available data and customized cartographic design specifications (or off-the-shelf graphics such as DRGs). The current Finder versions make use of the Google Maps API to access more current base maps of various styles that incur no development or maintenance costs. Similar statements can be made in reference to other functions, such as the Google Geocoding API that provides the ability to conduct intelligent spatial searches without the need to develop or maintain a digital gazetteer or complex search functions. Software is thus easier, cheaper, and faster to build and maintain.

By offering an alternative to functions and maps developed in-house, commercial APIs have impacted the demand for specific competencies within the office. Manual methods of cartography are now wholly obsolete. And while map design continues to be important, more emphasis is placed on the technology components of database design, automated maintenance processes, and code development to support Web-based map and information delivery. Much of the basic cartography is taken care of already. Cartography, like GIS, has become more IT-focused over the last decade and will continue to be so in coming years.

18.3.4 APIs and Customization

While Web-mapping technology allows for user interaction and customization, true customizability still has a long way to go. Modern Web-mapping frameworks and APIs make it relatively easy to incorporate additional layers of content (e.g., counties and PLSS features) in a map interface; this is the basic definition of a "mashup" after all. But modifying existing map symbology, incorporating localized spellings, or including place names that do not exist in a stock gazetteer are not easily achieved within this framework. In most Web-mapping applications, customizability has rather narrow, pre-defined limits. The commercial map tiles delivered by Google and other APIs are raster images, not live vector data, and they contain immutable, pre-rendered representations of features. Changes in map symbology, and in feature selection and generalization, occur only at fixed zoom levels and only according to rules imposed a priori by the data provider. Likewise, the ancillary layers superimposed on these tiles may give the appearance of live data, but often are also pre-rendered maps delivered in image format and offering limited user customizability.

Ultimately the Web offers the potential for "deep interactivity" – bi-directional information exchange between agents where the exchange transforms the functioning of the entities involved (Lindborg 2008). This means customized, on-the-fly cartographic renderings tailored to the specific needs and special queries of the user. Achieving deep interactivity requires a stored set of rules and procedures governing such mapping tasks as feature selection, generalization, labeling and symbology, coupled with an underlying dataset containing the data that drives these rules and procedures. Properly implemented, deep interactivity promises the user the ability to receive, in real-time, answers to custom queries formatted in ways that conform to the user's special requirements. Technologically there is still some distance to go, but it can be argued that this capability is the real value proposition of Web mapping.

In the meantime an additional challenge that needs to be addressed is the impact of the Web on traditional cartographic design. In traditional cartography, maps are designed as composite objects. That is, the layers of content composing the map sandwich, the symbology assigned to these layers, and all associated annotation and marginal content are blended together to create a visually satisfying and effective result. Web-based mapping, especially when relying on pre-rendered base maps, is at odds with this design philosophy since it typically makes use of a variety of content sources that have not been designed to be combined in ways that produce an effective graphical result (Cartwright 2008). There is no guarantee that the various layers will have the same level of generalization, that features will interact properly with each other, or that competing symbology will not obfuscate the map's message. As the Web continues to grow in popularity as a mapping medium, these issues will come increasingly to the fore, but for now they remain obstacles to effective design and communication.

18.4 Conclusion

In this chapter we have discussed the evolution of the SCO's efforts to employ Web-mapping technology to support its cataloging and information dissemination mission. The geospatial data themes of importance to the SCO user community, including aerial imagery, geodetic control and PLSS data, have remained fairly constant over time, while the technology used to deliver this information has gone through several significant changes. The major steps in this evolution include paper maps, digital media such as CDs, static Web pages, interactive Web maps and, finally, Web-mapping applications built on open source software and commercial map APIs.

As we have tried to demonstrate here, the evolution of the SCO's information delivery media has been stimulated by both technological change and the changing needs of the user community. The success of the SCO's efforts in the realm of Web-based mapping owes much to ongoing efforts to understand and respond to new user requirements and align system functionality with user needs. According to this user-centered design (UCD) approach, user requirements are at the forefront of the design process. The benefits of designing systems in this way include higher productivity, greater user satisfaction, increased demand and use, and improved return on investment (Haklay 2010).

At the same time the new technological means has impacted service delivery and data and software maintenance workflows and cycles, in both positive and negative ways. On the negative side, there is still much work to be done improving the customizability of Web-based maps to allow this medium to achieve its full potential for information exchange and delivery. Likewise, additional effort needs to be focused on cartographic design to ensure Web-based maps communicate information accurately and effectively. On the positive side, modern Web-mapping methods have facilitated broader access to data and information, and have contributed to reduced system development and maintenance costs. As awareness of these methods continues to grow, geospatial technology and applications are increasingly being incorporated into traditionally non-spatial disciplines (Goodchild and Janelle 2010; Bodenhamer et al., 2010). In particular, the availability of Web-map APIs and services, and the relative ease with which these can be integrated into online applications, offers the almost unlimited ability to combine data and functions in novel and unique ways to serve an increasingly broad spectrum of users.

Acknowledgements The projects summarized in this chapter build on the efforts of two previous Wisconsin State Cartographers, Arthur L. Ziegler and Ted Koch, and numerous SCO staff members and students. Brenda Hemstead and AJ Wortley in particular have been closely involved, as have numerous student assistants who have helped build, test and maintain the applications described here. The SCO's efforts in Web-based delivery of geospatial data have been greatly facilitated by funding from the Federal Geographic Data Committee (http://www.fgdc.gov/grants) and the Ira and Ineva Reilly Baldwin Wisconsin Idea Endowment (http://www.provost.wisc.edu/baldwin/). Partners on the Baldwin Endowment project include the Department of Geography at the University of Wisconsin-Madison, the Arthur H. Robinson Map Library, and the University of Wisconsin Digital Collections Center. Editorial assistance for this chapter was provided by Brenda Hemstead, AJ Wortley, Jim Lacy, and Michael Bricknell. All attempts have been made to outline the SCO's history and activities as accurately as possible based on staff memories and office publications. Any errors or omissions are the responsibility of the authors.

References

Bodenhamer DJ, Corrigan J, Harris RM (eds) (2010) The spatial humanities: GIS and the future of humanities scholarship. Indiana University Press, Bloomington

Cartwright W (2008) Delivering geospatial information with Web 2.0. In: Peterson MP (ed) International perspectives on maps and the internet. Springer, Heidelberg, pp 11–30

Cloud Computing (2011) In Citizendium. http://en.citizendium.org/wiki/Cloud_computing. Accessed Feb 2011

Federal Geographic Data Committee (FGDC) (1999) Z39.50 application profile for geospatial metadata or "GEO". http://www.fgdc.gov/standards/projects/GeoProfile. Accessed Feb 2011

Goodchild MF, Janelle DG (2010) Toward critical spatial thinking in the social sciences and humanities. GeoJournal 75:3–12

Haklay M (ed) (2010) Interacting with geospatial technologies. Wiley-Blackwell, Chichester

Lindborg P (2008) Reflections on aspects of music interactivity in performance situations. Canadian Electroacoustic Community, 10.4. http://cec.concordia.ca/econtact/10_4/lindborg_interactivity.html. Accessed Feb 2011

Software as a service (2011) In: Wikipedia http://en.wikipedia.org/wiki/SaaS. Accessed Feb 2011

Veregin H, Gorman PC, Stoltenberg J, Wortley AJ, Bricknell M (2010) An open source web application for historic air photo display and distribution in Wisconsin. In: Proceedings, AutoCarto 2010, Orlando

Chapter 19
WebGIS Systems for Planetary Data Access at the PDS Geosciences Node

J. Wang, D.M. Scholes, and K.J. Bennett

Abstract NASA's Planetary Data System (PDS) is a geographically distributed system with five discipline and three support nodes that work together to archive and distribute scientific data from NASA's planetary missions. The Geosciences Node focuses on data for Earth's Moon, Mercury, Venus, and Mars. As of September 2010, the Geosciences Node holdings consist of about 50 TB of data, with the addition of about 1 TB per month. To support search and retrieval of data from the Geosciences Node archives, two online systems, Analyst Notebooks (ANs) and Orbital Data Explorers (ODEs) have been developed. Both ANs and ODEs have WebGIS systems embedded to enhance their map-based search and display capabilities. An ESRI® ArcGIS Server was used to build maps and publish web services. JavaScript application programming interface (API) was used extensively to access those map services and build web interface with interactive maps. This paper first reviews the existing planetary online map systems used in the planetary community. Then the challenges of planetary GIS are discussed. The development of WebGIS for ANs and ODEs with ArcGIS Server and JavaScript API is introduced. This paper also discusses data sharing through planetary Web service including WMS/WFS based on Open Geospatial Consortium (OGC) standards. A solution is presented in this paper to solve the map projection issue of planetary Web services. Free client tools used to test the OGC WMS/WFS services are covered at the end of this paper.

J. Wang (✉)
Department of Earth and Planetary Sciences, Washington University in St. Louis, 1 Brookings Drive, CB 1169, St. Louis, MO 63130, USA
e-mail: wang@wunder.wustl.edu

19.1 Introduction

NASA's Planetary Data System (PDS, Slavney et al., 2011) is a geographically distributed system with five discipline and three support nodes that work together to archive and distribute scientific data from NASA's planetary missions. The Geosciences Node is one of the science discipline nodes, focusing on data for Earth's Moon, Mercury, Venus, and Mars. As of September 2010, the Geosciences Node holdings consist of about 50 TB of data, with the addition of about 1 TB per month. Those data come from different missions including orbital missions such as Mars Global Surveyor (MGS), Mars Reconnaissance Orbiter (MRO), ESA's Mars Express, Mercury MESSENGER (Mercury Surface, Space Environment, Geochemistry and Ranging) mission, and Lunar Reconnaissance Orbiter (LRO) program, as well as the surface missions such as the Mars Exploration Rovers (MER) Spirit and Opportunity, Mars Phoenix Lander, LCROSS (Lunar Crater Observation and Sensing Satellite) program, and Apollo landed missions. The Geosciences Node Web site is visited by people from all over the world to search and download data of interest. In the 12 months preceding September 2010, roughly 50 GB of data were downloaded daily (personal contact with Dr. Raymond E. Arvidson).

With more international planetary missions planned, large amounts of data will be available in the coming decades. The worldwide demand for the planetary data requires research to find efficient ways to organize, present, share, and distribute the large amount of data to the science community and general public users. Planetary sciences have greatly benefitted from the advancements in Geographic Information Systems (GIS) over the past two decades. GIS tools are used to view data, process data, and generate geology or thematic maps. Free desktop GIS tools applicable to the planetary science data include JMARS (Java Mission-planning and Analysis for Remote Sensing) from the Mars Space Flight Facility at Arizona State University (Gorelick et al., 2003), as well as NASA World Wind (World Wind 2010) from NASA Ames Research Center. Traditional desktop GIS has limitations related to sharing and distributing the data among users. The application of WebGIS makes it possible to publish large amounts of planetary GIS data to the science community and general public through the internet. Several planetary WebGIS applications have been developed using various platforms. For example, The United States Geological Survey (USGS)'s PIGWAD (Planetary Interactive G.I.S.-onthe-Web Analyzable Database) has been available since 1999 using Environmental Systems Research Institute's (ESRI) ArcIMS (Hare and Tanaka 2000, 2002). An OSU Mars WebGIS system (Li et al., 2007) was built in 2004 at The Ohio State University. This system used a similar ESRI® ArcIMS technique to display the MER mission landing site maps and to distribute some of the derived data products. MARSOWEB was created by NASA Ames research center to support landing sites selection for future missions to Mars. Google has both desktop application and internet versions to allow users to view the Mars and Moon maps generated from PDS data. Additionally, Open Geospatial Consortium (OGC) Web Map

Service (WMS) is used at websites such as USGS Map-A-Planet at the PDS Imaging Node, as well as the Web Map Servers, OnMars and OnMoon, at Jet Propulsion Laboratory (Plesea et al., 2007). USGS also provides the Gazetteer of Planetary Nomenclature through OGC WMS or Web Feature Service (WFS) (Akins et al., 2010).

The Geosciences Node has developed two online systems, Analyst Notebooks (AN) and Orbital Data Explorers (ODE), to support search and retrieval of surface and orbital data, respectively. These two systems complement the above mentioned tools and Web systems, and will be augmented based on user comments. For the landed missions, Spirit, Opportunity, Phoenix, LCROSS, and Apollo Analyst Notebooks have been developed to allow the user to "replay" the Mars/Moon surface missions by integrating maps, images, sequence information, relevant data, and documentation. For orbital data, the Node has developed Mercury, Moon, and Mars ODEs with map and text based searches and displays of data housed within the PDS. Both ANs (Stein et al., 2010; Wang et al., 2010a) and ODEs (Bennett et al., 2008; Wang et al., 2009, 2010b, c) have WebGIS systems embedded to facilitate online maps to enhance the map-based search and retrieval capabilities. The following paper will first discuss the challenges in the planetary GIS. Then, it will introduce the development of WebGIS for ODEs and ANs using ArcGIS server and JavaScript application programming interface (API) techniques. This paper will also talk about planetary Web services and the related issue with map projection. Free client tools used to test the OGC WMS/WFS services will be covered at the end of this paper.

19.2 Challenge in the Planetary GIS

There are many challenges when implementing GIS solutions in the planetary science. First, users may have difficulty following GIS standards and manipulating various data into different formats and planetary coordinate systems. For example, thousands of older images and maps from multiple missions and instruments are referenced in a wide range of disparate coordinate systems, such as the positive-west system. It is a challenge to integrate, register, and analyze these data sets with the ones using the positive-east system. The data needs to be standardized in a GIS supported formats and common coordinate systems. To support this effort, the data also need to be compatible with each other and to be compatible among different commercial or free GIS packages. PDS Imaging Node is working with USGS to provide a Unified Planetary Coordinates (UPC) database to the planetary science community (Akins et al., 2009). A Planetary Image Locator Tool (Pilot) was developed at the Imaging Node to serve the UPC database. The unified system will eliminate the potential confusion among different users and let users to search, correlate, combine, and compare data from multiple sources.

Secondly, a gap still exists between Earth-based GIS and planetary GIS. One particular example is the planetary map projections. For most of the commercial

GIS software and free GIS tools, it is often impossible to find a European Petroleum Survey Group (EPSG) code to match the projection of planetary data. Hare et al., (2006) have proposed to improve support for planetary coordinate reference systems (CRS) within existing OGC standards and geospatial applications. As the demand for planetary GIS data is increasing, commercial GIS companies have gradually added support for planetary coordinate systems. ESRI has included the geographic coordinate systems for planets of the solar system in the past few years. However, there is no predefined projection coordinate system for planets other than Earth in the ESRI software.

Finally, how to distribute and share the data is another concern. Traditional desktop GIS might have full GIS functionalities, but is inefficient when sharing the data. They also have high learning curve for non-GIS users among the planetary science community. The geospatial service-oriented architecture (SOA) could be the way to share planetary data. By providing planetary Web services such as WMS, WFS, and Web Coverage Service (WCS) services, one can access data via different levels. WMS can be used to access a static map image. To access geospatial features, one can use WFS. Users can obtain the actual image data through WCS. The rapid development in Web technology also enables user to have sleek controls by manipulating available Web services through open source toolkits.

Several software packages have the ability to process geospatial data, create maps and provide Web services for data sharing. The commercial Web mapping packages include Intergraph GeoMedia® WebMap, ESRI® ArcIMS, ArcGIS Server, and ERDAS APOLLO. ESRI and Intergraph Corporation are GIS software leaders. ERDAS has reputation in photogrammetric data capture and remote sensing image processing. ERDAS strengthened its capability in Web Feature Sever and Image Web Server by purchasing the IONIC Software and Earth Resource Mapping Ltd (ER Mapper) in 2007. Some free open source platforms include GeoServer, Mapserver of the University of Minnesota (UMN), MapGuide, and Mapnik. GeoServer is a Java-based free software server that supports OGC standards WMS/WFS services for map creation and data sharing (Geoserver.org 2010). It has integrated OpenLayers, a free mapping library, to make map generation quick and easy. UMN Mapserver (Mapserver 2010) was developed in the mid-1990s. It is maintained by international developers. Other open source tools, such as CartoWeb and Chameleon, are built on MapServer for developing Web Mapping applications. MapGuide open source is a Web-based platform supporting Web mapping applications and geospatial Web services (MapGuide 2010). Mapnik is a free toolkit written in C++ for both desktop and Web development (Mapnik 2010).

Among these platforms, ESRI® ArcGIS Server was chosen for our application because of a university license, and the experience of the developer with ESRI® ArcGIS. As an industrially standardized GIS platform, ArcGIS Server provides easy-to-use online software development kits (SDKs) for building customized WebGIS applications. Development options include .NET Web Application Developer Framework (ADF), Java Web ADF, and a variety of REST-based APIs such as Flex, JavaScript, and Microsoft Silverlight. In addition, ArcGIS Server is OGC compliant product. The services generated through ArcGIS Server benefit the data sharing among most OGC supported tools and APIs.

19.3 Development of WebGIS Systems

The PDS Geosciences Node has developed ANs and ODEs for planetary data search and downloading. The Apollo AN was developed to "replay" the landed missions by integrating documentation, maps, and images. The Mercury, Moon, and Mars ODEs were developed to search, display, and download orbital data across multiple orbital missions. Currently, the Apollo AN, and all the ODEs have WebGIS embedded in their systems. Work is underway to incorporate WebGIS function within the MER and Phoenix AN to enhance their online mapping capabilities. The WebGIS in ANs and ODEs were designed to fit the SOA model. The online map systems not only provide a platform with mapping capabilities and data representations functions, but also have search and downloading capabilities enabled for easy access of the multi-mission data. The Web interfaces are open to the science community and the general public accessible at the PDS Geosciences Node Web, http://geo.pds.nasa.gov/.

19.3.1 System Overview

A simple version of Mars ODE was initially developed to search the Mars orbital data. Its online map was developed based on ArcGIS Server 9.2 and .Net technique. Search results were plotted on a MGS MOLA basemap. Users can choose the map information such as product centers, latitude/longitude bounding boxes, and approximate product footprints to be displayed on the basemap. Additional search by area function was also provided. An updated WebGIS system for the Mars, Mercury, and Moon ODEs based on recent versions of ArcGIS Server and JavaScript API was released in late 2010. Apollo AN was built in 2009 in memory of the 40th Anniversary of Apollo mission, as well as to support the LRO and future lunar explorations. The WebGIS of Apollo AN was developed as a prototype using an ESRI® ArcGIS Server 9.3.1 and JavaScript API technique. The general procedure to build the WebGIS system for ODEs or ANs is shown in the following diagram.

As shown in Fig. 19.1, first, metadata of the products from multiple missions and instruments are standardized by a back-end processor into a standard format and a unified coordinate system. Standardized metadata are then loaded into a relational database, and product shapefiles are stored in an ESRI® Geodatabase. Maps are created based on the Geodatabase contents. And map services are generated compliant with the OGC standards. Finally, a WebGIS interface is developed to serve maps using JavaScript API. Users may visit the map service either through this WebGIS interface or using free or commercial GIS software packages.

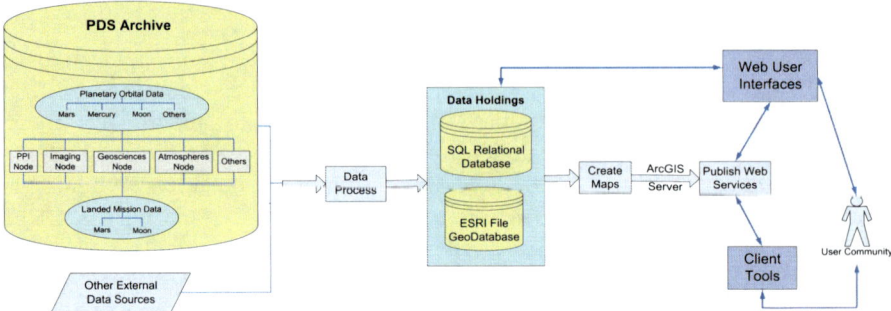

Fig. 19.1 General procedure to build the WebGIS system

The data preparation and Web mapping application using JavaScript API are described in the following section.

19.3.2 Data Preparation

As mentioned before, PDS holds various data sets from multiple missions, instruments, and planets. Data sets have disparate providers, data formats, coordinate systems (particularly older data sets), and projections, which makes it difficult to search, display, and correlate data from various instruments. For example, MGS Mars Orbiter Camera (MOC) used the areographic latitude and west positive longitude, while MGS MOLA data used the east positive, areocentric coordinates. It was inconvenient to register or overlay these two datasets, if they were not unified in the same coordinate systems.

To address these issues, before being loaded into a database, data was processed in a standard format and projected in the same coordinate system. For the orbital missions, product footprints from various data sets were either standardized by a background processor or directly acquired from the USGS UPC database. Quality of the footprint data was checked, especially for those records located at polar area or across the 0/180 longitude. Footprints were validated to eliminate the records with null geometry and repair records with errors such as incorrect ring ordering and self-intersections. Compared with the orbital mission, the surface mission landed at a small region required much higher resolution images in order to display more detailed information of the surface features. Most of the global basemaps were downloaded from USGS Web. Higher resolution images were registered to the lower resolution global basemap, and the rover traverse was registered to the basemap mosaics.

The data preparation provided the standardized data for the creation of maps and services. After Web services were published, JavaScript API was used to create an interactive user interface for Web mapping application.

19.3.3 JavaScript API for Web Mapping Applications

JavaScript was used intensively in the development of WebGIS for ANs and ODEs. JavaScript provides the client-side scripting ability to enhance user interfaces and develop dynamic Web sites. The following will introduce the implementation of JavaScript API in ODEs and ANs to support their WebGIS functionality.

First, the free ArcGIS JavaScript API was used to connect to a map service at an ArcGIS Server and load a cached or non-cached map service into an application as a layer. Several layers could be overlain together. Then, more functions can be implemented using JavaScript such as setting the initial extent of the map to an interesting area, deciding the order of layers to be displayed, adding a navigation toolbar with pan, zoom, and other navigation buttons, displaying the coordinates, as well as setting visible scales and map layouts. Figure 19.2 is a map display from the Mars ODE 3.0 interface, which shows the Mars Express High Resolution Stereo Camera (HRSC) image footprints draped on top of the MGS MOC image mosaics. In the left map control window, users can choose different layers to be displayed. Figure 19.3 presents an interactive map at Apollo 17 Landing Site, which has additional traverse and features plotted. Both interactive maps let users view digital maps centered at a point of interest. Users can further zoom into or out of a map with different levels of details presented.

The additional toolkit Dojo was used to enhance the Web mapping applications in both ANs and ODEs. Dojo (Geoweb Guru 2010) is a flexible JavaScript toolkit which allows users to choose desired pieces of the toolkit. Dojo includes three separate projects: Dojo, Dijits, and DojoX. Dojo was used to handle browser normalization, fix browser incompatibilities, as well as support drag and drop. For example, in the map control window of Fig. 19.2, users can change the display order by directly dragging and dropping the layer. Dijit was used to implement a pop-up information window such as the overview map and the identification result

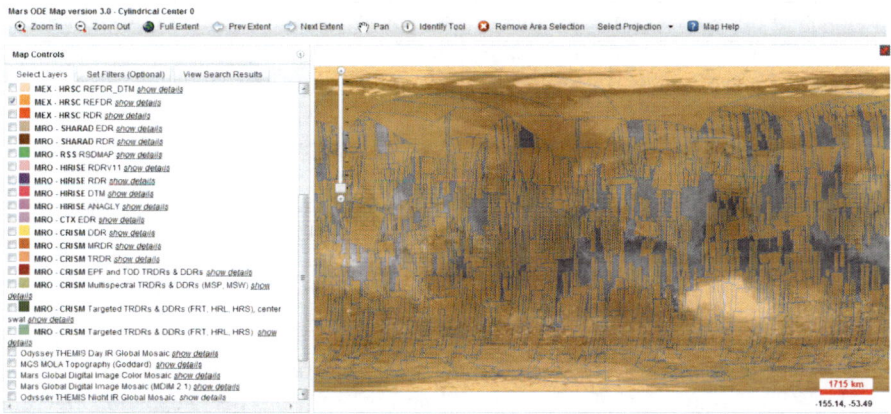

Fig. 19.2 Client access of the ODE Web interface

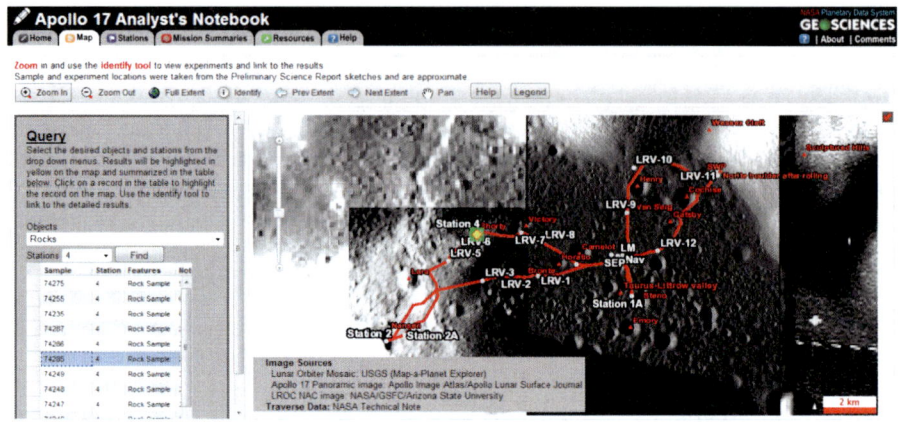

Fig. 19.3 Interactive map for Apollo 17 landing site

window when a user clicks on the map. DojoX was used to add a data grid widget to display the query results as shown in the left panel of Fig. 19.3.

In addition to the mapping and GIS functions, both ANs and ODEs support basic search function based on mission, instrument, product type, location, time, or product ID. ODEs provide additional function to search by product coverage. By selecting a coverage area of visible product layers through the map interface, data products with footprints intersecting the selected area are found and highlighted in the map.

19.4 Planetary Web Services

Planetary Web services are OGC standard web services including WMS/WFS/WCS used in the planetary community. WMS, WFS and WCS are OGC standard protocols released since 1999 for access of maps, features and coverage (OGC 2010). WMS, WFS, and WCS services define a simple HTTP interface and operations that enable interoperable access to the geographic data. The client generates the request and posts it to a Web feature server or a Web map server using HTTP. Results are returned at different levels, which depend on the kind of service being connected and requests being sent. WMS serves static map images in the form of raster tiles such as PNG, GIF, and JPG. Unlike WMS, WFS serves geographic features with accessible geometry and attributes information. In addition, it allows users to query on the published WFS service, and supports operations such as INSERT, UPDATE, and DELETE on geographic features using HTTP. Unlike WFS to return discrete geospatial features, WCS returns geospatial coverage data with attributes and detailed descriptions.

ODE has created WFS services of the product footprint data for easy sharing and distribution. WMS basemap services are also supported. Although still experimental, these services are currently available to the science community and the general public. WCS is under consideration to be used for publishing the actual image data in the future. The following section will discuss the solution of the projection issue faced when creating a WMS/WFS service for planetary data. Examples of free access of WMS/WFS services will also be covered.

19.4.1 Map Projection of the Planetary WMS/WFS

WMS or WFS are published together with the map services through ArcGIS Server. One of the key issues to access a planetary WMS/WFS service is the map projection. Failure to set a correct map projection when creating a WMS/WFS service may result in client software displays blank maps. As mentioned in the previous section, it is often impossible to find an EPSG code to match a planetary projection. When ArcGIS Server publishes Web services, the WMS/WFS services cannot pick the user pre-defined planetary projections correctly from the data set. The default setting of a WMS or WFS service is still using EPSG 4326 (based on WGS84) to define the CRS.

To solve this issue, using an external capabilities file is recommended in the ESRI documentation. This file is a connection between a WMS/WFS service and a WMS/WFS client, which can be edited to change the CRS of a WMS/WFS service. It works well with the existing EPSG code. However, the external capabilities file cannot fully support the planetary projections, if there is no definition of planetary projections at the Server site. Therefore, a set of user self-defined "EPSG" codes were used to define map projections for the planetary data. These "EPSG" codes were distinguished with the existing common-used ones to avoid any conflict. In order to set up these codes, a set of definitions were generated at the ArcGIS Server site. These definitions were used to define a planetary body with an ellipsoid or a sphere, datum, as well as other projection parameters. When a WMS/WFS service with these "EPSG" codes is published through ArcGIS server, the defined codes are picked and the projections are recognized by ESRI products and other GIS client tools.

ODEs provide WMS services of basemaps and WFS services of orbital product footprints in the simple cylindrical projection centered at 0 or 180 longitudes, as well as the polar stereographic projection at North or South Pole. For surface data, Apollo AN supports map services in simple cylindrical projection centered at 0 longitude, also called equirectangular, or plate carrée projection. If needed, other projections can be implemented by adding customized definitions.

19.4.2 Client Access of WMS/WFS

Once a WMS/WFS service is published, HTTP is used as the distributed computer platform to link the client application with a WMS/WFS service. The data retrieve process is shown in Fig. 19.4. First, a client application sends a request to the WMS/WFS service. Then, the service connects to the database stored in ArcGIS server and processes the request. This step is opaque to the client. Finally, response is returned to the client.

Several client tools were used in our test to access the WMS/WFS services, including the Web browser, and several commercial and free GIS packages. This paper will only focus on the web browser and some of the free clients. Examples are given in the following sections.

1. Web Browser

The simplest client of a WMS/WFS service could be a Web browser. Table 19.1 lists the example of URL strings used to access a WMS/WFS service. The URL string consists of a number of options and settings such as server name, path, service name, and service type being requested. For example, in the URL string for a WMS service, CRS = EPSG:202034 is the self-defined EPSG code for the moon south polar stereographic projection. WIDTH and HEIGHT are used to specify the size of an image, and the BBOX specifies the map extents. More specifications of the URL component can be found in OGC document (2004,

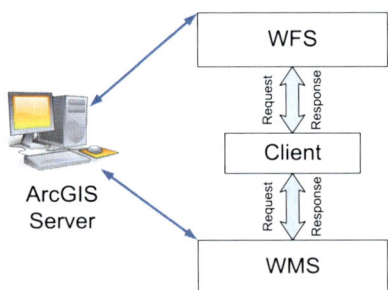

Fig. 19.4 Data retrieve using a WMS/WFS service

Table 19.1 Strings used to access a WMS/WFS service

Service	Access URL string
WMS	http://odeogcserver.rsl.wustl.edu/ArcGIS/services/moon_polar_south/moon_basemap_clem_uvvis_sp/MapServer/WMSServer?REQUEST=GetMap&SERVICE=WMS&VERSION=1.3.0&LAYERS=0&STYLES=&FORMAT=image/png&BGCOLOR=0xFFFFFF&TRANSPARENT=TRUE&CRS=EPSG:202034&BBOX=-1084194.45547945,-931021.62,1084076.00547945,931140.07&WIDTH=1020&HEIGHT=876
WFS	http://odeogcserver.rsl.wustl.edu/ArcGIS/services/moon_polar_north/moon_clem_hires_mdim_na/MapServer/WFSServer?request=GetFeature&TypeName=moon_polar_north_moon_clem_hires_mdim_na:moon_clem_hires_mdim_na&service=WFS

2005). The WMS service returns a static image in the Web browser as shown in Fig. 19.6a. WFS returns a number of records with geographic position and associated attributes listed in a XML file and presents the results in a Web browser.

2. Other Free Client Tools

The following examples illustrate results of the client access to the OGC standard WMS/WFS services published by Geosciences Node. Free WMS viewers like Gaia (http://www.thecarbonproject.com/gaia.php), NASA World Wind, and CARIS Easy View (http://www.caris.com/products/easy-view/) were used to test the WMS services. First, a client application send the GetCapabilities request to a Web Map Server to get the metadata of the service. Then, the GetMap function is used to request a map image from the Map Server. In Fig. 19.5, the 3D interactive Mars viewer of NASA World Wind presents the MGS MOLA topography data by connecting to its WMS service. Figure 19.6b displays a Clementine ultraviolet-visual (UVVIS) image mosaic in south polar stereographic projection using Gaia to connect to the WMS service.

Other open source WFS clients such as Gaia and UDig (http://udig.refractions.net/) were used to test the WFS services. Figures 19.7 and 19.8 are examples of the client access to WFS services with maps in simple cylindrical and polar stereographic projections. The returned results from the WFS services are added as a layer into a map through operations of GetCapabilities, DescribeFeatureType, and GetFeature. As shown in Fig. 19.8, users can remotely access the data and acquire its geographic position and associated attributes by the information tool in Gaia. The map layer can be further output in other format for geospatial analysis.

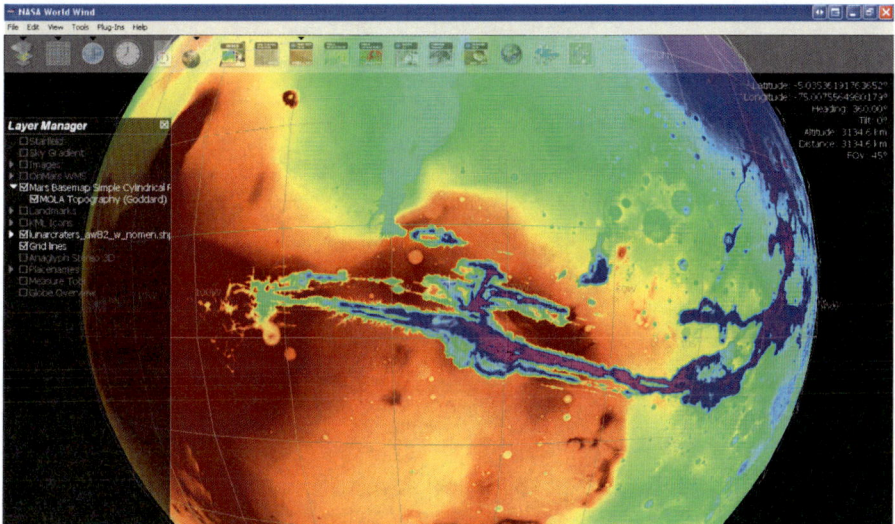

Fig. 19.5 Access WMS of MGS MOLA topography with NASA World Wind. (The Mars topographic map was produced by Goddard Space Flight Center and available to be downloaded from USGS ftp://pdsimage2.wr.usgs.gov/pub/pigpen/mars/mola/mola128_88Nto88S_Simp_clon0.zip)

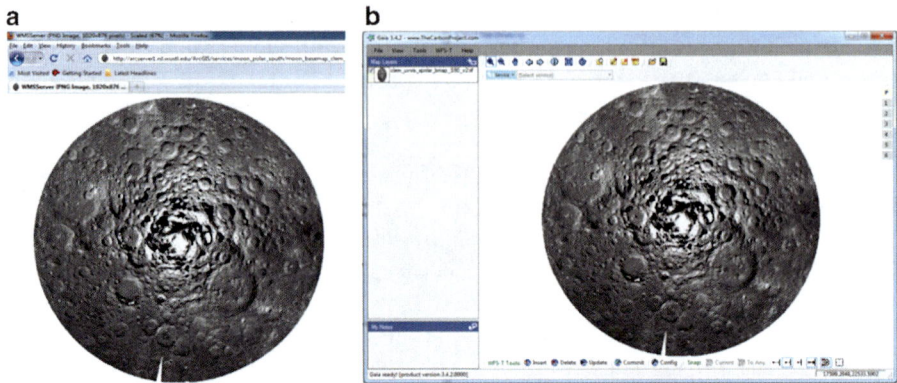

Fig. 19.6 Access WMS of Clementine UVVIS image mosaic (Clementine UVVIS image mosaic was produced by USGS and available at ftp://pdsimage2.wr.usgs.gov/pub/pigpen/moon/clementine/UVVIS_ULCN2005_Basemap_v2/mosaic_256ppd/. Subset of this data set was re-projected in south polar stereographic projection in this research.): (**a**) using a Web browser, (**b**) using Gaia

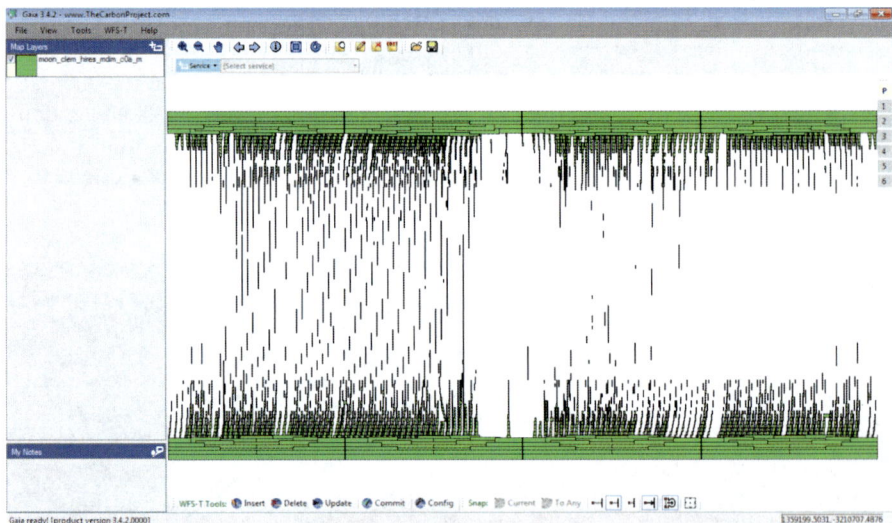

Fig. 19.7 Access WFS of Clementine HIRES MDIM footprint map with Gaia (data set is in simple cylindrical projection centered at 0 longitude)

19.5 Future Development

Both ODEs and ANs provide access to different mission data archives by integrating engineering data, science data, and documentation into interactive Web-accessible pages, which facilitate mission "replay" and support search and analysis across disparate data sources.

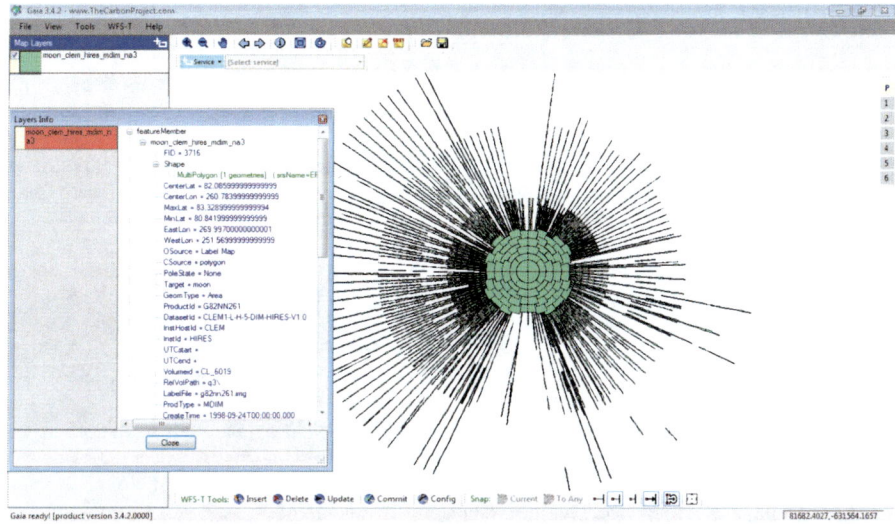

Fig. 19.8 Access WFS of Clementine HIRES MDIM footprint map with Gaia (data set is in north polar stereographic projection)

In addition to providing WFS services of product footprints, ODEs will provide the actual data such as images through WCS. With these services, the data sharing and distribution will be more efficient. The WebGIS in the Apollo AN was developed as a prototype for the landed Apollo missions. Similar techniques might be used to update the online map systems in Mars Spirit, Opportunity, and Phoenix ANs to help playing back the surface activities, but the possibility of using a non-ArcGIS server solution could also be considered. A Mars Science Laboratory (MSL) interactive online map archiving was created in 2011 for landing site safety and feasibility analysis before the MSL mission.

The PDS Geosciences Node is continuously improving the Analyst Notebooks and Orbital Data Explorers to improve functionality, improve user interfaces, and incorporate feedback from mission science teams and PDS users.

Acknowledgments Analyst Notebooks and Orbital Data Explorers are developed through funding provided by NASA's PDS Geosciences Node. Both tools are available from the Web site of PDS Geosciences Node. We would like to thank all of the colleagues working at the Geosciences Node for data archiving and valuable comments. Especially the comments on the WebGIS systems from Mr. Tom Stein and his proof reading of this paper are appreciated. Special thanks are given to Mr. Feng Zhou from the Geosciences Node and Ms. Melita from ESRI for their help to solve the user self-defined coordinate system issue when developing the WMS/WFS services.

The Geosciences Node welcomes questions and comments from the user community. Please send email to geosci@wunder.wustl.edu.

References

Akins S, Raub R, Hare T, Blue J (2010) Web feature service for planetary nomenclature. In: Abstract (2 pages) and poster, 41th lunar and planetary science conference, The Woodlands, Abstract #2251

Akins SW, Gaddis L, Becker K, Barrett J, Bailen M, Hare T, Soderblom L, Raub R (2009) Status of the PDS unified planetary coordinates database and the planetary image locator tool (PILOT). In: Abstract (2 pages) and poster, 40th lunar and planetary science conference, The Woodlands, Abstract #2002

Bennett KJ, Scholes D, Arvidson R, Slavney S, Guinness EA, Stein TC (2008) Accessing mars data using PDS geosciences node's orbital data explorer, LPSC XXXIX, Abstract #1379

Geoserver.org (2010) http://geoserver.org/display/GEOS/Welcome. Accessed 12 Jan 2012

Geoweb Guru (2010) Using dojo to enhance web mapping applications with the ArcGIS server, 2010. http://www.geowebguru.com/articles/168-using-dojo-to-enhance-web-mapping-applications-with-the-arcgis-server. Accessed 12 Jan 2012

Gorelick NS, Weiss-Malik M, Steinberg B, Anwar S (2003) JMARS: a multimission data fusion application, XXXIV lunar and planetary science conference, Houston, Abstract #2057

Hare TM, Tanaka KL (2000) PIGWAD – new functionality for planetary GIS on the Web. In: 31st lunar and planetary science conference, Houston, Abstract #1889

Hare TM, Tanaka KL (2002) PIGWAD – OpenGis and image technologies for planetary data analysis. In: XXXIII lunar and planetary science conference, Houston, Abstract #1365

Hare TM, Archinal B, Plesea L, Dobinson E, Curkendall D (2006) Standards proposal to support planetary coordinate reference systems in open geospatial web services and geospatial application. In: XXXVII lunar and planetary science conference, Houston, Abstract #1931

Li R, Di K, Wang J, Niu X, Agarwal S, Brodyagina E, Oberg E, Hwangbo JW (2007) A WebGIS for spatial data processing, analysis, and distribution for the MER 2003 mission. J Photogr Rem Sens 73(6):671–680

Mapserver (2010) http://www.mapserver.org. Accessed 12 Jan 2012

MapGuide Open Source (2010) http://mapguide.osgeo.org/. Accessed 12 Jan 2012

Mapnik (2010) http://mapnik.org. Accessed 12 Jan 2012

NASA World Wind (2010) http://worldwind.arc.nasa.gov/. Accessed 12 Jan 2012

OGC document 04–094, 2005. Web feature service implementation specification, version 1.1.0. http://www.opengeospatial.org/standards/wfs. Accessed 12 Jan 2012

OGC document 03-109r1 (2004) OGC Web map service interface, version 1.3.0. http://www.opengeospatial.org/standards/wms. Accessed 12 Jan 2012

Open Geospatial Consortium (2010) http://www.opengeospatial.org/standards. Accessed 12 Jan 2012

Plesea L, Hare TM, Dobinson E, Curkendall D (2007) Description of the JPL planetary web mapping server. ISPRS working group IV/7 extraterrestrial mapping, Advances in planetary mapping, Houston, Abstract

Stein TC, Arvidson RE, Heet TL, Wang J (2010) PDS Analyst's notebook: enriching planetary data archives by integrating mission data and documents. In: Abstract (2 pages) and poster, 41th lunar and planetary science conference, The Woodlands, Abstract #1414

Slavney S, Arvidson RE, Guinness EA, Stein TC (2011) PDS geosciences node data and services. In: Abstract (2 pages) and poster, 42th lunar and planetary science conference, The Woodlands, Abstract #1895

Wang J, Stein TC, Heet T, Scholes DM, Arvidson RE, Heil-Chapdelaine V (2010a) A WebGIS for Apollo analyst's notebook. In: Second international conference on advanced geographic information systems, applications, and services, DOI 10.1109/GEOProcessing.2010.20

Wang J, Bennett KJ, Scholes DM, Arvidson RE (2010b) Lunar data access through the orbital data explorer from the PDS geosciences node. Paper (6 pages) and poster, Global lunar conference, the 11th ILEWG conference on exploration and utilisation of the moon, Beijing, China, May 31–June 3 2010. Paper #GLUC-2010-1.7. A.P.1

Wang J, Bennett KJ, Scholes DM, Ward JG, Slavney S, Guinness EA, Arvidson RE (2010c) Updates to the orbital data explorer from the PDS geosciences node. In: Abstract (2 pages) and poster, 41th lunar and planetary science conference, The Woodlands, Abstract #2251

Wang J, Bennett KJ, Scholes D, Arvidson R, Ward JG, Slavney S, Guinness EA, Stein TC, Heil-Chapdelaine V (2009) Planetary data access through the orbital data explorer from the PDS geosciences node. In: Abstract (2 pages) and poster, 40th lunar and planetary science conference, The Woodlands, Abstract #1193

Index

A
Accessibility, 94, 96–98, 102, 109, 153
Aerial Photograph Catalog (APCatalog), 282, 284
Aerial photography, 280–282, 284, 288–290, 293, 294
AJAX *See* Asynchronous JavaScript and XML (AJAX)
Animated maps, 205–216
APCatalog *See* Aerial Photograph Catalog (APCatalog)
Asynchronous JavaScript and XML (AJAX), 4, 7–10, 62, 63, 107, 114, 124, 180, 183, 245, 265, 272, 276

C
Calendar, 109, 111
Canvas, 25, 27–29, 35, 238
Cartographic catalogs, 281–284
Cartography, 4, 23, 94, 100, 143–145, 147, 149, 152, 153, 160, 171, 173, 178, 203, 206, 212, 295, 296
Cloud computing, 294, 295
Commercial APIs, 61–70, 96, 280, 288, 290, 295
Commercial map APIs, 279, 288–291, 295
Continuously Operating Reference System (CORS), 287
ControlFinder, 286–290, 292
CORS *See* Continuously Operating Reference System (CORS)

D
Deep interactivity, 296
Delphi, 43, 244–245, 260

Digital Raster Graphics (DRGs), 286, 287, 290, 295
Django, 76, 79–80, 82
3D modeling, 29, 228
DRGs *See* Digital Raster Graphics (DRGs)

E
Education, 37–57, 142
Environmental Science Research Institute (ESRI), 7, 26, 27, 31, 32, 62, 76, 78, 79, 125–127, 129–131, 135, 241, 247, 249, 266, 286, 300, 302, 303, 307
eXtensible Markup Language (XML), 8, 11, 25–27, 39, 41, 42, 52, 55, 62, 79, 108, 111, 113, 125–128, 131–135, 145, 149, 152, 180, 199, 206–209, 213, 221, 229, 245, 266–268, 273, 276, 309
Extensible Stylesheet Language Transformation (XSLT), 143, 149–152

F
Federal Geographic Data Committee (FGDC), 284–286
Fedora Commons, 285, 290, 291
FGDC *See* Federal Geographic Data Committee (FGDC)
Flash, 25–30, 34, 35, 66, 124–127, 133, 135, 207, 276
Flex, 25, 26, 35, 123–136, 302
Flickr, 63–69, 159, 163, 266, 274

G
Generalization, 158, 159, 167, 171, 173, 178, 296

Geocoding, 7, 19, 20, 94, 95, 97, 98, 102, 108, 246, 270, 276, 290, 295
Geodetic control data, 281, 282, 286, 297
Geolocation, 37, 242, 246, 251
GeoMOOSE, 285, 288
Geospatial data, 144, 146, 280, 281, 286, 297, 302, 306
Geovisualization, 38, 44, 89
Globes, 144, 219–238
Google, 4–8, 15–19, 24, 30, 42, 63, 65–67, 69, 76, 77, 79, 108, 109, 117, 121, 124, 147, 161, 169, 170, 196–198, 201, 220, 222, 223, 225–228, 250, 252, 266, 268, 272–274, 276, 285, 290, 291, 296, 300
Google Geocoding API, 290, 295
Google Maps, 3–5, 7, 8, 13, 15–19, 37, 39, 43–45, 52, 63–66, 69, 93, 100, 101, 109, 113, 116, 117, 121, 124, 158, 159, 161, 170, 179, 182, 195–203, 241, 245, 246, 249, 251, 265, 266, 268–270, 273
Google Maps API, 7, 30, 38, 39, 44, 52–54, 63–66, 108, 113, 116, 120, 159, 161, 170, 197, 203, 249, 251, 265–276, 289, 290, 295
Graduated circles, 195–203

H
Height Modernization Program (HMP), 286
History teaching, 38
HMP *See* Height Modernization Program (HMP)

I
Icon aggregation, 158–160, 166–168, 170–172
Icon cluttering, 158–160, 168, 171, 173
Icon placement, 158–160, 166–168
IF *See* Information filtering (IF)
*i*MapInvasive, 73–89
Information filtering (IF), 159, 160, 171–173
Interactive mapping, 24, 196, 209, 245, 285
Invasive species, 74, 75, 80–84, 88, 89

J
JavaFX, 25, 27, 28, 35
JavaScript, 9, 10, 25–30, 34, 39, 42, 45, 52, 63–66, 75, 78, 93, 95, 98, 99, 103, 109, 125, 183, 197, 213, 235, 237, 238, 249, 251, 265–276, 285, 301, 302, 305
JavaScript API, 7, 45, 109, 208, 226, 228, 249, 266, 269, 301, 303–306

K
Keyhole Markup Language (KML), 18, 32, 52, 79, 87, 145, 147, 149, 152, 220–228, 266, 276, 287, 290, 294

L
Location-based-service (LBS), 19, 105, 106, 285
Location-based systems (LBS), 11

M
Map mashups, 4, 13, 14, 30, 37–57, 64–66, 157–173, 196, 242, 243, 250–254, 266
Mapping APIs, 4, 7, 17–20, 37, 77, 78, 91–103, 159, 180, 266
MapQuest, 4, 7, 17, 93
MapServer, 30, 31, 127, 129, 131, 183, 184, 208, 209, 214, 285, 286, 290, 302, 308
Map services, 15, 77, 97–99, 102, 127, 129, 133, 146, 178, 209, 242, 249, 251, 265–276, 279–297, 303, 305, 307
Map styles, 15, 17, 251
Map symbology, 296
Map tiles, 5, 9, 17, 101, 245, 290, 291, 296
Mashups, 4, 12, 18–20, 30, 38, 44, 52, 64, 65, 158, 159, 161, 163, 164, 166, 168–173, 196, 209, 242, 243, 249, 266, 295, 296
Mobile, 3, 7, 11, 12, 14, 15, 20, 28, 29, 34, 35, 83–84, 96, 106, 107, 109–111, 113, 116, 117, 143, 159, 182, 183, 209, 242, 247, 248, 252, 255, 266, 294
Mobile mapping, 11–12, 18–19
Multimedia mapping, 61–70, 266
Multipublishing, 177–191

N
National Aeronautics and Space Administration (NASA), 15, 46, 281, 300, 309
National Geodetic Survey (NGS), 282, 286
National Spatial Data Infrastructure (NSDI), 91, 92, 284

Index 317

NGS *See* National Geodetic Survey (NGS)
NSDI *See* National Spatial Data Infrastructure (NSDI)

O
OGC *See* Open Geospatial Consortium (OGC)
Online mapping, 3–12, 37, 93–95, 196, 241, 242, 266
Online mapping service, 7, 91, 93, 94, 100, 101, 265–276
Open architecture, 180, 182
Open Geospatial Consortium (OGC), 10, 39–41, 52, 96, 97, 102, 143, 145–149, 151–153, 180, 191, 208, 209, 211, 213, 214, 216, 285, 286, 302
  standards, 10, 31, 96, 97, 145, 146, 153, 180, 216, 285, 302, 303, 306, 309
  WMS/WFS, 46, 206, 285, 300–302, 306, 308, 309
OpenLayers, 30, 38, 45, 47, 76, 78, 87, 98, 103, 179, 180, 183, 184, 191, 208, 285, 288, 302
Open source, 7, 11, 15, 24, 27, 28, 30–32, 38, 73–89, 98, 108, 109, 124, 126, 127, 143, 148, 149, 183, 208, 209, 225, 235, 265–276, 280, 285, 286, 288, 289, 297, 302, 309
OrthoFinder, 285, 286, 288
Orthophotography, 281, 285
Outdoor activities, 178, 179, 186, 191

P
PDS *See* Planetary Data System (PDS)
People's Garden, 266, 267, 270, 273
Planetary Data System (PDS), 299–311
Planetary GIS, 300–302
Planetary Web services, 301, 302, 306–310
PLSS *See* Public Land Survey System (PLSS)
PLSSFinder, 288–290, 292, 293
PostGIS, 31–33, 39, 76, 79, 80, 88, 183, 184, 208, 213, 285, 286, 290, 291
PostgreSQL, 31–33, 39, 75, 76, 79, 80, 88, 108, 183, 208, 285, 286, 290
Psychological scaling method, 195
Public Land Survey System (PLSS), 280, 287, 289–291, 296, 297
Python, 75, 76, 79, 80, 88

R
Rich Internet Application (RIA), 25, 26, 28, 35, 123–125

S
SaaS *See* Software as a Service (SaaS)
Sample Flex Viewer (SFV), 125–128, 131, 133–135
Scalable Vector Graphics (SVG), 25, 27, 28, 30–35, 142, 143, 145, 148, 151, 152, 180, 184, 206–216
SCO *See* State Cartographer's Office (SCO)
Semantics, 92, 142, 143, 145, 146, 153, 168, 171, 173, 249, 260, 261
Service oriented architecture (SOA), 143, 216, 302, 303
SFV *See* Sample Flex Viewer (SFV)
Shapefiles, 31, 32, 62, 286, 287, 290, 294, 303
Silverlight, 25–28, 33–35, 302
SMIL *See* Synchronized Multimedia Integration Language (SMIL)
Social network interactions, 241–261
Software as a Service (SaaS), 294, 295
Spatio-temporal data, 205–207, 209, 211, 213, 214, 216, 253–255
Species tracking, 74, 82, 84
State Cartographer's Office (SCO), 279–297
SVG *See* Scalable Vector Graphics (SVG)
Synchronized Multimedia Integration Language (SMIL), 206, 207, 209–213, 215, 216

T
Task planning, 105–121
Technology, 4, 11, 15, 18–20, 24–29, 33, 35, 38, 62, 69, 70, 75, 76, 98, 99, 106, 123–125, 178, 206, 256, 260, 280–282, 286, 294–297, 302
Thematic maps, 4, 81, 89, 141–154, 196, 198, 300
Tile, 4–6, 9, 10, 15, 17, 97, 101, 102, 183, 184, 196, 245, 285, 286, 290, 291, 296, 306
Twitter, 18, 158, 241, 242, 244–257, 259–261

U
UCD *See* User-centered design (UCD)
UI *See* User interface (UI)

United States Department of Agriculture (USDA), 266, 267, 272, 274–276, 281, 289
University of Wisconsin, 285, 288
Usability, 52, 86, 93, 94, 96, 100, 102, 178, 179, 181–182, 190, 191
Usability design, 178, 179, 181–182, 191
USDA *See* United States Department of Agriculture (USDA)
User-centered design (UCD), 297
User interface (UI), 9, 27, 30, 34, 45, 111, 128–129, 178, 179, 181–183, 185, 186, 191, 210, 212–213, 272, 289, 304, 305, 311
US Geological Survey (USGS), 281, 282, 286, 289, 300, 301, 304, 309, 310

V
Vector, 10, 17–19, 23–35, 79, 87–88, 123, 145–147, 151, 152, 180, 183, 184, 189, 205, 207, 216, 296
Visualization, 19, 20, 29–31, 33, 39, 49, 52, 87, 144, 153, 158, 159, 181, 191, 206, 216, 243, 244, 246, 248, 249, 251, 255, 260, 269, 270

W
WCS *See* Web Coverage Service (WCS)
Web-based mapping applications, 62
Web cartography, 94, 145, 152
Web Coverage Service (WCS), 10, 11, 31, 145, 146, 302, 306, 307, 311
Web Feature Services (WFS), 11, 31–33, 45, 87, 97, 145, 146, 151, 180, 183, 184, 285, 286, 301, 302, 306–311
WebGL, 25, 27, 29, 35, 235
Web(-based) mapping, 24, 34, 38, 52, 62, 106, 180–182, 206, 245, 266, 280, 285–289, 291, 293, 295–297, 302, 304–306
Web(-based) maps, 18, 23, 30, 31, 178–181, 266, 293, 295, 297
Web Map Services (WMS), 7, 10–11, 13–20, 31, 33, 45, 46, 63, 78, 87, 97, 105, 109, 126, 128, 134, 135, 145, 146, 148, 153, 180, 183, 206–214, 216, 249, 285, 290, 301, 302, 306–310
Web Processing Service (WPS), 143, 145–147, 149–153
WFS *See* Web Feature Services (WFS)
WHAIFinder *See* Wisconsin Historic Aerial Image Finder (WHAIFinder)
WiscLINC *See* Wisconsin Land Information Clearinghouse (WiscLINC)
WiscMap, 286
Wisconsin Historic Aerial Image Finder (WHAIFinder), 289–293
Wisconsin Idea, 288
Wisconsin Land Information Clearinghouse (WiscLINC), 284, 286
Wisconsin State Cartographer's Office, 279–297
WMS *See* Web Map Services (WMS)
WPS *See* Web Processing Service (WPS)

X
XAML, 25, 26, 35
XML *See* eXtensible Markup Language (XML)
XSLT *See* Extensible Stylesheet Language Transformation (XSLT)

Y
YouTube, 63–69, 163, 169, 266, 274

Printed by Publishers' Graphics LLC USA
MO20120419-080
2012